Plant DNA Barcodes, Community Ecology, and Species Interactions

Plant DNA Barcodes, Community Ecology, and Species Interactions

Editors

W. John Kress
Morgan R. Gostel

MDPI • Basel • Beijing • Wuhan • Barcelona • Belgrade • Manchester • Tokyo • Cluj • Tianjin

Editors

W. John Kress
National Museum of
Natural History
Smithsonian Institution
Washington, DC, USA

Morgan R. Gostel
Botanical Research
Institute of Texas
Fort Worth, TX, USA

Editorial Office
MDPI
St. Alban-Anlage 66
4052 Basel, Switzerland

This is a reprint of articles from the Special Issue published online in the open access journal *Diversity* (ISSN 1424-2818) (available at: https://www.mdpi.com/journal/diversity/special_issues/Plant_DNA_Barcodes?listby=viewed&view=default).

For citation purposes, cite each article independently as indicated on the article page online and as indicated below:

LastName, A.A.; LastName, B.B.; LastName, C.C. Article Title. *Journal Name* **Year**, *Volume Number*, Page Range.

ISBN 978-3-0365-6043-4 (Hbk)
ISBN 978-3-0365-6044-1 (PDF)

Cover image courtesy of Kenneth Wurdack and W. John Kress, Smithsonian Institution.

Contents

About the Editors

W. John Kress

W. John Kress, PhD, is a Distinguished Scientist and Curator Emeritus in the National Museum of Natural History, Smithsonian Institution; Co-Chair for the Earth BioGenome Project; and visiting Scholar for Dartmouth College and The Arnold Arboretum of Harvard University. His research interests include systematic biology: systematics of Heliconia and other Zingiberales, especially with regard to reproductive biology, molecular variation, and cladistic relationships; generic concepts in the Zingiberales; phylogeny and evolution of the monocotyledons; evolutionary biology: plant reproductive biology, the evolution of breeding and pollination systems, and genetic diversity and speciation in tropical angiosperms; conservation biology: forest fragmentation, gene flow, and genetic variation in tropical and temperate plant species.

Morgan R. Gostel

Morgan R. Gostel, PhD, is a Research Botanist at the Botanical Research Institute of Texas and Director of the Global Genome Initiative for Gardens (GGI-Gardens). His research interests are based in plant diversity and evolution with a focus on systematics and taxonomy. He is also interested in advancing plant genomics, by supporting collaboration between the botanical garden and genome sequencing communities with GGI-Gardens.

Editorial

Plant DNA Barcodes, Community Ecology, and Species Interactions

W. John Kress [1,2,3,*] and Morgan R. Gostel [1,4]

[1]   National Museum of Natural History, Smithsonian Institution, P.O. Box 37012, Washington, DC 20013, USA; mgostel@brit.org
[2]   Department of Biological Sciences, Dartmouth College, Hanover, NH 03755, USA
[3]   The Arnold Arboretum, Harvard University, Boston, MA 02130, USA
[4]   Botanical Research Institute of Texas (BRIT), Fort Worth, TX 76107, USA
[*]   Correspondence: kressj@si.edu; Tel.: +1-(202)-372-7745

Citation: Kress, W.J.; Gostel, M.R. Plant DNA Barcodes, Community Ecology, and Species Interactions. *Diversity* **2022**, *14*, 453. https://doi.org/10.3390/d14060453

Received: 10 May 2022
Accepted: 1 June 2022
Published: 6 June 2022

Publisher's Note: MDPI stays neutral with regard to jurisdictional claims in published maps and institutional affiliations.

1. Introduction

The community of biologists has been eager to realize the promise of DNA barcodes since the concept of a rapid method for genetic identification of species was first proposed in 2003. As we approach twenty years of DNA barcoding, the application of these short, but highly variable sequences continue to increase and methods continue to be developed that utilize this ever-expanding resource for multiple fields of biology. The nearly ten million DNA barcodes for life on Earth available today provide a database that is especially useful for ecology and evolutionary biology. In particular, DNA barcodes provide a rapid resource to identify taxa; to quantify and understand species richness; and to determine community interactions in primary and secondary habitats. Many ecologists, who are concerned with the assembly and maintenance of species richness at local and regional scales, have driven empirical and conceptual advances in the field of community ecology. At the same time evolutionary biologists have focused on the description and classification of species diversity, factors controlling the origin and ancestry of biodiversity, and the network of interactions that connect evolutionary units through time and space.

Today, thanks to the ever-expanding and well-curated DNA barcoding resources now available, fundamental biological questions can be more rigorously addressed regarding community evolution, assembly, productivity, and species interactions across and among diverse habitats and organisms, including plants, animals, fungi, and microorganisms. DNA barcodes are now routinely used to discover new species, to determine phylogenetic patterns of community diversity, and to uncover the complexities of interactions in almost all domains of life to understand diets, symbioses, pollinator networks, and historically challenging biomes, such as below-ground soil and deep-water marine communities. This Special Issue of *Diversity* addresses the wide variety of applications of DNA barcodes, especially in plants. The eleven papers included in this Special Issue illustrate how the DNA barcode library continues to be expanded, the range of ecological and evolutionary questions that can be answered with DNA barcodes, and how plant-human interactions are better understood using DNA barcodes as a research tool.

2. Building the Plant DNA Barcode Library

The diversity of gene regions that serve as DNA barcodes continues to expand from the original cytochrome oxidase 1 mitochondrial sequences applied to many groups of animals. To date, no single gene region fits all lineages of life as a universal DNA barcode. For that reason, researchers continue to experiment and search for the most effective DNA barcode for specific clades on the Tree of Life and particular type or condition of tissues within organisms. In this Special Issue Dal Forno et al. [1] explored the application of DNA barcodes in both fresh and historical collections of lichen-forming basidiomycetes. Their results demonstrate that barcode sequences can effectively be generated from both fresh and historical collections more than 100 years old and that fungal ITS barcode sequences provide powerful resources for species delimitation in an integrative taxonomic framework., Omelchenko and colleagues [2] continued the quest to identify the most efficient regions for metabarcoding members of the grass family and suggest additional spacer regions from nuclear ribosomal DNA (e.g., ITS1 and ETS in addition to the more common ITS2 barcode) provide enhanced discriminatory power for species identification in mixed pollen samples of grass species. In a third paper Kenfack, Abiem and Chapman [3] tested the effectiveness and efficiency of applying the standard plastid plant barcode regions (rbcLa, matK and trnH-psbA) to over one hundred species of trees in a montane forest in Nigeria and concluded that the combination of rbcLa and matK is sufficient for species discrimination. As part of the Global Genome Initiative for Gardens, Gostel et al. [4] release in the Special Issue 2722 DNA barcode sequences from 174 families and 702 genera of land plants that represent taxa without previous barcode sequences in GenBank. Each of these papers represents a significant contribution to building the DNA barcode library for plants.

3. Using Plant DNA Barcodes to Understand Ecological Patterns and Evolutionary Processes

With the advancement of DNA barcodes as a reliable source for genetic species identification, biologists increasingly use this tool to track species interactions. Such studies are now being conducted across the globe in both temperate and tropical environments. The review by Gostel and Kress [5] in this Special Issue outlines the recent progress that has been made in these investigations as a result of novel computational and sequencing capacities, high-throughput barcoding methodologies, and the expansion of the global DNA barcode database.

Three additional papers in the Issue highlight how DNA barcodes help to uncover previously obscure interactions, e.g., mate location in highly complex tropical forests; the abundance and diversity of large mammalian herbivores and their woody plant food sources; and in plant-insect pollinator communities. In an attempt to uncover how Orthopterans (crickets, katydids, and grasshoppers) use acoustic cues to find mates, Palmer and colleagues [6] turned to plant DNA barcodes to test the specificity of food plants that could facilitate mate location in katydids on Barro Colorado Island in Panama. Their results showed that most katydids are generalist herbivores and food choice would most likely not facilitate mate location. In a semi-arid African savanna Freeman et al. [7] demonstrated the important role of megaherbivores in shaping vegetation across landscapes. Using data from plant DNA barcodes, they were able to ascertain that some habitats, which deter large mammalian herbivores, serve as refuges for plant species that otherwise are quite palatable to these animals. Finally, a comparison between plant metabarcoding and non-molecular methods of tracking plant-pollinator interactions demonstrated the advantages of a DNA barcode approach in determining the complexity of these communities [8].

4. Plant DNA Barcodes and Human Interactions

It should not be forgotten that one of the most important plant-animal interactions on the planet is between plants and humans. Three final papers in the Special Issue address the application of DNA barcodes to tracking medicinal plants, invasive species, and habitat conservation. Jamdade et al. [9] demonstrate that as the DNA barcode library is built for the flora of the United Arab Emirates the current plant DNA barcode regions provide sufficient markers for the safe usage, prevention of adulteration, and the regulation of medicinal plant trading. DNA barcode sequence data were employed by Yessoufou and Ambini [10] to build a molecular phylogeny of the 210 known naturalized alien woody plants in South Africa. Based on this phylogeny they demonstrated that the benefits humans obtain from an alien species had significant evolutionary signal, but that non-invasive species exhibited more benefits to humans than their introduced, invasive counterparts. Such phylogenetic metrics can also contribute to plant conservation. Pearl et al. [11] generated DNA barcodes for 366 species of plants in the heathland ecosystems in Queensland, Australia. The resulting measures of phylogenetic diversity found in these communities combined with other patterns of diversity suggested contrasting conservation and management implications for these historical "refugial environments".

This Special Issue of *Diversity* on "Plant DNA Barcodes, Community Ecology, and Species Interactions" provides a taste of the current variety of investigations and publications that are a result of the expansion of DNA barcodes in the biological sciences. It is hoped that the papers contained herein will inspire and encourage future applications of DNA barcoding to the exploration of ecological and evolutionary systems across the globe.

Author Contributions: W.J.K. and M.R.G. contributed equally to this publication. All authors have read and agreed to the published version of the manuscript.

Funding: This research received no external funding.

Acknowledgments: W.J.K. and M.R.G. are grateful to the authors who have contributed their work to this Special Issue.

Conflicts of Interest: The authors declare no conflict of interest.

References

1. Dal Forno, M.; Lawrey, J.D.; Moncada, B.; Bungartz, F.; Grube, M.; Schuettpelz, E.; Lücking, R. DNA barcoding of fresh and historical collections of lichen-forming Basidiomycetes in the genera *Cora* and *Corella* (Agaricales: Hygrophoraceae): A success story? *Diversity* **2022**, *14*, 284. [CrossRef]
2. Omelchenko, D.O.; Krinitsina, A.A.; Kasianov, A.S.; Speranskaya, A.S.; Chesnokova, O.V.; Polevova, S.V.; Severova, E.E. Assessment of ITS1, ITS2, 5′-ETS, and *trnL-F* DNA barcodes for metabarcoding of Poaceae pollen. *Diversity* **2022**, *14*, 191. [CrossRef]
3. Kenfack, D.; Abiem, I.; Chapman, H. The efficiency of DNA barcoding in the identification of Afromontane forest tree species. *Diversity* **2022**, *14*, 233. [CrossRef]
4. Gostel, M.R.; Carlsen, M.M.; Devine, A.; Barker, K.B.; Coddington, J.A.; Steier, J. Data Release: DNA barcodes of plant species collected for the Global Genome Initiative for Gardens (GGI-Gardens) II. *Diversity* **2022**, *14*, 234. [CrossRef]
5. Gostel, M.R.; Kress, W.J. The expanding role of DNA barcodes: Indispensable tools for ecology, evolution, and Conservation. *Diversity* **2022**, *14*, 213. [CrossRef]
6. Palmer, C.M.; Wershoven, N.L.; Martinson, S.J.; ter Hofstede, H.M.; Kress, W.J.; Symes, L.B. Patterns of herbivory in neotropical forest katydids as revealed by DNA barcoding of digestive tract contents. *Diversity* **2022**, *14*, 152. [CrossRef] [PubMed]
7. Freeman, P.T.; Ang'ila, R.O.; Kimuyu, D.; Musili, P.M.; Kenfack, D.; Lokeny Etelej, P.; Magid, M.; Gill, B.A.; Kartzinel, T.R. Gradients in the diversity of plants and large herbivores revealed with DNA barcoding in a semi-arid African savanna. *Diversity* **2022**, *14*, 219. [CrossRef]
8. Lowe, A.; Jones, L.; Witter, L.; Creer, S.; de Vere, N. Using DNA metabarcoding to identify floral visitation by pollinators. *Diversity* **2022**, *14*, 236. [CrossRef]
9. Jamdade, R.; Mosa, K.A.; El-Keblawy, A.; Shaer, K.A.; Harthi, E.A.; Al Sallani, M.; Al Jasmi, M.; Gairola, S.; Shabana, H.; Mahmoud, T. DNA barcodes for accurate identification of selected medicinal plants (Caryophyllales): Toward barcoding flowering plants of the United Arab Emirates. *Diversity* **2022**, *14*, 262. [CrossRef]
10. Yessoufou, K.; Estelle Ambani, A. Are introduced alien species more predisposed to invasion in recipient environments if they provide a wider range of services to humans? *Diversity* **2021**, *13*, 553. [CrossRef]
11. Pearl, H.; Ryan, T.; Howard, M.; Shimizu, Y.; Shapcott, A. DNA Barcoding to Enhance Conservation of Sunshine Coast Heathlands. *Diversity* **2022**, *14*, 436. [CrossRef]

Article

DNA Barcoding to Enhance Conservation of Sunshine Coast Heathlands

Hilary Pearl [1,*], Tim Ryan [2], Marion Howard [1], Yoko Shimizu [1] and Alison Shapcott [1]

[1] School of Science, Technology and Engineering, University of the Sunshine Coast, Sippy Downs 4556, Australia; marion.howard@research.usc.edu.au (M.H.); yoko.shimizu@des.qld.gov.au (Y.S.); ashapcot@usc.edu.au (A.S.)

[2] Queensland Herbarium, Mount Coot-Tha Rd, Toowong 4066, Australia; tim.ryan@des.qld.gov.au

[*] Correspondence: h_p051@student.usc.edu.au

Abstract: Conservation priorities and decisions can be informed by understanding diversity patterns and the evolutionary history of ecosystems, and phylogenetic metrics can contribute to this. This project used a range of diversity metrics in concert to examine diversity patterns in the Sunshine Coast heathlands, an ecosystem under intense pressure. The species richness and composition of 80 heathland sites over nine regional ecosystems of heathland on the Sunshine Coast were enhanced with phylogenetic metrics, determined by barcoding 366 heath species of the region. The resulting data were added to an existing phylogeny of regional rainforest species. The diversity metrics for sites and regional ecosystems were compared using univariate and multivariate statistics. The phylogeny from this study, and the low phylogenetic diversity of the heathlands, is consistent with the theory that heath species evolved on the fringes on a wider Australian rainforest flora. Distinctive heathland communities were highlighted, and the existence of geographically scattered, but compositionally similar, phylogenetically even sites points to a possible "refugial environment", characterised by moisture and instability. This suggests contrasting conservation implications: the protection of distinctive communities but also the management of the dynamic processes in other wet and alluvial "refugial environments". The potential for more focused conservation priorities is enhanced.

Keywords: heathlands; phylogenetic diversity; barcoding; phylogeny; conservation; refugia

Citation: Pearl, H.; Ryan, T.; Howard, M.; Shimizu, Y.; Shapcott, A. DNA Barcoding to Enhance Conservation of Sunshine Coast Heathlands. *Diversity* **2022**, *14*, 436. https://doi.org/10.3390/d14060436

Academic Editors: W. John Kress, Morgan Gostel and Michael Wink

Received: 28 March 2022
Accepted: 23 May 2022
Published: 29 May 2022

Publisher's Note: MDPI stays neutral with regard to jurisdictional claims in published maps and institutional affiliations.

1. Introduction

Conservation obligations are an imperative with the Convention on Biological Diversity calling for the 30% of land and sea areas of signatory countries to be protected by 2030 [1]. Overall, Australia has achieved the Aichi target of 17% of land area conserved [2]. However, conservation based on area alone is insufficient to reduce biodiversity loss, with calls for the consideration of factors, including ecological representation, genetic diversity, connectivity, endangered ecosystems, and species [2–6]. We need to know more about diversity patterns in ecosystems we are aiming to protect in order to inform conservation decisions.

Species richness (SR) is a standard measure of diversity, but incorporating metrics, such as genetic diversity, species composition, centres of endemism, and phylogenetic diversity, are advocated for identifying priority areas of biodiversity now and into the future [7–14]. Species composition patterns can identify areas of distinctiveness [15–17]. Phylogenetic diversity has been considered a useful surrogate for a diversity of traits, which provide ongoing material for evolutionary processes [14,18,19]. Phylogenetic diversity metrics have also been used in determining centres of diversity and distinctiveness [20] and in planning for conservation gains [21,22].

In addition, broader diversity measures may provide insight into community assembly processes, contributing to conservation prioritization. Areas of distantly related species, or "phylogenetic evenness", may indicate refugial areas of conservation significance, whilst areas of "phylogenetic clustering" could represent locations where species had been subject to environmental filtering, such as recently expanded communities [23,24]. Deep-past and biogeographical origins may impact on the observed phylogenetic structuring of communities [21,25–27].

However, phylogenetic metrics used in understanding community assembly may be used differently in conservation [14]. For example, low diversity "phylogenetically clustered" areas may suggest compelling and distinctive community assembly patterns, but these areas may then be viewed as unimportant for conservation. However, when considered together, patterns of diversity, including species richness and composition, along with phylogenetic diversity and structuring, could all contribute to understanding local ecological processes and history [28]. Can diversity metrics be used in concert to contribute to an understanding of the floristic history of a region and so inform the prioritization of conservation areas in a region? This study aimed to explore this question in a system under pressure in the south-east corner of Queensland, Australia.

For millennia, the heathlands of the Sunshine Coast have been a source of sustenance for indigenous people [29], as reflected in the many place names and indigenous words, such as "wallum" to describe *Banksia aemula* and associated plant communities [30,31]. The area is under development pressure as one of the fastest growing regions in Australia [32–34]. The heathlands are threatened by land clearing, forestry, urbanisation, inappropriate fire regimes, and climate change [35,36]. In the 1960s it was recognized that the heathland grew on low-nutrient soils, with terminology such as "depauperate" heath and sedge communities used to describe the "poverty" of these lowlands [36]. Indeed, it is theorised the Australian heath evolved in infertile, seasonally waterlogged soils on the fringes of rainforest, with sclerophylly being an adaptation to low nutrients [37–40]. Thus, it is predicted that the heathlands should have a lower phylogenetic diversity than the surrounding rainforest, but is this so? Previous phylogenetic work in the surrounding rainforests offers an opportunity to explore this [15–17,41].

Almost all heathland species in Australia are endemic to this continent and many are localised [35], including in the heaths of the south-east corner of Queensland [36,37,42]. A major centre of endemism and floristic distinctiveness, corresponding to the "wallum" landscapes, has been identified along this coastal fringe north of Brisbane [43]. This is echoed in the occurrences of other significant biota in these heaths, including the eastern ground parrot and the acid frogs [44,45]. These coastal heaths lie on a complex landscape of coastal sand plains and dunes (Holocene and Pleistocene), alluvium, decomposed sandstones, and volcanic basalts on hills and ranges [46,47]. Are the montane areas with heterogenous environments "phylogenetically even", reflecting local genetic studies [48] and the broader ecological findings of increased diversity in topographical heterogenous areas [49–51]?

There remain areas of heathlands that still reflect the pre-European, or "pre-clearing" conditions. In Queensland, vegetation communities are classified into "regional ecosystems" (RE). This framework, instigated in 1999, is used by private and government land managers for biodiversity and conservation assessment and management [52]. On the Sunshine Coast, the variation in the heath communities is reflected in nine heathland regional ecosystems. All have a predominant sclerophyllous shrub layer with a restiad or sedge ground-layer, growing either on wet and waterlogged substrates or on dry sands or rocky peaks [53]. The regional ecosystem classification, along with the beginnings of a barcode library for the local heath flora, provides an opportunity to assess the patterns of diversity and distinctiveness and the conservation significance of the heathland types.

Aims of This Study

With a goal towards exploring conservation priorities, this research plans to quantify the diversity and the distinctiveness of the Sunshine coast heathlands and to specifically address the questions:

1. Is there variation in species richness, phylogenetic diversity, and composition among heath regional ecosystems of the Sunshine Coast, which may be important for assessing conservation priorities?
2. Is there any evidence in the diversity metrics to inform on the ecological and evolutionary history of the heaths on the Sunshine Coast?
3. Do the species composition and phylogenetic metrics provide insights into the community assembly dynamics of the heath; are the regional ecosystems operating as distinct and discreet communities or is there overlap in species composition?
4. Is there evidence of sites of "refugia" or centres of diversity where are these located, and do they warrant consideration of enhanced protection?

2. Materials and Methods

2.1. Design and Data Collection

For this study, the Sunshine Coast heathlands were defined as extending from Cooloola in the north, 150 km south to Bribie Island, and west up to 30 km to Mapleton and the Glasshouse Mountains. These heathlands fall within the council areas of Gympie, Noosa, Sunshine Coast, and Moreton Bay (Figure 1). The current extent and protection status of the heathlands is shown in Supplementary Figure S1.

The Queensland regional ecosystem framework is based on bioregions (broadscale landscape patterns), land zones (based on geology, soils and landforms), and vegetation (described in terms of structure and floristics) and are locally mapped at a scale of 1:50,000 [52]. To capture the diversity across the heath and to ensure the representativeness of heath types, sampling was undertaken of the mapped nine regional ecosystem (RE) heath types [53]. The Queensland Herbarium CORVEG database contained an existing dataset of thirteen sites from five regional ecosystems [54]. This dataset was expanded so that each regional ecosystem was sampled over at least six sites, although the limited extent of RE 12.12.10 meant that only three sites were sampled (Table 1). Sites were chosen which had been unaffected by fire for at least 12 months to reduce this confounding factor. There are also areas along the coast with special significance or were sites of "conservation battles" in the past [34], and some of these "iconic" areas were included to assess their diversity. To address the stratified sampling for other aspects of this project, a further 67 sites were sampled from across the Sunshine Coast study area, resulting in data for a total of 80 sites over nine regional ecosystems.

Site were selected using ArcGis 10.5: [55]: shape files of the Regional Ecosystem mapping of remnant vegetation for south-east Queensland [56] were overlaid with the Protected Area Status [57] to enable selection of sites within the Protected Area Estate. As site data were being added to the CORVEG database, collection methods were consistent with the Queensland Herbarium CORVEG techniques with a standard proforma and plot size of 1000 m^2 [58]. Data collection occurred during late winter, spring, and early summer, from July 2016 to December 2018. At each site, GPS bearings (10 m accuracy) were taken, and a complete native species list was recorded. The comprehensiveness of species lists was maximised by visiting each site twice, with at least one site visit aimed to be during the spring flowering peak [59]. Orchids were noted but not included in species lists, as their cryptic nature meant they were unable to be recorded consistently; likewise, ferns and bryophytes were noted but their inclusion was beyond the scope of this study. Species were identified on site using a field herbarium prepared and developed by comparing samples with Queensland Herbarium samples and by using field handbooks and keys [60,61]. Any species not clearly identified on site were collected for later verification by Queensland Herbarium botanists.

Table 1. Description and diversity metrics of the heathland regional ecosystems on the Sunshine Coast, Queensland.

Regional Ecosystem	Short Description	Biodiversity Status	No. Sites	SR	PD	MPD	MNTD	NRI	NTI
12.2.9	*Banksia aemula* low open woodland on dunes and sand plains, usually deeply leached soils	No concern at present	12	121	3632.73 *L	208.48	32.94	0.63	2.91 *H
12.2.12	Closed heath on seasonally waterlogged sand plains	Of concern	22	151	4258.21 *L	209.17	33.10	0.09	2.24 *H
12.2.13	Open or dry heath on dunes and beaches	Endangered	6	61	2376.66 *L	203.59	45.79	2.41 *C	1.80 *H
12.3.13	Closed heathland on seasonally waterlogged alluvial plains usually near coast	No concern at present	7	89	3004.52 *L	210.71	38.66	−0.76	2.36 *H
12.3.14	*Banksia aemula* low woodland on alluvial plains usually near coast	Of concern	7	95	3153.69 *L	209.56	39.56	−0.16	1.94 *H
12.5.9	Sedgeland to heathland in low lying areas on complex of remnant Tertiary surface and Tertiary sedimentary rocks	Of concern	6	114	3564.05 *L	208.40	39.14	0.60	1.50
12.8.19	Heath and rock pavement with scattered shrubs or open woodland on Cainozoic igneous hills and mountains	Of concern	9	106	3295.21 *L	207.88	36.92	0.88	2.36 *H
12.9-10.22	Closed sedgeland/shrubland on sedimentary rocks, generally coastal	Of concern	8	123	3849.04 *L	208.68	38.48	0.41	1.49
12.12.10	Shrubland of rocky peaks on Mesozoic to Proterozoic igneous rocks	Of concern	3	61	2743.24	205.66	63.85	1.52	−0.88

Regional Ecosystem, the species richness (SR), phylogenetic diversity (PD), mean phylogenetic distance (MPD), mean nearest taxon distance (MNTD), net relatedness index (NRI), and nearest taxon index (NTI) are given. Values that were significantly different from random ($p = 0.05$) are indicated with an asterisk. Superscript letters indicate whether these values are higher (*H) or lower (*L) than expected or whether they are significantly or clustered (*C). No Regional Ecosystems were significantly even. Biodiversity status is that given by Queensland Herbarium assessments in 2018 [53].

For each species, a herbarium voucher and DNA sample in silica consisting of approximately 5 g of healthy, fresh, clean leaf material (permit numbers WITK 17429716 and WITK 18628117) were collected with vouchers lodged at Queensland Herbarium or the University of the Sunshine Coast Herbarium. Taxonomy used was the same as for the Queensland Flora Census [62]. A total of 255 vouched DNA samples was collected.

A complete list of Sunshine Coast heath plants comprising 366 species was compiled for later analysis, which included the species encountered on site, along with other species listed in local guidebooks [60,61,63,64] and confirmed through the Australian Virtual Herbarium [65].

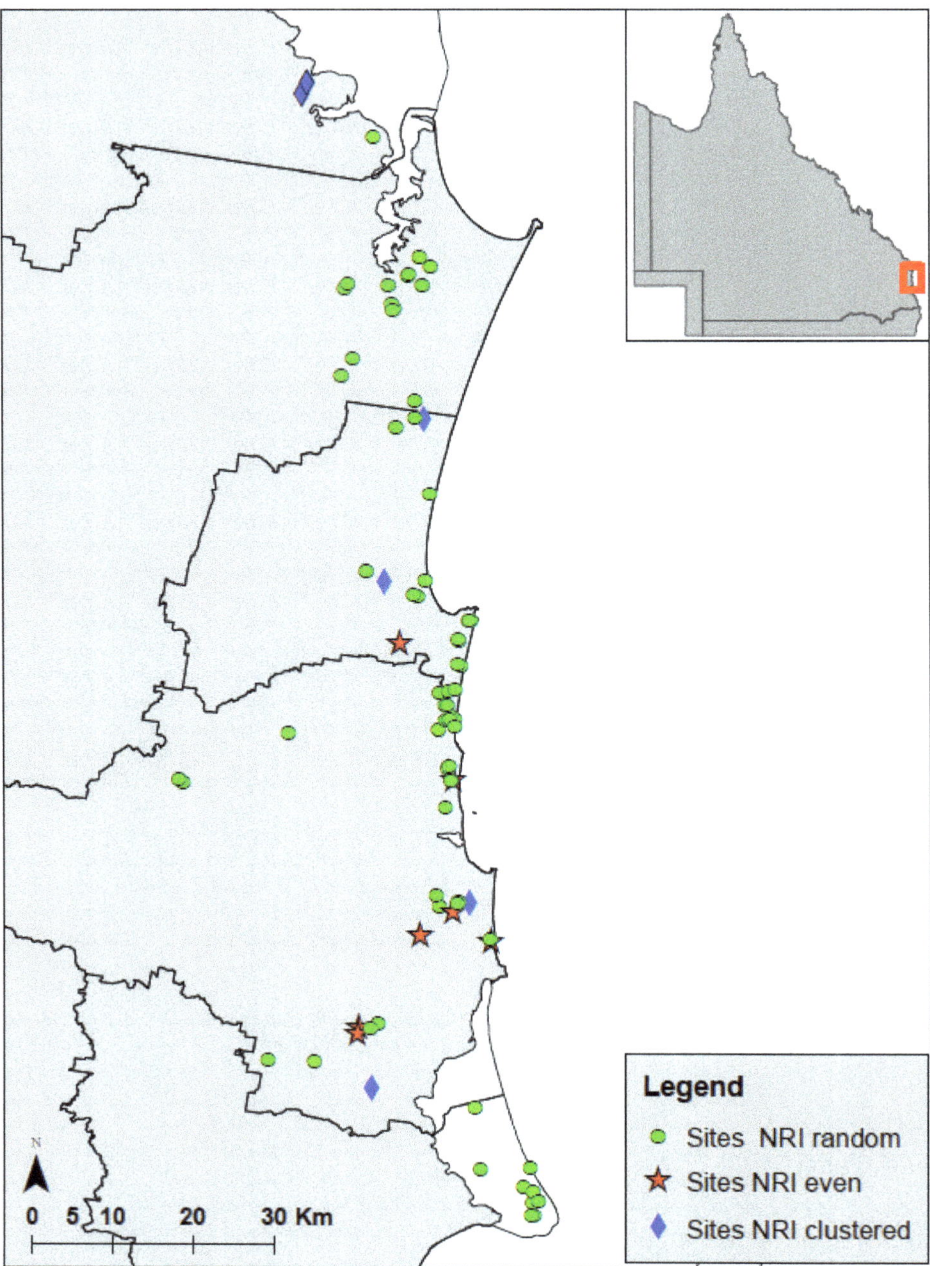

Figure 1. Map of Queensland [66], Australia indicating the Sunshine Coast study area. In the expanded box, the Sunshine Coast area showing the location of the 80 sites. Sites are marked by their NRI results: phylogenetically clustered, random, or even.

2.2. DNA Barcoding and Sequence Alignment

The set of silica dried samples used for this study comprised 255 samples and 130 samples from other collections made from a variety of previous collections undertaken by the Shapcott lab and held at the University the Sunshine Coast [16,17,67]. DNA was extracted from 385 samples following the methods used by Shapcott [17]. The PCR amplification and sequencing of three accepted plastid DNA barcode markers, rbcL, matK, and psbA-trnH, used established methods [68]. The PCR product was purified with exosap, and forward and reverse primers were used along with the Big Dye Terminator v3.1 cycle sequencing kit (ThermosFisher Scientific, Waltham, MA, USA) in a cycle sequence reaction to attach dyes in preparation for sequencing [68]. This was followed by a sephadex purification and rehydration with HiDi formamide to prepare samples for sequencing on an AB3500 Genetic Analyser (Applied Biosystems, Foster City, CA, USA) at the University of the Sunshine Coast. Any unsuccessful samples were reprocessed. Contigs were made using the forward and reverse sequences in Geneious version 10.2.6 (Biomatters, Auckland, New Zealand) (https://www.geneious.com (accessed on 27 March 2022) and were edited for accuracy and checked for quality and length. Contigs were exported to consensus sequences under the following quality control guidelines: a HQ score of a minimum of 65%, a sequence length of a minimum of 300 base pairs, and a minimal number of ambiguous base calls. Alignments of rbcL were completed using MUSCLE and the matK alignment was performed using MAFFT, in Geneious. The psbA-trnH makers were aligned using SATe [69]. All alignments were examined and manually adjusted to correct for homologies. Preliminary Trees were constructed in Geneious 10.2.6 for each marker to check the phylogenetic placement of species and any species that were clearly incorrectly placed on the tree were discarded, either as being contaminated DNA or a misidentification. In rare instances, sequences of less than HQ 65%, or less than 300 base pairs, were kept where they were placed correctly on the phylogenetic tree and there was no alternative sequence to use. Some samples were sequenced again for one or more loci to improve quality. For each plant species, at least two makers were used to construct the "barcode". Missing sequences were retrieved from the public database GENBANK (www.ncbi.nlm.nih.gov/genbank/ (accessed on 27 March 2022)). In the few instances where no markers were procured for the species, a congener was used.

To further improve the robustness of the phylogeny, the data for the 366 heath species were aligned with an existing dataset of south-east and central Queensland rainforest species using the same three gene markers [16]. The final alignments for rbcL, matK, and psbA-trnH were trimmed and concatenated to create a three gene alignment for the heath and rainforest species of south-east Queensland, resulting in a dataset for 1576 species.

2.3. Phylogenetic Reconstruction

To enforce taxonomic relationships based on the global dataset from the Angiosperm Phylogeny Group III [70], a constraint tree using the R20120829 tree was built in Phylomatic version 3 [71]. In Mesquite [72], the tree was edited so each family was a polytomy and that the barcode data could then be used to infer the relationships of the species below the level of family, consistent with the methods of Shapcott [17].

The 3-marker alignment for the 1576 heath and rainforest species and constraint tree were uploaded to the CIPRES portal [73], and the RAxML-HPC2onXSEDE tool was used with a mixed partition model to search for the best scoring ML tree. This was repeated for eight runs and the best tree with the smallest likelihood score was selected. This tree was rooted in Geneious and dated in PATHd8 [74], with the ages of nine orders found using fossil dates and an age for angiosperms of 250 mya after the methods of Shapcott [16,17] to produce a final dated tree with calibrated molecular branch lengths for the 1576 heath and rainforest species. This tree was pruned in PICANTE in R to produce a dated tree for the 366 regional heath species of the Sunshine Coast [75].

2.4. Diversity Measures and Analyses

Summary data of the numbers of species (SR), genera (GR), and families (FR) were generated for the Sunshine Coast heath community of 366 species. Each of the species from the Sunshine Coast heath taxa was represented by a barcode identifier displayed on the dated heath phylogeny and these were used to create community lists of species found in each regional ecosystem, based on the field work data. Species were assigned to each of the nine regional ecosystems on a presence or absence basis, to be able to make broad comparisons of the heath regional ecosystems in terms of diversity measures. Additionally, community lists were developed for each of the 80 sites on a presence or absence basis to enable diversity measures to be generated for each site, to be used to statistically investigate variation between sites grouped by Regional Ecosystems.

The dated phylogeny, the complete Sunshine Coast heath community file, the community lists for individual sites and the individual regional ecosystems were used to derive phylogenetic metrics, and all analyses used R software [76]. Phylogenetic diversity (PD), mean phylogenetic distance (MPD), and mean nearest taxon (MNTD) were calculated for each regional ecosystem, as well as for each site using PICANTE [24,75]. A randomised null model, using the whole Sunshine Coast heath taxa and shuffling the taxa labels across the tips of the phylogeny, was used to calculate the probability of the phylogenetic diversity measures deviating significantly from random distributions. PICANTE calculates a standardised size effect (ses) and this figure multiplied by -1 gives the net relatedness ness index (NRI) for MPD and a nearest taxon index (NTI) for MNTD. A NRI has a value of 0 for a completely random community, increases as the community becomes more clustered, and decreases as a community becomes more even, with the NTI following a similar pattern [23]. The NRI and NTI were tested for significance using a randomised null model in PICANTE [75]. All these diversity measures were obtained for each of the nine regional ecosystems and for each of the 80 sites.

Individual site diversity measures were used to test for differences in PD, SR, FR, GR, MNTD, MPD, NRI, and NTI as well as between structural data (maximum and minimum heights and percentage cover of vegetation layers) between sites grouped by regional ecosystems, using the Kruskal–Wallis test followed by a Dunn's post hoc test with a Bonferroni correction in the "stats" package and PMCNR packages [77]. Relationships between SR, GR, FR, and PD were tested using Spearman's rank correlation tests in the "stats" package [76]. The significance of the observed frequency of phylogenetically even and phylogenetically clustered sites in each regional ecosystem was tested using Pearson's chi-squared test in the gmodels package [78].

To investigate patterns and similarities among site communities, the presence/absence matrices of species composition were used to calculate pairwise dissimilarity matrices using Vegdist and the Bray-Curtis method in the Vegan package [79]. A dissimilarity matrix between sites was calculated based on PD using Unifrac, a measure of phylogenetic distance between sites, within PICANTE [80,81].

These distance matrices were used in non-metric multidimensional scaling (NMDS) to visualise relationships among sites using Vegan [79]. Northings and eastings data for each site were used to calculate geographic distance matrices using Vegdist and the euclidian method in Vegan. All the dissimilarity matrices were tested for correlation using the Mantel test and the Spearman method in Vegan [82].

To visualise and assist the interpretation of the regional ecosystem data, labelled phylogenetic trees were produced using the iTOL program [83].

3. Results

3.1. Phylogenetic Position

The Sunshine Coast heath taxa list comprised 366 species, excluding ferns and orchids, covering 26 orders, 73 families, and 201 genera. Ten of these families contained 10 or more species, and included Myrtaceae (42 species), Cyperaceae (34 species), Poaceae (29 species), Fabaceae (27 species), Ericacae (22 species), Mimosaceae and Proteaceae (19 species each),

Restionaceae and Rutaceae (12 species each), and Laxmanniaceae (10 species). Over the 80 sites of data collection, 280 species were encountered, which represented 76.6% of the Sunshine coast heath taxa of 366 species. These 280 species came from 157 genera and 56 families, with some of the families more frequently encountered on sites than others; Myrtaceae, Cyperaceae, Proteaceae, Xanthorrhoeaceae, and Fabaceae were each encountered on more than 70 of the 80 sites.

Whilst the south-east and central Queensland rainforest and heath species share some orders, the heath species are not represented by any unique orders. The heath taxa of 366 species are grouped within the larger south-east Queensland and central Queensland rainforest and heath phylogeny (Figure 2). Some orders, such as the Magnoliales, Pandanales, and Piperales, are not represented within the heath, many orders were poorly represented, including the Laurales, Solanales, and the Sapindales, whilst orders such as the Poales (including the families Restionaceae and Cyperaceae) and the Ericales are richly represented.

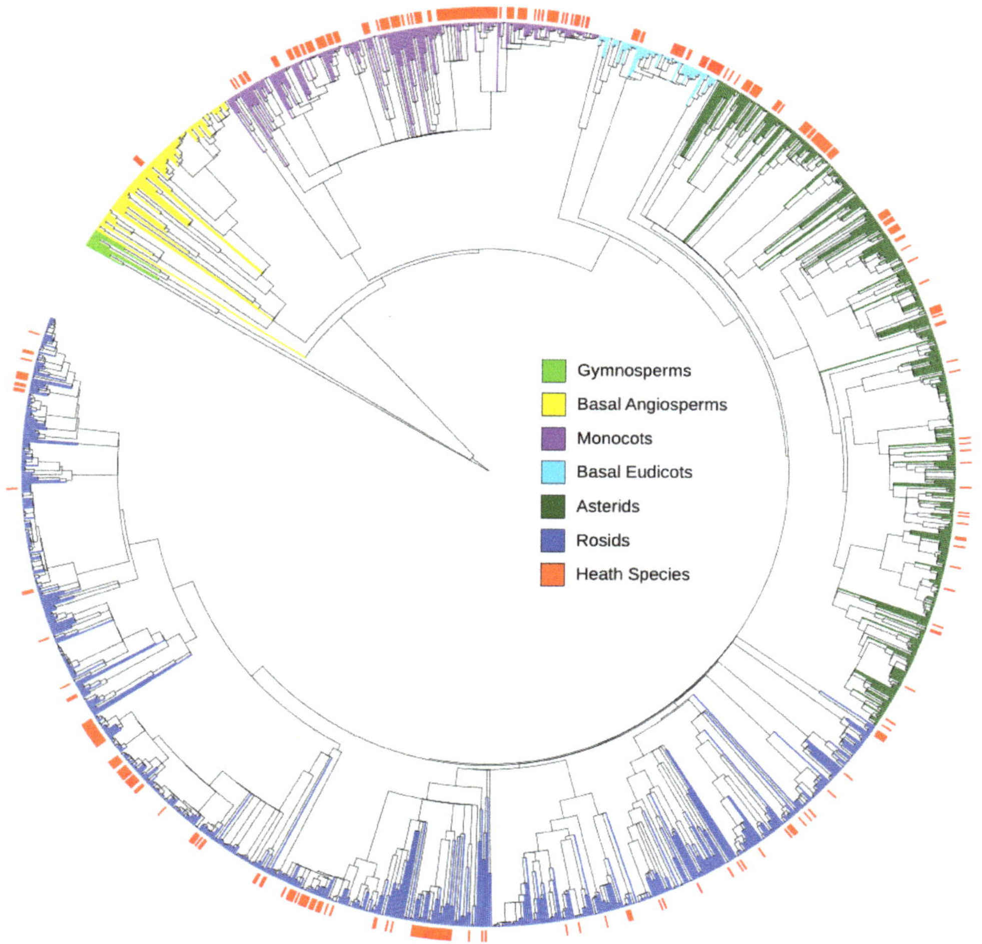

Figure 2. Phylogenetic position of the Sunshine Coast heath species (indicated in red) within the wider south-east Queensland and central Queensland rainforest phylogeny.

3.2. Species Richness and Phylogenetic Diversity Metrics

While the total combined species richness of the regional ecosystems ranged from 61 for RE 12.12.10 (3 sites) to 151 species for RE 12.2.12 (22 sites in total), reflecting in part the differential sampling across regional ecosystems, the mean species richness per site for each regional ecosystem varied from 28.7 to 43.3 species/1000 m^2 and was not significantly different (Table 2). The phylogenetic diversity of the individual nine regional ecosystems ranged from 2376 to 4258, with a total PD for the Sunshine Coast heath taxa of 8156 (Table 1). All regional ecosystems had a lower PD than expected in the context of the Sunshine Coast heath regional species pool ($p < 0.05$), except for the shrublands of rocky peaks, RE 12.12.10 (Table 1). The endangered dry open heath on dunes, RE 12.2.13 was the only heath type identified as "phylogenetically clustered" with a significant NRI ($p < 0.05$), with all other regional ecosystems identified as "phylogenetically random" (Table 1, Figure 3). Six of the nine regional ecosystems had a mean nearest taxon index (NTI) higher than expected by chance ($p < 0.05$), meaning that the species in these communities are more closely related at the terminal nodes (Table 1). Furthermore, when sites were grouped by regional ecosystems and compared, the NRI was found to be significantly higher (more clustered) in the heaths of the dunes compared to the alluvial closed heath: the NRI in RE 12.2.9 and in RE 12.2.13 were higher compared with RE 12.3.13 (Kruskal Wallis chi-squared = 20.912, $p = 0.0074$) (Table 2). There were no other significant differences in PD, MPD, MNTD and NTI, SR, GR, or FR between sites grouped by regional ecosystem (Table 2).

Table 2. Summary of mean diversity values of the Sunshine Coast heath sites grouped by regional ecosystem.

RE	No. Site	SR	GR	FR	PD	MPD	MNTD	NRI	NTI
12.12.10	3	30.3 (4.2)	27.0 (3.7)	18.3 (3.1)	1616.9 (126.7)	206.8 (0.7)	72.3 (7.5)	0.61 (0.20) ab	0.48 (0.51)
12.2.12	22	34.5 (12.1)	30.7 (10.0)	17.5 (5.0)	1754.6 (402.0)	211.1 (3.8)	68.7 (14.5)	−0.45 (0.94) ab	0.67 (0.64)
12.2.13	6	36.3 (2.7)	32.0 (1.4)	17.8 (1.0)	1771.3 (71.7)	206.0 (2.2)	56.1 (6.5)	0.87 (0.60) a	1.48 (0.52)
12.2.9	12	30.1 (11.2)	26.1 (9.2)	15.5 (4.3)	1570.0 (390.6)	205.5 (6.9)	70.8 (25.9)	0.92 (1.29) a	0.99 (0.77)
12.3.13	7	31.0 (7.9)	27.7 (6.7)	17.0 (3.7)	1652.2 (344.6)	213.5 (2.6)	66.2 (9.3)	−1.08 (0.74) b	0.90 (0.89)
12.3.14	7	28.7 (4.3	26.4 (4.0)	14.9 (2.9)	1464.3 (251.3)	208.3 (8.4)	60.4 (15.5)	0.09 (1.79) ab	1.47 (1.15)
12.5.9	6	43.3 (7.0)	37.3 (5.8)	21.2 (3.8)	1923.3 (302.0)	208.2 (5.2)	55.2 (5.8)	0.41 (1.62) ab	1.28 (0.79)
12.8.19	9	31.2 (16.5)	27.9 (14.2)	16.7 (8.7)	1564.9 (756.5)	206.6 (9.6)	61.9 (9.3)	−0.09 (1.34) ab	1.24 (1.10)
12.9-10.22	8	42.5 (12.3)	37.1 (11.2)	21.0 (5.9)	1991.1 (514.2)	208.9 (5.1)	62.9 (10.2)	−0.12 (1.05) ab	0.48 (1.16)
		NS	NS	NS	NS	NS	NS	KW = 20.9 $p = 0.007$	NS

For each regional ecosystem (RE), the species richness (SR), genus richness (GR), family richness (FR), phylogenetic diversity (PD), mean phylogenetic distance (MPD), mean nearest taxon distance (MNTD), net relatedness index (NRI), and nearest taxon index (NTI) are given. Values that were significantly different in Kruskal–Wallis tests ($p < 0.05$) are indicated with letters. Means sharing the same letter are not significantly different. Please note, the higher the NRI or NTI, the more closely the taxa in the community are related. Standard deviations are shown in brackets.

Across the 80 sites, diversity, as measured by species richness, varied widely, from 6–58 species per 1000 m^2, whilst genus richness ranged from 6–51 genera per 1000 m^2, suggesting few instances of congenic species in each site (Table 3). Indeed, species richness was positively correlated with genus richness (Spearman rho = 0.986, $p < 0.001$) and family richness (Spearman rho = 0.9256, $p < 0.001$) as well as PD (Spearman rho = 0.994, $p < 0.001$). Moreover, 19 of the individual sites had a lower PD than expected by chance: across the individual 80 sites, PD ranged from 443 to 2585, with no sites having a higher PD than expected ($p < 0.05$) (Table 3). The NTI showed that 17 sites contained species significantly more clustered on the terminal nodes than expected by chance ($p < 0.05$) (Table 3). Based on the NRI, seven of the 80 sites were "phylogenetically even" ($p < 0.05$), and these sites were found in RE 12.2.12, 12.3.13, and 12.3.14, ecosystems of alluvial or seasonally waterlogged areas (Table 3). Six sites were "phylogenetically clustered" ($p < 0.05$), and both these "clustered" and "even" sites were scattered across the Sunshine

Coast (Figure 1). Phylogenetically clustered sites were found in regional ecosystems 12.2.9, 12.3.14, 12.8.19, and 12.5.9, with the Banksia aemula heath of the dunes (12.2.9) found to have more phylogenetically clustered sites than expected by chance, and the closed heathland on alluvial plains (12.3.13) found to contain more phylogenetically even sites than expected ($\chi^2 = 28.76$, $p = 0.026$).

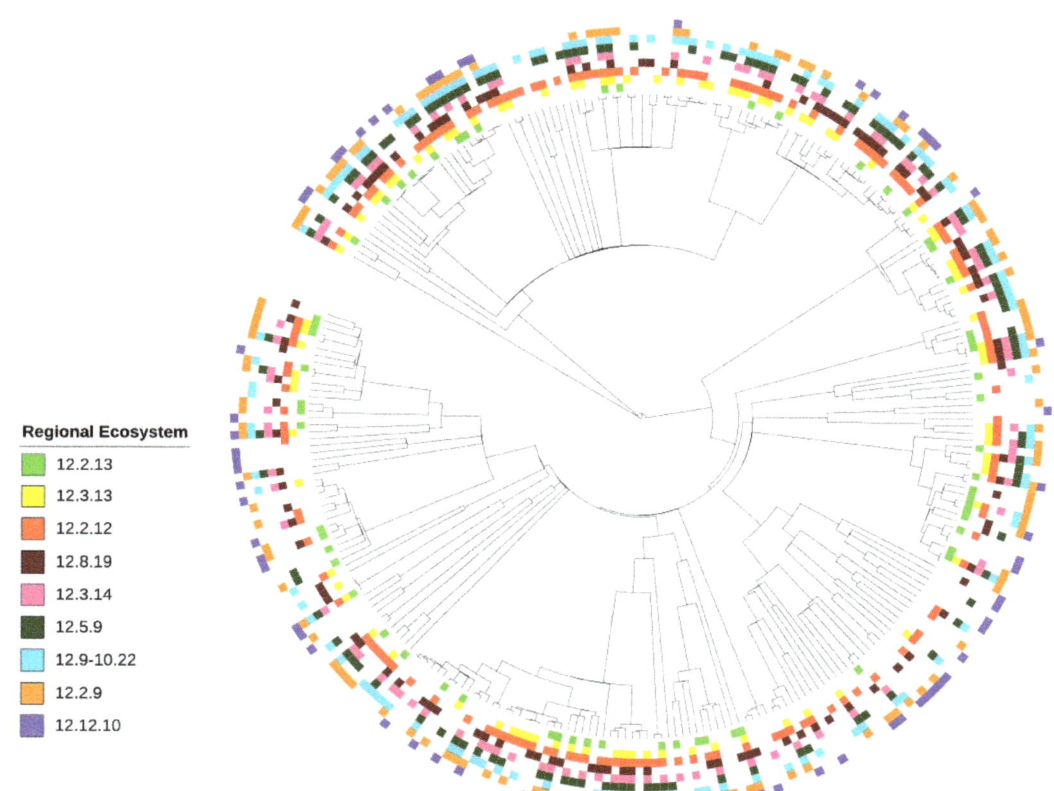

Figure 3. Phylogenetic position of the Regional Ecosystems represented against the phylogenetic tree of the Sunshine Coast heath site taxa.

Table 3. Phylogenetic metrics of each of the 80 study sites, grouped by regional ecosystem.

Regional Ecosystem	Site	SR	GR	FR	PD	MPD	MNTD	NRI	NTI
RE 12.2.9 **Banksia aemula, low open woodland on dunes and sand plains, usually deeply leached soils**	15248	34	32	19	1866.9	212.0	69.1	−0.72	0.41
	15250	9	9	8	824.5	214.5	143.7	−0.41	−0.7
	15621	20	17	9	1061.0 *L	189.5	60.9	3.56 *C	1.90 *H
	16488	28	25	16	1687.7	207.2	76.7	0.42	0.27
	16491	25	22	13	1433.6	211.8	75.0	−0.49	0.65
	BI29-63	37	33	18	1774.9	208.9	58.9	0.13	1.12
	BI29-65	16	15	11	1122.2	197.5	91.2	1.8	0.51

Table 3. *Cont.*

Regional Ecosystem	Site	SR	GR	FR	PD	MPD	MNTD	NRI	NTI
	COO29-44	40	34	19	1817.0 *[L]	200.9	52.7	2.54 *[C]	1.68
	COO29-62	47	40	20	2081.9	205.9	49.9	1.11	1.69 *[H]
	LCOO29-43	40	34	19	1881.3	204.5	56.3	1.46	1.25
	MCNP29-60	28	23	15	1458.6	207.4	61.0	0.5	1.49
	NNP29-61	37	29	19	1830.7	205.7	54.2	1.13	1.63 *[H]
RE 12.2.12 Closed heath on seasonally waterlogged sand plains	15228	28	26	12	1449.0	205.7	66.9	0.8	0.97
	15252	29	27	19	1827.5	212.9	80.4	−0.87	−0.14
	16443	17	16	11	1262.6	213.2	101.2	−0.59	−0.15
	16450	23	21	12	1264.4	205.6	65.0	0.74	1.49
	16493	24	22	14	1514.1	216.2	91.2	−1.3	−0.46
	BI212-50	38	34	16	1778.2 *[L]	204.5	58.4	1.38	1.15
	BI212-51	28	24	17	1648.7	215.5	76.3	−1.4	0.32
	BI212-53	45	37	21	1878.1 *[L]	207.7	51.3	0.57	1.61
	COO212-24	44	39	21	2052.1	209.1	64.5	0.12	0.17
	COO212-45	22	19	12	1341.1	214.1	78.7	−0.9	0.61
	COO212-46	57	47	25	2585.6	213.0	57.2	−1.43	0.29
	ES212-10	17	17	11	1160.5	215.8	94.5	−0.94	0.26
	ES212-5	39	35	21	2045.6	212.1	70.7	−0.79	−0.12
	ES59-40	23	22	13	1467.9	213.2	83.9	−0.71	0.22
	KMCP212-31	48	41	21	2053.3 *[L]	209.7	51.7	−0.13	1.45
	KMCP212-7	58	51	28	2528.4	213.5	51.6	−1.66 *[E]	1.04
	MCNP212-39	41	39	24	2243.9	217.2	65.5	−2.38 *[E]	0.2
	MCNP212-42	27	23	12	1359.7 *[L]	208.3	59.4	0.18	1.73 *[H]
	ME212-2	27	24	16	1445.6	205.1	70.4	0.85	0.84
	NNS212-33	41	38	19	1930.6	212.0	59.2	−0.8	0.9
	PEP212-27	36	33	18	1793.7	209.6	61.1	−0.11	0.99
	PEP212-49	47	40	22	1969.6 *[L]	211.0	52.7	−0.57	1.3
RE 12.2.13 Open or dry heath on dunes and beaches	MHD213-3	32	30	17	1761.5	208.8	64.0	0.09	0.98
	MHD213-35	38	34	19	1846.3	208.8	54.6	0.14	1.51
	NNP213-13	38	32	17	1686.3 *[L]	205.4	45.9	1.04	2.35 *[H]
	NNP213-14	38	33	18	1764.1	204.1	52.6	1.43	1.78 *[H]
	NNP213-32	34	32	17	1706.3	203.8	60.9	1.34	1.15
	NNP213-47	38	31	19	1863.2	205.0	58.6	1.19	1.1
12.3.13 Closed heathland on seasonally waterlogged alluvial plains, usually near coast	16454	22	22	14	1316.3	211.3	69.1	−0.38	1.32
	BSA313-4	29	26	17	1588.2	214.9	73.3	−1.25	0.43
	COO313-17	25	21	13	1277.7 *[L]	208.6	53.8	0.19	2.22 *[H]
	MRNP313-16	35	33	21	1939.0	215.3	74.0	−1.63 *[E]	−0.14
	MRNP313-59	26	22	13	1380.0	213.8	66.4	−1.03	1.17
	PV313-37	35	32	21	1959.8	215.3	74.1	−1.60 *[E]	−0.15
	TNP313-36	45	38	20	2104.3	215.2	52.8	−1.85 *[E]	1.49
12.3.14 Banksia aemula, low woodland on alluvial plains, usually near coast	15622	24	22	12	1405.8	196.6	81.7	2.65 *[C]	0.21
	BSA314-38	28	28	17	1610.3	215.9	76.0	−1.53 *[E]	0.27
	BSA314-6	31	29	17	1702.5	215.6	66.0	−1.54 *[E]	0.9
	BSA314-8	28	24	14	1385.5 *[L]	215.2	48.0	−1.37	2.38 *[H]
	MRNP314-15	23	21	10	962.8 *[L]	197.3	38.0	2.27 *[C]	3.34 *[H]
	MRNP314-57	35	31	17	1530.9 *[L]	206.7	52.5	0.63	1.89 *[H]
	MRNP314-58	32	30	17	1652.0	211.1	60.9	−0.45	1.28
12.5.9 Sedgeland to heathland in low lying areas on complex of remnant Tertiary surface and Tertiary sedimentary rocks	COO59-19	46	40	23	2094.7	209.3	57.7	0.05	0.89
	COO59-25	39	36	20	1891.7	212.6	59.9	−0.97	1.01
	COO59-26	32	28	16	1499.8 *[L]	212.1	54.4	−0.69	1.89 *[H]
	COO59-56	51	45	26	2264.9	208.7	59.0	0.19	0.37
	LC59-37	49	40	24	2147.3	208.1	55.7	0.34	0.96
	LC59-48	43	36	18	1641.3 *[L]	198.3	44.2	3.55 *[C]	2.54 *[H]

Table 3. *Cont.*

Regional Ecosystem	Site	SR	GR	FR	PD	MPD	MNTD	NRI	NTI
12.8.19 **Heath and rock pavement** **with scattered shrubs or** **open woodland on Cainozoic** **igneous hills and mountains**	13962	6	6	4	443.7 *[L]	203.2	66.1	0.34	2.46 *[H]
	MB819-12	12	11	7	721.5 *[L]	197.7	74.0	1.22	1.73 *[H]
	MCNP819-40	44	38	23	2234.2	213.6	68.9	−1.35	−0.32
	MCNP819-41	38	33	20	1916.6	213.4	62.2	−1.14	0.73
	ME212-29	45	41	24	2121.7	211.6	58.4	−0.76	0.84
	ME819-1	34	31	21	1824.7	213.4	72.1	−1.07	0.13
	ME819-11	42	37	21	1961.2	212.5	52.4	−1	1.66
	ME819-30	48	42	25	2286.6	208.2	56.6	0.35	0.84
	WHM819-9	12	12	5	573.8 *[L]	185.4	46.4	2.60 *[C]	3.11 *[H]
12.9-10.22 **Closed sedgeland/shrubland** **on sedimentary rocks,** **generally coastal**	COO910-20	49	45	20	1975.0 *[L]	210.7	46.9	−0.48	1.97 *[H]
	COO910-54	45	40	23	2087.8	205.2	64.8	1.32	0.03
	COO910-55	50	44	26	2458.4	211.8	69.3	−0.94	−0.88
	COO910-70	44	41	22	2025.4	210.1	60.0	−0.3	0.67
	ES910-33	53	41	27	2402.4	210.2	60.5	−0.4	0.08
	ES910-71	50	44	23	2255.8	210.9	59.8	−0.59	0.37
	NNS910-34	16	12	8	818.2 *[L]	198.1	59.6	1.67	2.34 *[H]
	NNS910-72	33	30	19	1906.3	214.5	82.6	−1.29	−0.74
12.12.10 **Shrubland of rocky peaks on** **Mesozoic to Proterozoic** **igneous rocks**	MNP1210-30	27	25	15	1483.3	206.2	69.9	0.69	0.91
	MNP1210-66	29	25	19	1632.1	207.7	80.7	0.38	−0.08
	SP1210-31	35	30	21	1735.4	206.6	66.3	0.76	0.61

For each site, the species richness (SR), genus richness (GR), family richness (FR), phylogenetic diversity (PD), mean phylogenetic distance (MPD), mean nearest taxon distance (MNTD), net relatedness index (NRI), and nearest taxon index (NTI) are given. Values that were significantly different from random ($p = 0.05$) are indicated with an asterisk. Letters indicate whether these values are higher (*[H]) or lower (*[L]) than expected or whether they are significantly even (*[E]) or clustered (*[C]).

3.3. Species and Phylogenetic Composition

Sunshine Coast heath sites varied in their species composition with some groupings based on regional ecosystems apparent in the NMDS analysis (Figure 4). The heaths of the old volcanic, Mesozoic to Proterozoic rocky peaks (RE 12.12.10) displayed similarities in species composition between sites. In contrast, the heaths of the rocky pavements of Cainozoic igneous rocks, RE 12.8.19, were divided in two, with the Glasshouse Mountains clustering as a group and the sites further north on Mt Coolum and Mt Emu forming a separate grouping (Figure 4). There were strong similarities in species composition between the heaths of the dry dunes, RE 12.2.13, and RE12.2.9, whilst the remainder of the heath types, which are heathlands of alluvial or seasonally waterlogged areas, showed considerable overlap in species composition (Figure 4). Phylogenetic composition showed a similar patterning to the vegetation composition, but it is striking that the heath of the dry dunes, RE 12.2.13 and 12.2.9, appears more phylogenetically similar to the heaths of the dry rocky peaks, 12.12.10 and 12.8.19, than do the other heath types (Figure 4). Mantel tests indicated a significant strong and positive correlation ($r = 0.8563$, $p < 0.001$) between species composition and phylogenetic distance, whereas geographic distance between sites was very weakly correlated with phylogenetic distance ($r = 0.1319$, $p = 0.006$) and species composition ($r = 0.1284$, $p = 0.006$).

Whilst the "even" and "clustered" sites were scattered across the region (Figure 1), the even sites showed distinct similarities in terms of species composition and phylogenetic distance (Figure 4).

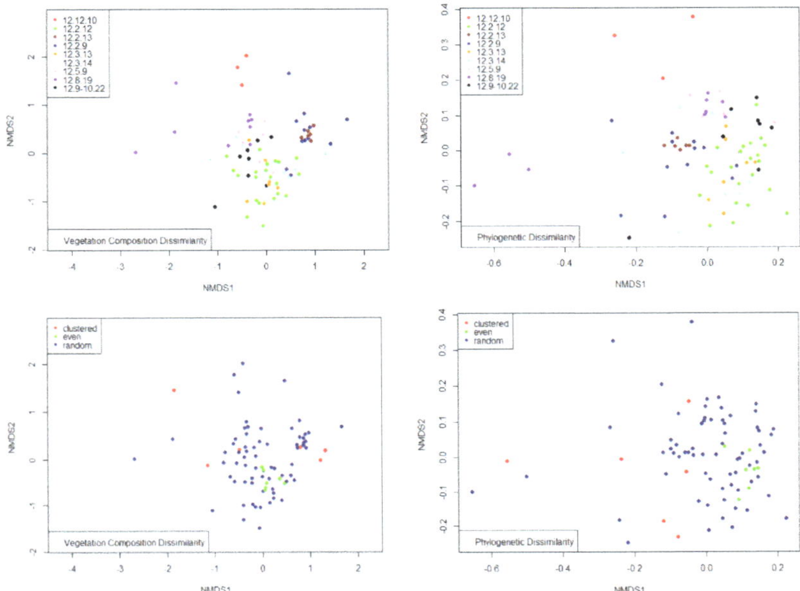

Figure 4. Comparison of non-metric multidimensional scaling (NMDS) analyses for comparing relationships among 80 Sunshine Coast heathland sites. Vegetation composition using Bray–Curtis dissimilarity is shown on the (**left**) and phylogenetic Unifrac dissimilarity is on the (**right**). Sites have been coloured by their regional ecosystems (**above**) and by their NRI significance (**below**).

4. Discussion

4.1. Value of a Range of Diversity Metrics

Phylogenetic and species composition metrics from this study have provided deeper insight into the diversity and composition of the Sunshine Coast heathlands than species richness alone. This has resulted in a more nuanced characterization of community distinctiveness and ongoing ecological processes [84]. In this study, no regional ecosystem on the Sunshine Coast was significantly different in terms of species richness, family richness, genus richness, or phylogenetic diversity. It was expected from earlier studies in heath over south-eastern Australia that the wetter alluvial regional ecosystems of heath would be significantly lower in species richness than the drier heath regional ecosystems [37,85,86], raising the contention that these wetter systems are not a priority for conservation. However, in some coastal heath regions of New South Wales, no species richness differences were found between wet and dry heaths but, as was found in this study, compositional differences were apparent [87]. So, what do the compositional and phylogenetic data from this study suggest?

4.2. Data Consistent with Theory of Evolutionary History of Heath Flora

The phylogenetic tree from this study, and the lack of unique orders within the heath, is consistent with the theory that sclerophyll and heath vegetation evolved in the low nutrient and/or waterlogged areas within the extensive Australian Mesozoic rainforest [37]. Phylogenetic data in other studies, as well as the fossil record, have provided support for rainforest being the ancestral state of Australian flora, with the sclerophyllous component becoming more common after the Oligocene, with increasing climatic variability, seasonality, and aridity [38,88,89]. This resulted in the radiation of sclerophyll flora, including the Myrtaceae, Proteaceace, Ericaceae, and Casuarinaceae [90,91], all typical components of the heath flora. The overall phylogenetic diversity of the heath is considerably lower than found in studies of corresponding rainforest areas [16,17,41]. All the heath ecosystems, except one, had a low phylogenetic diversity, yet they also contained ancient lineages such as Proteaceae and Restionaceae. These data support the possible impact of deep-past processes on the phylogenetic patterns seen in the heath; it may not only be a result of present-day assembly processes, such as environmental filtering [26]. The NTI for each regional ecosystem showed a stronger (higher) clustering than the NRI, which has been suggested as indicating more recent diversification events [92], further supporting recent speciation in the heath. This has implications in terms of conservation, as maximizing phylogenetic diversity alone, without historical interpretation, can have negative outcomes [27].

4.3. Phylogenetic Clustering

The Sunshine Cast heath data found phylogenetic clustering (a high NRI) on one of the nine regional ecosystems, in the dry, high, wind-buffeted, leached sand dunes, and along with a high NTI (clustered), which provides evidence for single clusters of taxa on the species pools, as opposed to several clusters evenly distributed around the tree [93]. This clustering, also found in global dune plant communities, leads to the conclusion that environmental filtering is at play [92]. This is supported by this regional ecosystem (RE 12.2.13) having a distinct community composition but overlapping in composition with a structurally taller system (RE 12.2.9), both growing in highly nutrient leached sands [94]. No other heath regional ecosystems were found to have a clustered NRI, but six had a clustered NTI, suggesting several clusters evenly distributed around the tree [93]. Phylogenetic clustering, due to the presence of closely related species, could indicate environmental filtering [23] and may be disadvantageous in terms of species competition. Yet, in long term heath studies in New South Wales, phylogenetic clustering became more pronounced over time since fire, in contrast to the expectation that increased competition would inhibit the coexistence of species with high niche overlap [95]. Furthermore, it is argued that closely related species may not necessarily compete more closely than distantly related ones, that filtering may not be the dominant mechanism, and that facultative interactions may play a part in this [14,96,97]. The co-occurrence of closely related species may be beneficial, for example, when facilitated by mutualistic exchange, such as through mycorrhizal symbiosis shared by co-occurring members of a clade, increasing the host's tolerance to environmental change [98]. Mycorrhizal strategies have been suggested as drivers of heath assemblages in Western Australia and Brazil, with well-known strategies used by the distinctive heath families Ericaceae, Orchidaceae, and Myrtaceae [99,100]. For the Sunshine Coast heaths, whose phylogeny is consistent with the general theory of sclerophyll and heath derivation, there are indicators that environmental filtering is a driver of their assembly but hints that facilitation and mutualism may be factors contributing to their community assembly and their general pattern of phylogenetic clustering.

4.4. Community Assembly

The diversity and composition patterns of the Sunshine Coast heaths are also intriguing against the background of contrasting theories of communities moving towards a climax assemblage [101], as opposed to random, temporary, and fluctuating assemblages of individual species responding to stochastic events [102]. Furthermore, this is a rejection of these two extreme views: ecosystem constraints and individual variability constantly interact, with the relative importance of each unsolved [103]. In the Sunshine Coast heaths this study found some distinct compositional communities as in the Glasshouse Mountains (RE 12.8.19), the montane heaths of the old volcanics (RE 12.12.10), and the dry sands (RE 12.2.9 and RE 12.2.13), with each of these communities also having their own unique species. However, this study also found the overlapping composition of the other five ecosystems, all found on the wetter sands, sediments, and alluviums. In this context, the observations of an early ecologist ring true—that the "wallum" vegetation of the coastal lowlands are a dynamic ecosystem with no climax and are subject to continual disturbance in the form of fire, wind, cyclones, periodic drought, storms, and flooding [36]. The vegetation was described as a continuum, as an oscillation of ecotones between floristic reservoirs, and consisting of flora of predominantly wide potential environment, with an infusion of species with much narrower environmental requirements [36]. Likewise, heaths of the Sydney sandstones have been described as dynamic "mosaics" over time [104] and that for Australian heathlands in general it is difficult to designate any one species as dominant, with proportions changing with microhabitat [105]. This mix of overlapping regional ecosystems in the wet and alluvial areas appears to be a dynamic system, where there is movement of species within the constantly changing environment, subject to extremes of moisture, fire, wind, and drainage. In terms of long-term conservation, it suggests the facilitation of this movement may be essential, and connectivity between these protected areas needs to be considered.

4.5. A Refugial Environment?

The phylogenetically even sites were all located in the wet and alluvial ecosystems, and despite being scattered across the coast, were similar in terms of species composition and phylogenetic distance. Conservation priorities have focused on refugial areas, which have facilitated survival of biota for millennia and are likely to do so into the future [106–109]. Phylogenetically even sites may be indicative of refugial areas [17,110], although it is essential to be mindful of historical evolutionary processes [12,27]. Moisture has been a factor that contributes to sites acting as refugia, with dated core samples from the wetlands of nearby Stradbroke Island suggesting the wetlands have acted as refugia from regional drying for over the last 100,000 years [111]. Waterways, mesic habitats, and riparian areas have been linked to relictual taxa [49,89,112–114]. Refugia act as buffers to extreme conditions, and it is important to understand the evolutionary history of sites and the processes being protected by them. It could be argued that these wet and alluvial heath areas, with overlapping composition, are dynamic "refugial environments" and conserving these areas is protecting these ongoing processes of change. It is possible that they are not so much fixed in space as reflecting the concept of "shifting refuges", driven partly by stochastic events [115].

5. Conclusions

The Sunshine Coast heathlands are ecosystems under pressure from development, and this study aimed to use diversity measures, in concert, to inform conservation priorities. The phylogeny from this study is consistent with the theory that heath evolved on the fringes of a wider Australian rainforest flora, with the phylogenetic diversity being low in comparison with the surrounding rainforest flora. Whilst the heath regional ecosystems on the Sunshine Coast could not be discriminated by SR, GR, FR, or PD, composition and NRI highlighted the distinctive heath communities of the coastal high dunes and the montane areas. The wet and alluvial heaths of the coastal lowlands displayed an overlapping composition but also contained the phylogenetically even sites, possibly pointing to a "refugial environment" characterised by stochastic events and instability. This suggests contrasting conservation implications: the conservation of distinctive communities but also conservation of the dynamic processes in the wet and alluvial "refugial environments". With this enhanced understanding, further examination of the remnant extent and protection status of each regional ecosystem of heath is warranted.

Supplementary Materials: The following are available online at https://www.mdpi.com/article/10.3390/d14060436/s1. Figure S1: Map of the study area showing the areas of heath that have been cleared (in black), the areas of heath currently protected in National Parks (in green), and the remaining areas of heath (in red) tenured privately, or in lower levels of protection. The large area of National Park at the top of the map is the Cooloola section of the Great Sandy National Park. The major population centers are in the mid to southern coastal sections of the map.

Author Contributions: Conceptualization, H.P. and A.S.; methodology, A.S., T.R., H.P., M.H. and Y.S.; investigation, H.P. and M.H.; software, A.S., H.P., M.H. and Y.S.; formal analysis, H.P. and A.S.; resources, H.P., A.S. and T.R.; data curation, H.P., T.R., M.H. and A.S.; writing—original draft preparation, H.P.; writing—review and editing, H.P., A.S., T.R. and Y.S.; supervision, A.S., T.R. and Y.S.; project administration, A.S. and H.P.; funding acquisition, H.P. and A.S. All authors have read and agreed to the published version of the manuscript.

Funding: This research was funded by Holsworth Wildlife Research Endowment—Equity Trustees Charitable Foundation & the Ecological Society of Australia, and also by the the Friends of the Maroochy Bushland Botanic Garden. Specimen collection and fieldwork on protected areas was conducted with permit numbers WITK 17429716 and WITK 18628117 issued by the Department of Environment and Heritage Protection, Queensland.

Institutional Review Board Statement: Not applicable.

Data Availability Statement: Site data in this study are openly available in the Queensland Herbarium CORVEG database [54]. Voucher specimens are stored at Queensland Herbarium and the University of the Sunshine Coast herbarium. All other data are held in the University of the Sunshine Coast research repository and can be made available following publication of the PhD thesis of the lead author.

Acknowledgments: Queensland Herbarium curators and technical staff are thanked for their assistance. The fieldwork assistance of Ann Moran and Allan Ward was deeply appreciated. Rachele Wilson, Brittany Elliott, and USC laboratory and technical staff are thanked for their training and support.

Conflicts of Interest: The authors declare no conflict of interest. The funders had no role in the design of the study; in the collection, analyses, or interpretation of data; in the writing of the manuscript or in the decision to publish the results.

References

1. Convention on Biological Diversity. First Draft of the post-2020 Global Biodiversity Framework. In Proceedings of the Open Ended Working Group on the Post 2020 Global Biodiversity Framework, Third Meeting, Online, 23 August–3 September 2021; Available online: https://www.cbd.int/doc/c/abb5/591f/2e46096d3f0330b08ce87a45/wg2020-03-03-en.pdf (accessed on 27 March 2022).
2. Department of Agriculture Water and the Environment. *Australia's Sixth National Report to the Convention on Biological Diversity 2014–2018*; Department of Agriculture Water and the Environment, Commonwealth of Australia: Canberra, Australia, 2020.
3. Watson, J.E.M.; Watson, J.E.M.; Simmonds, J.S.; Narain, D.; Ward, M.; Maron, M.; Maxwell, S.L. Talk is cheap: Nations must act now to achieve long-term ambitions for biodiversity. *One Earth* **2021**, *4*, 897–900. [CrossRef]
4. Maron, M.; Juffe-Bignoli, D.; Krueger, L.; Kiesecker, J.; Kumpel, N.F.; Kate, K.; Milner-Gulland, E.J.; Arlidge, W.N.S.; Booth, H.; Bull, J.W.; et al. Setting robust biodiversity goals. *Conserv. Lett.* **2021**, *14*, e12816. [CrossRef]
5. Maron, M.; Simmonds, J.S.; Watson, J.E.M. Bold nature retention targets are essential for the global environment agenda. *Nat. Ecol. Evol.* **2018**, *2*, 1194–1195. [CrossRef] [PubMed]
6. Maxwell, S.L.; Cazalis, V.; Dudley, N.; Hoffmann, M.; Rodrigues, A.S.L.; Stolton, S.; Visconti, P.; Woodley, S.; Kingston, N.; Lewis, E.; et al. Area-based conservation in the twenty-first century. *Nature* **2020**, *586*, 217–227. [CrossRef] [PubMed]
7. Thornhill, A.H.; Mishler, B.D.; Knerr, N.J.; Gonzalez-Orozco, C.E.; Costion, C.M.; Crayn, D.M.; Laffan, S.W.; Miller, J.T. Continental-scale spatial phylogenetics of Australian angiosperms provides insights into ecology, evolution and conservation. *J. Biogeogr.* **2016**, *43*, 2085–2098. [CrossRef]
8. Orme, C.D.L.; Davies, R.G.; Burgess, M.; Eigenbrod, F.; Pickup, N.; Olson, V.A.; Webster, A.J.; Ding, T.S.; Rasmussen, P.C.; Ridgely, R.S.; et al. Global hotspots of species richness are not congruent with endemism or threat. *Nature* **2005**, *436*, 1016–1019. [CrossRef]
9. Fleishman, E.; Noss, R.F.; Noon, B.R. Utility and limitations of species richness metrics for conservation planning. *Ecol. Indic.* **2006**, *6*, 543–553. [CrossRef]
10. Brooks, T.M.; Cuttelod, A.; Faith, D.P.; Garcia-Moreno, J.; Langhammer, P.; Perez-Espona, S. Why and how might genetic and phylogenetic diversity be reflected in the identification of key biodiversity areas? *Philos. Trans. R. Soc. B-Biol. Sci.* **2015**, *370*, 20140019. [CrossRef]
11. Forest, F.; Crandall, K.A.; Chase, M.W.; Faith, D.P. Phylogeny, extinction and conservation: Embracing uncertainties in a time of urgency. *Philos. Trans. R. Soc. B-Biol. Sci.* **2015**, *370*, 20140002. [CrossRef]
12. Laity, T.; Laffan, S.W.; Gonzalez-Orozco, C.E.; Faith, D.P.; Rosauer, D.F.; Byrne, M.; Miller, J.T.; Crayn, D.; Costion, C.; Moritz, C.C.; et al. Phylodiversity to inform conservation policy: An Australian example. *Sci. Total Environ.* **2015**, *534*, 131–143. [CrossRef]

13. Gonzalez-Orozco, C.E.; Pollock, L.J.; Thornhill, A.H.; Mishler, B.D.; Knerr, N.; Laffan, S.; Miller, J.T.; Rosauer, D.F.; Faith, D.P.; Nipperess, D.A.; et al. Phylogenetic approaches reveal biodiversity threats under climate change. *Nat. Clim. Chang.* **2016**, *6*, 1110–1114. [CrossRef]
14. Veron, S.; Saito, V.; Padilla-Garcia, N.; Forest, F.; Bertheau, Y. The Use of Phylogenetic Diversity in Conservation Biology and Community Ecology: A Common Base but Different Approaches. *Q. Rev. Biol.* **2019**, *94*, 123–148. [CrossRef]
15. Shapcott, A.; Liu, Y.N.; Howard, M.; Forster, P.I.; Kress, W.J.; Erickson, D.L.; Faith, D.P.; Shimizu, Y.; McDonald, W.J.F. Comparing Floristic Diversity and Conservation Priorities across South East Queensland Regional Rain Forest Ecosystems Using Phylodiversity Indexes. *Int. J. Plant Sci.* **2017**, *178*, 211–229. [CrossRef]
16. Howard, M.G.; McDonald, W.J.F.; Forster, P.I.; Kress, W.J.; Erickson, D.; Faith, D.P.; Shapcott, A. Patterns of Phylogenetic Diversity of Subtropical Rainforest of the Great Sandy Region, Australia Indicate Long Term Climatic Refugia. *PLoS ONE* **2016**, *11*, e0153565. [CrossRef]
17. Shapcott, A.; Forster, P.I.; Guymer, G.P.; McDonald, W.J.F.; Faith, D.P.; Erickson, D.; Kress, W.J. Mapping Biodiversity and Setting Conservation Priorities for SE Queensland's Rainforests Using DNA Barcoding. *PLoS ONE* **2015**, *10*, e0122164. [CrossRef] [PubMed]
18. Faith, D.P. Conservation evaluation and phylogenetic diversity. *Biol. Conserv.* **1992**, *61*, 1–10. [CrossRef]
19. Cadotte, M.W.; Tucker, C.M. Difficult decisions: Strategies for conservation prioritization when taxonomic, phylogenetic and functional diversity are not spatially congruent. *Biol. Conserv.* **2018**, *225*, 128–133. [CrossRef]
20. Forest, F.; Grenyer, R.; Rouget, M.; Davies, T.J.; Cowling, R.M.; Faith, D.P.; Balmford, A.; Manning, J.C.; Proches, S.; van der Bank, M.; et al. Preserving the evolutionary potential of floras in biodiversity hotspots. *Nature* **2007**, *445*, 757–760. [CrossRef]
21. Pollock, L.J.; Thuiller, W.; Jetz, W. Large conservation gains possible for global biodiversity facets. *Nature* **2017**, *546*, 141–144. [CrossRef]
22. Pollock, L.J.; Rosauer, D.F.; Thornhill, A.H.; Kujala, H.; Crisp, M.D.; Miller, J.T.; McCarthy, M.A. Phylogenetic diversity meets conservation policy: Small areas are key to preserving eucalypt lineages. *Philos. Trans. R. Soc. B-Biol. Sci.* **2015**, *370*, 20140007. [CrossRef]
23. Webb, C.O.; Ackerly, D.D.; McPeek, M.A.; Donoghue, M.J. Phylogenies and Community Ecology. *Annu. Rev. Ecol. Syst.* **2002**, *33*, 475–505. [CrossRef]
24. Swenson, N.G. Phylogenetic Analyses of Ecological Communites Using DNA Barcode Data. In *DNA Barcoding Methods and Protocols*; Kress, W.A., Erickson, D.L., Eds.; Springer: New York, NY, USA, 2012; pp. 409–419.
25. Gerhold, P.; Cahill, J.F.; Winter, M.; Bartish, I.V.; Prinzing, A. Phylogenetic patterns are not proxies of community assembly mechanisms (they are far better). *Funct. Ecol.* **2015**, *29*, 600–614. [CrossRef]
26. Gerhold, P.; Carlucci, M.B.; Proches, S.; Prinzing, A. The Deep Past Controls the Phylogenetic Structure of Present, Local Communities. *Annu. Rev. Ecol. Evol. Syst.* **2018**, *49*, 477–497. [CrossRef]
27. Costion, C.M.; Edwards, W.; Ford, A.J.; Metcalfe, D.J.; Cross, H.B.; Harrington, M.G.; Richardson, J.E.; Hilbert, D.W.; Lowe, A.J.; Crayn, D.M. Using phylogenetic diversity to identify ancient rain forest refugia and diversification zones in a biodiversity hotspot. *Divers. Distrib.* **2015**, *21*, 279–289. [CrossRef]
28. Tucker, C.M.; Cadotte, M.W.; Carvalho, S.B.; Davies, T.J.; Ferrier, S.; Fritz, S.A.; Grenyer, R.; Helmus, M.R.; Jin, L.S.; Mooers, A.O.; et al. A guide to phylogenetic metrics for conservation, community ecology and macroecology. *Biol. Rev.* **2017**, *92*, 698–715. [CrossRef] [PubMed]
29. Neal, R.; Stock, E. Pleistocene occupation in the southeast Queensland coastal region. *Nature* **1986**, *323*, 618–621. [CrossRef]
30. Griffith, S.J.; Bale, C.; Adam, P.; Wilson, R. Wallum and related vegetation on the NSW North Coast: Description and phytosociological analysis. *Cunninghamia* **2003**, *8*, 202–252.
31. Reed, A.W. *Aboriginal Place Names*; Reed Books: French Forests, NSW, Australia, 1970.
32. Australian Bureau of Statistics. Table 3.2 Australian Historical Population Statistics. 2019. Available online: https://www.abs.gov.au/statistics/people/population/historical-population/latest-release#data-download (accessed on 1 December 2021).
33. Green, E. *Piece by Piece: Conservation and Development on the Sunshine Coast 1960–2020*; Wildlife Preservation Society of Queensland, Sunshine Coast & Hinterland Inc.: Caloundra, QLD, Australia, 2021.
34. Wellington, T. *Noosa and Cooloola*; Beaut Books: Tinbeerwah, QLD, Australia, 2014.
35. Keith, D.; Lindenmayer, D.; Lowe, A.; Russell-Smith, J.; Barrett, S.; Enright, N.; Fox, B.; Guerin, G.; Paton, D.; Tozer, M.; et al. Heathlands. In *Biodiversity and Environmental Change: Monitoring, Challenges and Direction*; Lindenmayer, D., Burns, E., Thurgate, N., Lowe, A., Eds.; CSIRO Publishing: Collingwood, VIC, Australia, 2014.
36. Coaldrake, J.E. *The Ecosystem of the Coastal Lowlands ("Wallum") of Southern Queensland*; Commonwealth Scientific and Industrial Research Organisation: Melbourne, VIC, Australia, 1961.
37. Specht, R.L. The Sclerophyllous (Heath) Vegetation of Australia: The Eastern and Central States. In *Ecosystems of the World: Heathlands and Related Shurblands*; Specht, R.L., Ed.; Elsevier Scientific Publishing Company: Amsterdam, The Netherlands, 1979.
38. Byrne, M.; Steane, D.A.; Joseph, L.; Yeates, D.K.; Jordan, G.J.; Crayn, D.; Aplin, K.; Cantrill, D.J.; Cook, L.G.; Crisp, M.D.; et al. Decline of a biome: Evolution, contraction, fragmentation, extinction and invasion of the Australian mesic zone biota. *J. Biogeogr.* **2011**, *38*, 1635–1656. [CrossRef]

39. Barlow, B.A.; Clifford, H.T.; George, A.S.; Kanis, A.; McClusker, A. *Flora of Australia Volume 1*; AGPS: Canberra, Australia, 1981; Volume 1.

40. Hill, R.S. Origins of the southeastern Australian vegetation. *Philos. Trans. R. Soc. Lond. Ser. B Biol. Sci.* **2004**, *359*, 1537–1549. [CrossRef]

41. Scanlan, L.; McDonald, W.J.F.; Shapcott, A. Phylogenetic diversity and conservation of rainforests in the Sunshine Coast region, Queensland, Australia. *Aust. J. Bot.* **2018**, *66*, 518–530. [CrossRef]

42. Specht, R.L.; Clifford, H.T.; Arianoutsou, M.; Bird, L.H.; Bolton, M.P.; Forster, P.I.; Grundy, R.I.; Hegarty, E.E.; Specht, A. Structure, floristics and species richness of plant communities in southeast Queensland. *Proc. R. Soc. Qld.* **1991**, *101*, 27–78.

43. Young, P. How significant is the plant biodiversity of localised patches of heathy vegetation growing on low fertility soils on the hills and ranges and adjacent inland of southern Queensland? In *Faculty of Science, Health, Education and Engineering*; University of the Sunshine Coast: Sippy Downs, QLD, Australia, 2015.

44. Filer, A.; Beyer, H.L.; Meyer, E.; Van Rensburg, B.J. Distribution mapping of specialized amphibian species in rare, ephemeral habitats: Implications for the conservation of threatened "acid" frogs in south-east Queensland. *Conserv. Sci. Pract.* **2020**, *2*, e143. [CrossRef]

45. Baker, J.; Whelan, R.J.; Evans, L.; Moore, S.; Norton, M. Managing the Ground Parrot in its fiery habitat in south-eastern Australia. *Emu Austral Ornithol.* **2010**, *110*, 279–284. [CrossRef]

46. Willmott, W. *Rocks and Landscapes of the Sunshine Coast*, 2nd ed.; Geological Society of Australia, Queensland Division: Brisbane, QLD, Australia, 2007.

47. Pickett, J.W.; Thompson, C.H.; Kelley, R.A.; Romans, D. Evidence of high sea-levl during isotope stage-5C in Queensland, Australia. *Quat. Res.* **1985**, *24*, 103–114. [CrossRef]

48. Lamont, R.W.; Stokoe, R.L.; Shapcott, A. Ecological genetics of the wind-pollinated, tetraploid, *Allocasuarina emuina* L. Johnson (*Casuarinaceae*) from southeast Queensland reveals montane refugia for coastal heath during the last interglacial. *Aust. J. Bot.* **2012**, *60*, 718–734. [CrossRef]

49. Reside, A.E.; Welbergen, J.A.; Phillips, B.L.; Wardell-Johnson, G.W.; Keppel, G.; Ferrier, S.; Williams, S.E.; Vanderwal, J. Characteristics of climate change refugia for Australian biodiversity. *Austral Ecol.* **2014**, *39*, 887–897. [CrossRef]

50. Crisp, M.D.; Laffan, S.; Linder, H.P.; Monro, A. Endemism in the Australian flora. *J. Biogeogr.* **2001**, *28*, 183–198. [CrossRef]

51. Keppel, G.; Mokany, K.; Wardell-Johnson, G.W.; Phillips, B.L.; Welbergen, J.A.; Reside, A.E. The capacity of refugia for conservation planning under climate change. *Front. Ecol. Environ.* **2015**, *13*, 106–112. [CrossRef]

52. Nelder, V.J.; Butler, D.W.; Guymer, G.P. *Queensland's Regional Ecosystems: Building and Maintaining a Biodiversity Inventory, Planning Framework and Information System for Queensland Version 2.0*; Queensland Herbarium; Queensland Department of Environment and Science: Brisbane, QLD, Australia, 2019.

53. Queensland Herbarium. *Regional Ecosystem Description Database (REDD). Version 10.1 (March 2018)*; DSITI; Queensland Herbarium: Brisbane, QLD, Australia, 2018.

54. Queensland Herbarium. *Queensland CORVEG Database, Version 5/2019*; Department of Environment and Science; TERN AEKOS: Brisbane, QLD, Australia, 2012. Available online: http://aekos.org.au/collection/qld.gov.au/corveg (accessed on 24 September 2020).

55. Esri Inc. *ArcGIS Version 10.5*; Esri Inc.: Redlands, CA, USA, 2016.

56. Queensland Herbarium. *Biodiversity Status of Pre-Clearing and 2015 Remnant Regional Ecosystems—Version 10.0*; Department of Environment and Science; Queensland Government: Brisbane, QLD, Australia, 2017.

57. Queensland Parks and Wildlife Service. *Protected Areas of Queensland*; Department of Environment and Science; Queensland Government: Brisbane, QLD, Australia, 2017.

58. Nelder, V.J.; Wilson, B.A.; Dillewaard, H.A.; Ryan, T.S.; Butler, D.W. *Methodology for Survey and Mapping of Regional Ecosystems and Vegetation Communities in Queensland. Version 4.0*; I.T.a.I. Queensland Department of Science; Queensland Herbarium: Brisbane, QLD, Australia, 2017.

59. Elphick, C.S. How you count counts: The importance of methods research in applied ecology. *J. Appl. Ecol.* **2008**, *45*, 1313–1320. [CrossRef]

60. Leiper, G.; Glazebrook, J.; Cox, D.; Rathie, K. *Mangroves to Mountains*, 2nd ed.; Queensland, Society of Growing Australian Plants (Queensland Region) Inc.: Logan River Branch, QLD, Australia, 2017.

61. Harrold, A. *Wildflowers of the Noosa-Cooloola Area*; Noosa Parks Association Inc.: Noosa Heads, QLD, Australia, 1994.

62. Bostock, P.D.; Holland, A.E. Census of the Queensland Flora 2017. 2017. Available online: https://data.qld.gov.au/dataset/census-of-the-queensland-flora-2015 (accessed on 22 January 2016).

63. Carr, A. *A Field Guide to Native Plants of Bribie Island and Nearby Coastal South-East Queensland*; Caboolture Daytime Branch, Native Plants Queensland: Caboolture, QLD, Australia, 2018.

64. MacRae, I.C. *Wildflowers of Bribie Island*; Bribie Island Environmental Protection Association Inc.: Bribie Island, QLD, Australia, 1996.

65. AVH. The Australasian Virtual Herbarium. 2018. Available online: https://avh.chah.org.au/ (accessed on 1 September 2020).

66. Department of Natural Resources, Mines and Energy. Mainland-Queensland. 2007. Available online: http://qldspatial.information.qld.gov.au/catalogue/custom/search.page?q=%22Mainland%20-%20Queensland%22 (accessed on 28 October 2021).

67. Etherington, R.; Shapcott, A. Do habitat fragmentation and fire influence variation of plant species composition, structure and diversity within three regional ecosystems on the Sunshine Coast, Queensland, Australia? *Aust. J. Bot.* **2014**, *62*, 36–47. [CrossRef]
68. Kress, W.J.; Erickson, D.L.; Jones, F.A.; Swenson, N.G.; Perez, R.; Sanjur, O.; Bermingham, E. Plant DNA barcodes and a community phylogeny of a tropical forest dynamics plot in Panama. *Proc. Natl. Acad. Sci. USA* **2009**, *106*, 18621–18626. [CrossRef] [PubMed]
69. Yu, J.; Holder, M.T.; Sukumaran, J.; Mirarab, S.; Oaks, J. SATé Version 2.2.7. 2013. Available online: http://phylo.bio.ku.edu/software/sate/sate.html (accessed on 23 April 2020).
70. Bremer, B.; Bremer, K.; Chase, M.W.; Fay, M.F.; Reveal, J.L.; Soltis, D.E.; Soltis, P.S.; Stevens, P.F.; Anderberg, A.A.; Moore, M.J.; et al. An update of the Angiosperm Phylogeny Group classification for the orders and families of flowering plants: APG III. *Bot. J. Linn. Soc.* **2009**, *161*, 105–121. [CrossRef]
71. Webb, C.O.; Donoghue, M.J. Phylomatic: Tree assembly for applied phylogenetics. *Mol. Ecol. Notes* **2005**, *5*, 181–183. [CrossRef]
72. Maddison, W.P.; Maddison, D.R. Mesquite: A Modular System for Evolutionary Analysis. Version 3.61. Available online: http://www.mesquiteproject.org (accessed on 1 May 2020).
73. Miller, M.A.; Pfeiffer, W.; Schwartz, T. Creating the CIPRES Science Gateway for inference of large phylogenetic trees. In Proceedings of the Gateway Computing Environments Workshop (GCE) 2010, New Orleans, LA, USA, 14 November 2010.
74. Britton, T.; Anderson, C.L.; Jacquet, D.; Lundqvist, S.; Bremer, K. Estimating divergence times in large phylogenetic trees. *Syst. Biol.* **2007**, *56*, 741–752. [CrossRef] [PubMed]
75. Kembel, S.W.; Cowan, P.D.; Helmus, M.R.; Cornwell, W.K.; Morlon, H.; Ackerly, D.D.; Blomberg, S.P.; Webb, C.O. Picante: R tools for integrating phylogenies and ecology. *Bioinformatics* **2010**, *26*, 1463–1464. [CrossRef] [PubMed]
76. R Core Team. R Project for Statistical Computing. 2020. Available online: https://www.r-project.org/ (accessed on 1 May 2021).
77. Pohlert, T. The Pairwise Multiple Comparison of Mean Ranks Package (PMCMR). 2014. Available online: https://cran.r-project.org/web/packages/PMCMR/index.html (accessed on 1 May 2021).
78. Warnes, G.R.; Bolker, B.; Lumley, T.; Johnson, R.C. Gmodels: Various R Programming Tools for Model Fitting. 2018. Available online: https://cran.r-project.org/web/packages/gmodels/index.html (accessed on 1 May 2021).
79. Oksanen, J.; Blanchet, F.G.; Friendly, M.; Kindt, R.; Legendre, P.; McGlinn, D.; Minchin, R.; O'Hara, B.; Simpson, G.; Solymos, P.; et al. Vegan: Community Ecology Package. 2020. Available online: https://cran.r-project.org/web/packages/vegan/index.html (accessed on 1 May 2021).
80. Lozupone, C.; Hamady, M.; Knight, R. UniFrac—An online tool for comparing microbial community diversity in a phylogenetic context. *BMC Bioinform.* **2006**, *7*, 371. [CrossRef]
81. Lozupone, C.; Knight, R. UniFrac: A new phylogenetic method for comparing microbial communities. *Appl. Environ. Microbiol.* **2005**, *71*, 8228–8235. [CrossRef]
82. Diniz, J.A.F.; Soares, T.N.; Lima, J.S.; Dobrovolski, R.; Landeiro, V.L.; Telles, M.P.D.; Rangel, T.F.; Bini, L.M. Mantel test in population genetics. *Genet. Mol. Biol.* **2013**, *36*, 475–485. [CrossRef]
83. Letunic, I.; Bork, P. Interactive Tree Of Life v2: Online annotation and display of phylogenetic trees made easy. *Nucleic Acids Res.* **2011**, *39*, W475–W478. [CrossRef]
84. Noss, R.F. Indicators for monitoring biodiversity—A hierarchical approach. *Conserv. Biol.* **1990**, *4*, 355–364. [CrossRef]
85. Specht, R.L.; Specht, A. Species richness of sclerophyll (heathy) plant -communities in Australia—The influence of overstory cover. *Aust. J. Bot.* **1989**, *37*, 337–350. [CrossRef]
86. Specht, R.L. Heathlands. In *Australian Vegetation*; Groves, R.H., Ed.; Cambridge University Press: Cambridge, UK, 1994; pp. 321–344.
87. Myerscough, P.J.; Clarke, P.J.; Skelton, N.J. Plant coexistence in coastal heaths: Floristic patterns and species attributes. *Aust. J. Ecol.* **1995**, *20*, 482–493. [CrossRef]
88. Sniderman, J.M.K.; Jordan, G.J.; Cowling, R.M. Fossil evidence for a hyperdiverse sclerophyll flora under a non-Mediterranean-type climate. *Proc. Natl. Acad. Sci. USA* **2013**, *110*, 3423–3428. [CrossRef] [PubMed]
89. Byrne, M.; Murphy, D.J. The origins and evolutionary history of xerophytic vegetation in Australia. *Aust. J. Bot.* **2020**, *68*, 195–207. [CrossRef]
90. Crisp, M.; Cook, L.; Steane, D. Radiation of the Australian flora: What can comparisons of molecular phylogenies across multiple taxa tell us about the evolution of diversity in present-day communities? *Philos. Trans. R. Soc. Lond. Ser. B-Biol. Sci.* **2004**, *359*, 1551–1571. [CrossRef]
91. Carpenter, R.J.; Macphail, M.K.; Jordan, G.J.; Hill, R.S. Fossil evidence for open, Proteaceae-dominated heathlands and fire in the Late Cretaceous of Australia. *Am. J. Bot.* **2015**, *102*, 2092–2107. [CrossRef]
92. Brunbjerg, A.K.; Cavender-Bares, J.; Eiserhardt, W.L.; Ejrnaes, R.; Aarssen, L.W.; Buckley, H.L.; Forey, E.; Jansen, F.; Kattge, J.; Lane, C.; et al. Multi-scale phylogenetic structure in coastal dune plant communities across the globe. *J. Plant Ecol.* **2014**, *7*, 101–114. [CrossRef]
93. Vamosi, J.C.; Magallon, S.; Mayrose, I.; Otto, S.P.; Sauquet, H. Macroevolutionary Patterns of Flowering Plant Speciation and Extinction. *Annu. Rev. Plant Biol.* **2018**, *69*, 685–706. [CrossRef]
94. Specht, R.L. Plant communities of North Stradbroke Island: Development of structure and species richness. *Proc. R. Soc. Qld.* **2011**, *117*, 181–191.
95. Letten, A.D.; Keith, D.A.; Tozer, M.G. Phylogenetic and functional dissimilarity does not increase during temporal heathland succession. *Proc. R. Soc. B-Biol. Sci.* **2014**, *281*, 20142102. [CrossRef]

96. Cadotte, M.W.; Davies, T.J.; Peres-Neto, P.R. Why phylogenies do not always predict ecological differences. *Ecol. Monogr.* **2017**, *87*, 535–551. [CrossRef]

97. Munkemuller, T.; Gallien, L.; Pollock, L.J.; Barros, C.; Carboni, M.; Chalmandrier, L.; Mazel, F.; Mokany, K.; Roquet, C.; Smycka, J.; et al. Dos and don'ts when inferring assembly rules from diversity patterns. *Glob. Ecol. Biogeogr.* **2020**, *29*, 1212–1229. [CrossRef]

98. Prinzing, A.; Ozinga, W.A.; Brandle, M.; Courty, P.E.; Hennion, F.; Labandeira, C.; Parisod, C.; Pihain, M.; Bartish, I.V. Benefits from living together? Clades whose species use similar habitats may persist as a result of eco-evolutionary feedbacks. *New Phytol.* **2017**, *213*, 66–82. [CrossRef] [PubMed]

99. Lambers, H.; Brundrett, M.C.; Raven, J.A.; Hopper, S.D. Plant mineral nutrition in ancient landscapes: High plant species diversity on infertile soils is linked to functional diversity for nutritional strategies. *Plant Soil* **2011**, *348*, 7–27. [CrossRef]

100. Zemunik, G.; Turner, B.L.; Lambers, H.; Laliberte, E. Increasing plant species diversity and extreme species turnover accompany declining soil fertility along a long-term chronosequence in a biodiversity hotspot. *J. Ecol.* **2016**, *104*, 792–805. [CrossRef]

101. Clements, F.E. Nature and structure of the climax. *J. Ecol.* **1936**, *24*, 252–284. [CrossRef]

102. Gleason, H.A. The Individualistic Concept of the Plant Association. *Am. Midl. Nat.* **1939**, *21*, 92–110. [CrossRef]

103. Loreau, M. The Ecosystem: Superorganism, or Collection of Individuals? In *Unsolved Problems in Ecology*; Dobson, A., Holt, R.D., Tilman, D., Eds.; Princeton University Press: Princeton, NJ, USA, 2020; pp. 218–224.

104. Keith, D. Mosaics in Sydney heathland vegetation: The roles of fire, competition and soils. *CALMSci. Suppl.* **1995**, *4*, 199–206.

105. Specht, R.L. Conservation: Australian Heathlands. In *Ecosystems of the World: Heathlands and Related Shrublands*; Specht, R.L., Ed.; Elsevier Scientific Publishing Company: Amsterdam, The Netherlands, 1981.

106. Keppel, G.; Van Niel, K.P.; Wardell-Johnson, G.W.; Yates, C.J.; Byrne, M.; Mucina, L.; Schut, A.G.T.; Hopper, S.D.; Franklin, S.E. Refugia: Identifying and understanding safe havens for biodiversity under climate change. *Glob. Ecol. Biogeogr.* **2012**, *21*, 393–404. [CrossRef]

107. Keppel, G.; Ottaviani, G.; Harrison, S.; Wardell-Johnson, G.W.; Marcantonio, M.; Mucina, L. Towards an eco-evolutionary understanding of endemism hotspots and refugia. *Ann. Bot.* **2018**, *122*, 927–934. [CrossRef]

108. Morelli, T.L.; Barrows, C.W.; Ramirez, A.R.; Cartwright, J.M.; Ackerly, D.D.; Eaves, T.D.; Ebersole, J.L.; Krawchuk, M.A.; Letcher, B.H.; Mahalovich, M.F.; et al. Climate-change refugia: Biodiversity in the slow lane. *Front. Ecol. Environ.* **2020**, *18*, 228–234. [CrossRef] [PubMed]

109. Rossetto, M.; Kooyman, R. Conserving Refugia: What Are We Protecting and Why? *Diversity* **2021**, *13*, 67. [CrossRef]

110. Kooyman, R.; Rossetto, M.; Cornwell, W.; Westoby, M. Phylogenetic tests of community assembly across regional to continental scales in tropical and subtropical rain forests. *Glob. Ecol. Biogeogr.* **2011**, *20*, 707–716. [CrossRef]

111. Tibby, J.; Barr, C.; Marshall, J.C.; McGregor, G.B.; Moss, P.T.; Arnold, L.J.; Page, T.J.; Questiaux, D.; Olley, J.; Kemp, J.; et al. Persistence of wetlands on North Stradbroke Island (south-east Queensland, Australia) during the last glacial cycle: Implications for Quaternary science and biogeography. *J. Quat. Sci.* **2017**, *32*, 770–781. [CrossRef]

112. Hopper, S.D.; Gioia, P. The Southwest Australian Floristic Region: Evolution and conservation of a global hot spot of biodiversity. *Annu. Rev. Ecol. Evol. Syst.* **2004**, *35*, 623–650. [CrossRef]

113. Schut, A.G.T.; Wardell-Johnson, G.W.; Yates, C.J.; Keppel, G.; Baran, I.; Franklin, S.E.; Hopper, S.D.; Van Niel, K.P.; Mucina, L.; Byrne, M.; et al. Rapid Characterisation of Vegetation Structure to Predict Refugia and Climate Change Impacts across a Global Biodiversity Hotspot. *PLoS ONE* **2014**, *9*, e0082778. [CrossRef]

114. Buerki, S.; Callmander, M.W.; Bachman, S.; Moat, J.; Labat, J.N.; Forest, F. Incorporating evolutionary history into conservation planning in biodiversity hotspots. *Philos. Trans. R. Soc. B-Biol. Sci.* **2015**, *370*, 20140014. [CrossRef]

115. Reside, A.E.; Briscoe, N.J.; Dickman, C.R.; Greenville, A.C.; Hradsky, B.A.; Kark, S.; Kearney, M.R.; Kutt, A.S.; Nimmo, D.G.; Pavey, C.R.; et al. Persistence through tough times: Fixed and shifting refuges in threatened species conservation. *Biodivers. Conserv.* **2019**, *28*, 1303–1330. [CrossRef]

Article

DNA Barcoding of Fresh and Historical Collections of Lichen-Forming Basidiomycetes in the Genera *Cora* and *Corella* (Agaricales: Hygrophoraceae): A Success Story?

Manuela Dal Forno [1,2,*], James D. Lawrey [3], Bibiana Moncada [4,5,6], Frank Bungartz [7,8,9], Martin Grube [10], Eric Schuettpelz [2] and Robert Lücking [4,5]

[1] Botanical Research Institute of Texas, Fort Worth, TX 76107, USA
[2] Department of Botany, Smithsonian Institution, National Museum of Natural History, 10th St. & Constitution Ave. NW, Washington, DC 20560, USA; schuettpelze@si.edu
[3] Department of Biology, George Mason University, Fairfax, VA 22030, USA; jlawrey@gmu.edu
[4] Botanischer Garten, Freie Universität Berlin, Königin-Luise-Straße 6-8, 14195 Berlin, Germany; bibianamoncada@gmail.com (B.M.); r.luecking@bo.berlin (R.L.)
[5] Negaunee Integrative Research Center, The Field Museum, 1400 South Lake Shore, Chicago, IL 60605, USA
[6] Biología y Licenciatura en Biología, Universidad Distrital Francisco José de Caldas, Cra. 4 No. 26D-54, Torre de Laboratorios, Herbario, Bogotá D.C., Colombia
[7] Biodiversity Knowledge Integration Center & School of Life Sciences, Arizona State University, Tempe, AZ 85287-4108, USA; frank.bungartz@gmail.com
[8] Charles Darwin Foundation, Puerto Ayora 200305, Santa Cruz Island, Galápagos, Ecuador
[9] Instituto Nacional de Biodiversidad, Quito 170506, Ecuador
[10] Institut Für Pflanzenwissenschaften, Karl-Franzens-Universität Graz (University of Graz), Holteigasse 6, 8010 Graz, Austria; martin.grube@uni-graz.at
* Correspondence: mdalforno@brit.org

Citation: Dal Forno, M.; Lawrey, J.D.; Moncada, B.; Bungartz, F.; Grube, M.; Schuettpelz, E.; Lücking, R. DNA Barcoding of Fresh and Historical Collections of Lichen-Forming Basidiomycetes in the Genera *Cora* and *Corella* (Agaricales: Hygrophoraceae): A Success Story? . *Diversity* **2022**, *14*, 284. https://doi.org/10.3390/d14040284

Academic Editors: W. John Kress and Morgan Gostel

Received: 24 February 2022
Accepted: 6 April 2022
Published: 10 April 2022

Publisher's Note: MDPI stays neutral with regard to jurisdictional claims in published maps and institutional affiliations.

Abstract: Lichens collected worldwide for centuries have resulted in millions of specimens deposited in herbaria that offer the potential to assess species boundaries, phenotypic diversification, ecology, and distribution. The application of molecular approaches to historical collections has been limited due to DNA fragmentation, but high-throughput sequencing offers an opportunity to overcome this barrier. Here, we combined a large dataset of ITS sequences from recently collected material and historical collections, obtained through Sanger, 454, or Illumina Sequencing, to test the performance of ITS barcoding in two genera of lichenized Basidiomycota: *Cora* and *Corella*. We attempted to generate new sequence data for 62 fresh specimens (from 2016) and 274 historical collections (collected between 1888 and 1998), for a final dataset of 1325 sequences. We compared various quantitative approaches to delimit species (GMYC, bPTP, ASAP, ABGD) and tested the resolution and accuracy of the ITS fungal barcoding marker by comparison with a six-marker dataset. Finally, we quantitatively compared phylogenetic and phenotypic species delimitation for 87 selected *Cora* species that have been formally described. Our HTS approach successfully generated ITS sequences for 76% of the historical collections, and our results show that an integrative approach is the gold-standard for understanding diversity in this group.

Keywords: ASTRAL; biological collections; cryptic species; diversification; fungal barcoding; fungarium; historical specimens; natural history collections; phenotype

1. Introduction

Natural history collections are invaluable resources for assessing biodiversity and studying the evolution, biology, ecology, morphology, anatomy, chemistry, and genetics of species [1–3]. They have primarily been used for biodiversity, taxonomy, and evolutionary research, but recent efforts, including those employing machine learning, have substantially broadened their use, allowing, for example, the evaluation of shifts associated with climate change [1,4–6]. In mycology, historical records have been used, for instance, to analyze changes in the fruiting date of mushrooms and how that relates to climate change [7], and to track phytopathogenic fungi (review by [8]). For lichenology in particular, a unique use of herbarium specimens has been to compare concentrations of different minerals in contemporary versus historical samples to assess how these levels may have changed over time [9–11].

High-throughput sequencing (HTS) has revolutionized collections-based research, as the commonly fragmented nature of DNA from historical (archival) specimens generally poses a challenge in Sanger sequencing [12,13]. This problem can be overcome by HTS approaches, not only allowing amplicon sequencing from less-degraded samples but also the assembly of partial or whole genomes in many cases [14,15].

Historical lichen collections, a term utilized here to describe material of at least two decades, have only been used in a limited number of genetic studies. Sohrabi et al. [16] were the first to successfully obtain a sequence acquired from a 75-year-old herbarium specimen of *Aspicilia aschabadensis* (J. Steiner) Mereschk. Subsequently, Redchenko et al. [17] sequenced multiple archival *Caloplaca* specimens, including the current record for oldest lichen specimen sequenced (a material from 1859) and Bendiksby et al. [18] sequenced specimens of *Staurolemma omphalarioides* (Anzi) P.M. Jørg. and Henssen of up to 100 years old. More recently, studies have started to utilize HTS to obtain DNA sequences from historical lichen collections. With two-step PCR and multiple primer combinations, Kistenich et al. [19] aimed to amplify a 900 base-pair mtSSU fragment from 56 specimens of eight different species, varying in age up to 125 years-old. They showed that consensus Ion Torrent sequences presented overall better quality than those generated by Sanger sequencing. Gueidan et al. [20] utilized PacBio amplicon sequencing and successfully recovered target sequence data from 89 of 96 samples (88.5%), with the studied samples being up to 25 years old. More recently, Gueidan and Li [21] furthered their studies to include 384 specimens collected between 1966 and 2020, quadrupling their sampling and including older samples with equivalent success rates (86.5%).

The first molecular study involving a historical basidiolichen collection was that of Schmull et al. [22], who successfully sequence the ITS1 region of a 30-year-old specimen of *Dictyonema*, describing a new, potentially hallucinogenic lichen, *Dictyonema huaorani* Dal Forno, Schmull, Lücking, and Lawrey. The first application of HTS for basidiolichen collections has been relatively recent: Lücking et al. [23] described the new species *Cora timucua* Dal Forno, Kaminsky, and Lücking, known only from archival specimens collected in Florida. The authors were able to generate sequences with both Sanger and Illumina sequencing from a sample from 1901, demonstrating that even Sanger sequencing can still be a valuable tool in attempting to acquire DNA sequences from archival specimens, as also shown by other studies [16,18,19]. Indeed, Sanger sequencing has been used successfully to obtain sequence data from historical collections of non-lichenized Basidiomycota [24,25].

The lichenized basidiomycete genus *Cora* Fr. presents a notable example of the usefulness of ITS in fungal barcoding for delimiting species and assessing biodiversity, as proposed by Schoch et al. [26]. *Cora* forms part of the *Dictyonematinae* subtribe, a lichenized lineage of five genera with different morphologies related to the non-lichenized mushroom genus *Arrhenia* Fr. [27–29]. Within this subtribe, *Cora* and the related genus *Corella* Vain. are distinguished by a foliose thallus with a compact surface, whereas the other genera (*Acantholichen* P.M. Jørg., *Cyphellostereum* D.A. Reid, *Dictyonema* C. Agardh ex Kunth) are filamentous or microsquamulose [28]. The taxonomy of foliose basidiolichens has suffered from the typical problems of traditional revisions based largely on herbarium collections that fail to recognize important field characters, such as consistency, lobe arrangement, color, and substrate (Figure 1). In addition, emphasis was historically placed on mycological characters, such as hymenophore anatomy and basidiospores, to establish species boundaries [30]. As a result, for a long time, only a single foliose species was recognized, named *Dictyonema pavonium* (Sw.) Parmasto and subsequently *D. glabratum* (Spreng.) D. Hawksw. [31], with all other previously proposed taxa, including those in the genus *Corella*, as synonyms [30].

Figure 1. Comparison of fresh samples in their natural habitat (**a**,**b**) versus dried and deposited in the herbarium (**c**,**d**)—(**a**,**c**): *Cora* spec-023 (Dal Forno 2042), (**b**,**d**): *Cora cyphellifera* (Dal Forno 1808).

Molecular approaches led to the realization that these foliose lichens represented more than one species and also supported the separation of the genera *Cora* and *Corella* [27,28]. A first broad sampling using the ITS barcoding locus resulted in an estimated 116 species of *Cora* and ten of *Corella* [32]. Although constituting a dramatic increase in species count, this estimate was still considered conservative: quantitative species recognition methods suggested up to 170 species based on the same data, and a novel prediction method, which takes into account unsampled regions and habitat suitability, estimated more than 450 species [32]. Soon after, with much increased sampling, Lücking et al. [33] distinguished 189 species. It is possible to challenge these findings based on potential problems in ITS barcoding, such as improper assessment of intragenomic variation or the occurrence of multiple ITS copies in the genome as a result of hybridization and introgression or gene duplication, possibly leading to taxonomic inflation or the recognition of artifactual taxa [34,35]. On the other hand, ITS has also been shown to lack resolution in recently evolving species complexes, including both non-lichenized and lichenized fungi [35–39], potentially counterbalancing issues with wrongly assessed ITS variation in terms of species counts but introducing additional inaccuracy.

In the genera *Cora* and *Corella* and in subtribe *Dictyonematinae* in general, the topology of ITS-based phylogenies has been found to be highly congruent with those of other markers, such as nuLSU and RPB2 [28,40], suggesting that ITS resolves these lineages accurately. Analysis of intragenomic variation of the ITS in *Cora inversa* Lücking and B. Moncada using a 454 pyrosequencing approach, did not demonstrate potential gene duplication or hybrid ITS arrays; instead, the variation detected stemmed almost entirely from sequencing errors and had no effect on accurate species delimitation when using a phylogenetic approach [41]. A similar result has been reported for the *Rhizoplaca melanophthalma* species complex, in which the observed ITS variation did not interfere with species discrimination in that group [42].

Phenotypes in phylogenetically delimited species of *Cora* are highly consistent with the underlying molecular data [33] and even photobiont haplotypes showed a high level of congruence with the ITS-based phylogeny of the associated mycobionts [43]. Therefore, in these lichenized Basidiomycota, ITS barcoding appears to provide highly accurate assessments of species richness.

Unfortunately, these earlier studies were biased towards geographic areas from which fresh material could be readily sampled. Natural history collections provide access not only to a much broader geographic range, but also to specimens that may have been sampled in regions that are now heavily altered or with their original habitats destroyed, as shown by the example of the possibly extinct *Cora timucua* [23,44]. Inserted within a broad phylogenetic and phenotypic framework, herbarium collections of *Cora* and *Corella* can thus provide unique opportunities to expand taxon and specimen sampling for these genera, to extend our understanding of their biology, and to test assumptions about geographical distributions of species currently only documented by recently collected material. The utility of historical collections in this regard is, however, dependent on the quality of the sequences obtained from these samples.

The United States National Herbarium (USA) at the Smithsonian National Museum of Natural History (SI-NMNH) is home to one of the largest lichen collections worldwide, with over 250,000 specimens. It contains one of the largest and most diverse collections globally of the subtribe *Dictyonematinae*, with over 400 specimens having broad geographical and temporal representation and unique morphologies. From these collections, we analyzed all *Cora* and *Corella* specimens, for a total of 274 samples with collection dates ranging from 1888 to 1998, complementing our already large dataset of 856 recently collected specimens from 18 countries (Figure 2) and representing the largest study of historical lichen collections using molecular approaches including HTS to date focusing on a single genus (most samples belong to *Cora*).

Our objectives for this study included: (1) testing the success rate of ITS barcoding from historical samples of *Cora* and *Corella* and comparing Sanger versus Illumina sequencing success; (2) expanding the existing ITS-based phylogeny with newly generated sequences and assessing the number of phylogenetically delimited species using various quantitative approaches (GMYC, bPTP, ABGD, ASAP) to test our earlier prediction of more than 450 species in this group; (3) further assessing potential intragenomic ITS variation by comparing Sanger and Illumina data from the same specimens; (4) testing the performance of ITS relative to a six-marker dataset using a subset of terminals; and (5) exploring the level of potential cryptic speciation in *Cora* and *Corella* by testing for consistency between molecular and morphological data.

Figure 2. Map showing country-based availability of sequenced contemporaneous (fresh) and historical (herbarium) samples of *Cora* (and *Corella*) throughout the Americas (generated with mapchart.net).

2. Materials and Methods

2.1. Newly Generated Sequence Data from Fresh and Historical Collections

2.1.1. Sampling

Our entire dataset included 1130 unique samples (Table S1), 62 were new for this study (or "fresh specimens" from 2016) and 274 represented "historical specimens" (collected between 1888 and 1998, i.e., at least more than two decades ago). The majority of the contemporary samples (remaining 794 samples in the dataset) were collected during field trips between 2007 and 2018 throughout Mexico, Central and South America, and the Caribbean. The historical samples analyzed here are all deposited at the US Herbarium of the Smithsonian National Museum of Natural History. Contemporary samples are deposited at multiple herbaria (mainly B, CDS, CR, F, GMUF, UDBC, US; but also at CUVC, EB-BUAP, FAUC, KRAM, LPB, QCNE, UPR).

2.1.2. DNA Extractions

DNA was isolated from lichen fragments of approximately 1 cm^2 using the PowerSoil®-htp 96 Well Soil DNA Isolation Kit (MO BIO, USA, now the DNeasy® PowerSoil® HTP 96 kit,

Qiagen, Germantown, MD, USA) following Marotz et al. [45]. Prior to DNA extraction, each individual lichen fragment was cleaned in 0.85% NaCl and subsequently in sterile water.

2.1.3. PCR, ITS Sanger Sequencing and Assembly

PCR amplification and Sanger sequencing followed the protocols established by Dal Forno et al. [28,40]. Newly generated sequences were organized and assembled in Geneious Prime 2 January 2020. We considered as an "accepted sequence" any piece longer than 100 bp (before any trimming) that matched the target genus.

2.1.4. PCR, ITS Illumina Sequencing and Bioinformatic Analyses

Foliose members of *Dictyonematinae* (*Cora* and *Corella*) were processed with protocols from the Earth Microbiome Project (EMP, https://earthmicrobiome.org/protocols-and-standards/ accessed on 5 July 2017). For the ITS Illumina Amplicon Protocol, primers ITS1F [46] and ITS2 [47] were ordered with Illumina constructs, so there is no need for additional PCR steps to attach Illumina barcodes, following Caporaso et al. [48]. Each sample was amplified with PCR in triplicate [49], with reactions combined after gel electrophoresis. DNA concentrations were measured with Qubit^{TM} dsDNA Broad Range Assay Kit (Invitrogen^{TM}, Waltham, MA, USA) and 240 ng of each PCR product was pooled into a single tube for each plate (amplicon pool). Each pool contained 96 samples plus two negatives. The quantity of negative sample added to each amplicon pool was the average used volume per plate (ranging from 11 µL to 30 µL). The post-PCR DNA yields from fresh specimens ranged from 16.7 to 54.4 ng/µL and from 1.9 to 50.48 ng/µL for historical samples. Amplicon pools were then cleaned with UltraClean PCR Clean-Up Kit (MO BIO, Carlsbad, CA, USA) following the manufacturer's protocol, but followed by one or more additional pool clean ups with AMPure XP beads (Beckman Coulter, Brea, CA, USA; modification from EMP protocol needed, given that only utilizing the suggested clean up kit would not be sufficient to remove primer dimers, unincorporated DNTPs, etc. when visualized in a 1% gel). The two pools (one for each plate) were then combined and a qPCR with a KAPA Library Quantification Kit for Illumina$^{®}$ platforms was performed (Kapa Biosystems, Wilmington, MA, USA). A 2-nM dilution was made and finally sequenced in a MiSeq Kit v2 500 cycles (paired end 2 × 250 bp; Illumina, Inc., San Diego, CA, USA) in house at the Laboratories of Analytical Biology (LAB) of the SI-NMNH (Washington, DC, USA).

Samples were de-multiplexed and processed bioinformatically using QIIME2 [50]. Sequences were de-noised using DADA2 [51]. Taxonomy was assigned with the UNITE dynamic fungal classifier [52] with the "feature-classifier classify-sklearn" option in QIIME2. Newly generated tables were compiled into a single file with Amplicon Sequence Variation (ASV) reads, number of reads per ASV and per sample, number of times that ASV was observed across samples, and taxonomy. All analyses were run on the Smithsonian Institution High Performance Computing Cluster [53] or locally.

2.1.5. Microfluidics PCR, Multi-Marker Illumina Sequencing and Bioinformatic Analyses

For a subset of samples, a microfluidics PCR followed by Illumina sequencing methodology was performed following Gostel et al. [54,55], but adapted to commonly used primer combinations in lichens. This approach and its application in lichenology will be fully treated in a separate publication; but, in summary, this is a PCR-based target enrichment procedure that allows a large number of samples and primer combinations (including genetic markers for both myco- and photobionts and associated microbiome) to be amplified in a single step due to a series of microtubes that intersect, forming a central matrix of thousands of reaction chambers [55]. It is here applied for the first time in lichens. For this current study, five fungi markers were selected (EF3, mitLSU, ITS, nuLSU, nuSSU), amplified with a Microfluidic PCR on a Fluidigm Access Array at the Center for Conservation Genomics (CCG) at the Smithsonian's National Zoo (Washington, DC, USA) and sequenced on an Illumina MiSeq Kit v3 (600 cycles) at the LAB (SI-NMNH). Samples were de-multiplexed and de-noised using QIIME2 [50], and then assigned taxonomy with a custom database with reference sequences for each of the loci sequenced.

2.2. Phylogenetic Analyses

2.2.1. ITS-Based Phylogeny

A total of 1325 ITS sequences were assembled to infer relationships within the *Acantholichen-Corella-Cora* clade (Table S1). *Acantholichen* was included as it represents the sister clade of *Corella*, with *Acantholichen* and *Corella* together sister to *Cora* [28]. Of the 1325 sequences, 28 represented *Acantholichen*, 54 *Corella*, and the remaining 1243 the genus *Cora*. We used a previously published alignment focusing on the genus *Cora* [33] as template and added further and newly generated sequences with MAFFT 7 "–add" [56] using the online server [https://mafft.cbrc.jp/alignment/server/add_sequences.html]. The resulting alignment was manually inspected and, after running an initial phylogenetic analysis (see below for settings), terminals were sorted in a phylogenetic order and the alignment reinspected. This approach was repeated several times, until no further alignment inconsistencies were detected. Ambiguously aligned regions were found to be short and were detected only in a broader context between distantly related taxa, whereas closely related taxa within supported subclades did not show notable alignment ambiguities, as assessed through the Guidance2 Server [57] (http://guidance.tau.ac.il/) for smaller subsets of the data. Therefore, ambiguously aligned regions were not removed. The final alignment contained 1325 sequences and was 916 bases long after terminal trimming (File S1). Of these 1325 sequences, 965 were generated with Sanger, 330 with Illumina, and 30 with 454-pyrosequencing approaches. Sequence length (after trimming) varied between 86 and 724 bases (File S1). Short sequences generally corresponded to the ITS1 region obtained through Illumina and 454 pyrosequencing, but some short Sanger sequences covered only the ITS2 region.

We employed RAxML 8 [58] on the CIPRES Science Gateway [59] to reconstruct a maximum likelihood phylogeny based on the ITS data, using the universal GTRGAMMA model and bootstrap pseudo-replicates automatically determined through the RAxML Black Box on the CIPRES server. The resulting tree was visualized in FigTree 1.44 (http://tree.bio.ed.ac.uk/software/figtree/). Using this tree, species-level clades (i.e., species hypotheses) were delimited ad hoc using a combination of stem branch length, support, ecology (substrate, habitat) and distribution of the corresponding terminals, without taking into consideration their morphology.

2.2.2. Six-Marker Phylogeny

Six genetic markers (EF3, mtLSU, ITS, nuLSU, nuSSU, and RPB2) were utilized to infer a species (coalescent) tree in ASTRAL (Accurate Species TRee ALgorithm) III [60] that encompassed a total of 147 terminals representing 83 species (File S2). Each of these loci had the following number of samples and alignment length: EF3 (33 taxa, 421 bp); mtLSU (62 taxa, 649 bp); ITS (147 taxa, 816 bp); nuLSU (139 taxa, 1471 bp); nuSSU (69 taxa, 1268 bp); and RPB2 (11 taxa, 1033 bp). The ITS alignment was separately subjected to ML analysis as described above for the full dataset (File S3).

To compare the resulting ITS-based ML tree with the six-marker ASTRAL tree, generalized Robinson–Foulds distances were calculated between the ASTRAL and the ITS tree in R Studio with the TreeDist package [61,62], following the tutorial by Smith (https://ms609.github.io/TreeDist/articles/Generalized-RF.html accessed on 1 July 2021).

2.3. Quantitative Species Delimitation

2.3.1. Single-Locus Tree-Based Methods

We predicted species limits utilizing the Generalized Mixed Yule Coalescent (GMYC) approach as described by Fujisawa and Barraclough [63] using the R package splits (default settings, including single threshold). We first generated an ultrametric tree in BEAST v1.10.4 [64–66], using the following settings according to Lücking et al. [32]: (a) strict molecular clock; (b) GTR substitution model with base frequencies estimated and Gamma and invariant sites with six Gamma categories; (c) speciation through a Yule process with the "yule.birthRate" prior set to an exponential distribution with 4.0 as mean; (d) the "ucld.mean" prior set to an exponential distribution with 0.001 as mean and all other priors with default values; and (e) 100,000,000 MCMC generations.

We also inferred putative species boundaries applying the Poisson Tree Process (PTP), including its Bayesian implementation (bPTP) [67] on the bPTP Web Server (https://species. h-its.org/ptp/ accessed on 15 January 2022), with the following settings: 500,000 MCMC generations, 100 Thinning, and 0.25 Burn-in.

2.3.2. Distance-Based Methods and Barcoding Gap Analyses

For the barcoding gap analysis, we used a subset of the ITS sequence data limited to the genus *Cora*, retaining only terminals with less than 30% missing data, spanning most or all of the ITS1 and ITS2 region (716 terminals; File S4). This subset corresponded to 175 species-level taxa plus one subspecies delimited ad hoc (see above). We computed a distance matrix using the Kimura 2-parameter model implemented in BioEdit 7 through DNADIST 3.5 [68,69]. Within-species and between-species distances were then assessed by arranging the terminals according to the ITS-based ML phylogeny (see above) and comparing the distances between subsequent terminals, using our ad hoc species hypotheses as grouping variables. We computed a threshold distance value by comparing within-species and between-species distances, selecting the value that best discriminated between the two, retaining a maximum of within-species and a minimum of between-species pairs. The resulting threshold was then used to visualize distance patterns in the full two-dimensional distance matrix between all terminals, enabling the assessment of the predefined species-level clades into three categories: (1) species-level clades well-delimited by the threshold distance; (2) species-level clades with internal variation frequently greater than the threshold distance; and (3) species-complexes in which hypothesized species-level clades were not well-resolved by the threshold distance.

We further employed Automatic Barcode Gap Discovery (ABGD) [70] and Assemble Species by Automatic Partitioning (ASAP) [71] on the same dataset of 716 *Cora* terminals (File S4), using the corresponding web servers [https://bioinfo.mnhn.fr/abi/ public/abgd/abgdweb.html; https://bioinfo.mnhn.fr/abi/public/asap/asapweb.html accessed on 15 January 2022]. For ABDG, we used the following settings: Kimura (K80) 2-parameter model, Pmin = 0.001, Pmax = 0.1, Steps = 10, X (relative gap width) = 1.5, 1.0. 0.5, 0.1 (stepwise), Nbins = 20. For ASAP, we used settings as follows: Kimura (K80) 2-parameter model, Split groups below this probability = 0.01, Keep 10 best scores, Use fixed seed value = −1, Highlights results between the genetic distances = 0.05 and 0.05.

2.3.3. ITS-Based BLAST Performance

We used a subset of 758 terminals with (near) complete ITS sequences as reference (File S5, "BLAST reference subset"), similar to the aforementioned *Cora* subset but also including *Acantholichen* and *Corella*, to assess BLAST performance of four datasets: (1) the ITS1 region only (298 bases, individual length (173–)200–220 bp) of the *Cora* species in the BLAST reference subset (716 terminals; File S6); (2) the ITS region of all original terminals with less than 10% gaps (including Illumina ASVs, 1217 terminals; File S7); (3) the ITS2 region only (344 bases, individual length 231–263 bp) of this BLAST reference subset (File S8); and (4) a short, subterminal (49 bp from the end of the ITS4 primer) piece of the ITS2 region (85 bases, individual length (43–)50–64 bp) of this BLAST reference subset (File S9). The analysis was performed using the local BLAST function in BioEdit 7 [69,72] with the following settings: E value = 1.0, maximum number of hits to report = 50, gap opening/extending penalty = 0, mismatch penalty = −3, match reward = 1 (see also [73,74]).

BLAST results for each of the scenarios were evaluated as follows. First, we computed the variation of total scores for hits (same species-level clade), self-hits (same terminal), and misses (different species-level clade) for each scenario. Then, we computed the minimum score for hits versus the maximum score for misses for each query, determining to what level a "BLAST gap" existed for each scenario. Finally, we computed the ratio of minimum hit score versus maximum miss score to calculate the proportion of three ratios per query: (1) >1.25, minimum hit score at least 25% higher than maximum miss score; (2) >1.10, minimum hit score at least 10% higher than maximum miss score; and (3) >1.00, minimum hit score higher than maximum miss score.

2.4. Phenotypic Assessment

In order to test the hypothesis that phylogenetically defined taxa in *Cora* are phenotypically undifferentiated and may reflect taxonomic inflation, we analyzed 87 formally described and sequenced species of *Cora* [33], plus two outgroup taxa in the genus *Corella* (Table S2). We established a one-sequence-per-taxon ITS tree, selecting the type sequence for each species when available and using the alignment from Lücking et al. [33] as a template (Table S2). After deleting the non-relevant sequences and keeping only the 89 target sequences (File S10), the alignment was subjected to RAxML 8.2.0 [58] analysis on a local CPU to reconstruct the best-scoring maximum likelihood tree under the universal GTRGAMMA model; bootstrapping was performed with 1000 pseudo-replicates.

For the 87 selected *Cora* species (and the two *Corella*), we assembled a matrix for a total of 20 characters, including one ecological character (preferred substrate; row 1), 11 morphological, anatomical, and chemical characters (rows 2–12, based on largely on previous studies [32,33,40], onward simply referred as "phenotypical characters"), and eight chorological characters, defined as biogeographic regions (rows 13–20; Table 1; scores on Table S3). We then defined phenotypes by combining the character scores for the 11 phenotypical characters (columns 2–12 in Table S3) into a character state string for each species. Phenotypes were considered identical if they agreed in the complete string between two species. The number of phenotype characters was kept intentionally low and limited to those that were constant within a lineage to avoid false positives in terms of different phenotypes (the number of potential phenotypes increases exponentially with the number of characters). Our aim was to test if even a very low number of characters would result in a sizable number of distinct phenotypes. A high level of cryptic speciation (or potential taxonomic inflation) could be expected if the number of distinct phenotypes resulting from the combination of these 11 characters was substantially lower than the number of lineages distinguished (87).

For each possible pair of species, we computed molecular phylogenetic distance using the Kimura 2-parameter model implemented in BioEdit 7 through DNADIST 3.5 [68,69]. In parallel, we computed phenotypic distance by calculating total character state difference over all 11 phenotypical characters (columns 2–12 in Table S3).

To assess potentially cryptic speciation, we defined four types of crypticity (Table 2). To avoid confusion with the use of the terms "cryptic" and "crypsis" in zoology, here we propose to add the prefix "phylo-" before cryptic when referring to a phylogeny-based phenetic context. We defined "same" phenotype as total phenotype character distance = 0 (identical) and "similar" phenotype as total phenotype character distance = 1 (low). Taxa differing in more than one character or with a total character distance >1 were considered distinct.

Table 1. Ecological (1), phenotypical (11), and chorological (8) characters and states scored for the selected 87 species in the genus *Cora* (plus two *Corella*). Only the 11 phenotypical characters (Size through Bleeding pigment) were utilized for generating the character strings.

Character\State		0	1	2	3	4	5	?
1	Substrate [a]	—	rock	soil	ground	epiphytic	—	—
2	Size	—	small (lobes > 1 cm)	10édium (lobes 1–3 cm)	large (lobes > 3cm)	—	—	—
3	Sutures [b]	absent	short	long	—	—	—	—
4	Color [c]	—	grey	(grey-)brown	olive (-green)	green	aeruginous	—
5	Surface	even	granular-rugose	pitted	broadly undulate	narrowly undulate	—	—
6	Trichomes [d]	absent	present (felty)	present (setose)	present (strigose)	—	—	—
7	Margin	glabrous	pilose	granular	granular-pilose	—	—	—
8	Soredia	absent	present	—	—	—	—	—
9	Cortex	absent	viaduct-shaped	collapsed	compacted	prosoplecten-chymatous	paraplecten-chymatous	—
10	Papillae	absent	present	—	—	—	—	—
11	Hymenophore	—	rounded-confluent	concentric	resupinate-cyphelloid	cyphelloid	—	not observed
12	Bleeding pigment	absent	present	—	—	—	—	—
13	Central America	absent	present	—	—	—	—	—
14	Caribbean	absent	present	—	—	—	—	—
15	Galapagos	absent	present	—	—	—	—	—
16	Northern Andes	absent	present	—	—	—	—	—
17	Central Andes	absent	present	—	—	—	—	—
18	Southern Andes	absent	present	—	—	—	—	—
19	Eastern Brazil	absent	present	—	—	—	—	—
20	Asia	absent	present	—	—	—	—	—

[a] Soil means growing directly on bare soil, while ground means growing between terrestrial vegetation, e.g., over grasses. [b] Sutures refer to the lines apparent between two lobes, typically appearing as if the lobes have been sown together—*short*: only present along a small part of the lobes (see Figures 3I,K and 4G in [33]); *long*: present along most of the lobes (see Figures 5F and 7D,N in [33]). [c] Color when fresh or rewetted, not dried specimen. [d] Most samples are glabrous, but those with trichomes can be further divided—*felty* (rare, see Figure 8K,L in [33]): with short hairs formed by single hyphae; *setose* (most common type, see Figures 3L, 5O and 7K,L in [33]): with distinct hairs composed of at least partly agglutinate hyphae; *strigose* (few species, see Figures 4D,E and 10E,F in [33]): with conspicuous, long trichomes always composed of agglutinate hyphae.

In addition to the maximum likelihood phylogenetic tree based on the ITS barcoding locus, we also computed a cladogram based on all 20 characters using PAUP 4.0b10 [75,76]. To test the hypothesis that tree structure based on phenotype and distribution was significantly different from random (i.e., there was correlated structure in the ecological, phenotypical, and chorological data), we computed 100 random trees in PAUP based on the taxon set to simulate random distribution of character states relative to tree topology. For each tree, we calculated the following five indices based on phenotype character state distribution: parsimony tree length (TL; in steps), consistency index (CI), retention index (RI), rescaled retention index (RC), and homoplasy index (HI). The first and last indices (TL, HI) are proportional to noise in the phenotype data and inversely proportional to structure, whereas the other indices (CI, RI, RC) behave the opposite way.

Table 2. Definition of different types of crypticity within an evolutionary context and the corresponding terms used in other studies [34].

Proposed Term	Etymology	Phenotype Distance	Phylogenetic Relationship	Distribution
Eu-(phylo-)cryptic ("cryptic")	Greek: eús, eû = good	zero (same phenotype)	closely related or sister species	sympatric
Kapo-(phylo-)cryptic ("near-cryptic")	Greek: kápos = somewhat	very low (similar phenotype)	closely related or sister species	sympatric
Allo-(phylo-)cryptic ("semi-cryptic")	Greek: állos = other	zero or very low (same or similar phenotype)	closely related or sister species	allopatric
Pseudo-(phylo-)cryptic ("homoplasic")	Greek: pseudés = false	zero or very low (same or similar phenotype)	unrelated or distantly related	sympatric or allopatric

3. Results

3.1. Comparison of Sequence Performance between Fresh and Historical Collections

The two Illumina MiSeq runs yielded the following raw data:

Run 01 (Plates 1 and 2): 331.6 MB (I1 or Indexes), 3.35 GB (R1 or forward reads) and 3.94 GB (R2 or reverse reads) (samples: NSF-001–106 and US-001–086); 8417853 total sequences;

Run 02 (Plates 3 and 4): 289 MB (I1 or Indexes), 3.06 GB (R1 or forward reads) and 3.55 GB (R2 or reverse reads) (samples: US-087–278); 11250308 total sequences.

With amplicon sequencing in the Illumina system, we were able to successfully sequence 76.3% or 209 of the 274 historical specimens and 93.6% or 58 of the 62 fresh specimens. With Sanger sequencing, these numbers were substantially lower: 19% of the historical and 58.1% of the fresh specimens (Figure 3a). The difference was, thus, particularly marked for the historical specimens, with a 75% decrease in success between methods, compared to a 38% decrease for the fresh specimens. The pattern was largely consistent among the substrata where these lichens were growing (Figure 3b), although, for the rock substrate, Sanger fresh and historical success rates were more similar.

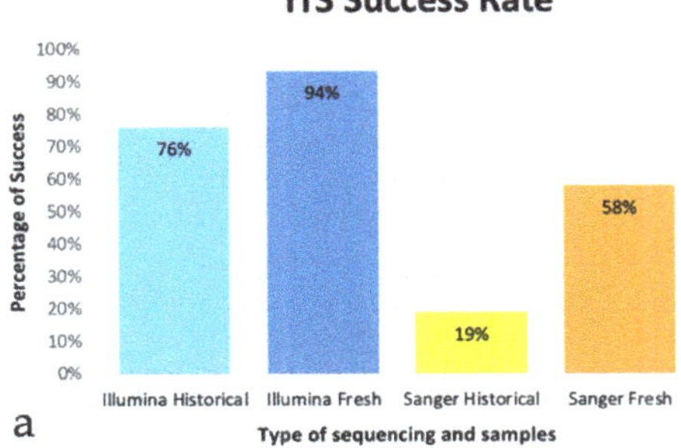

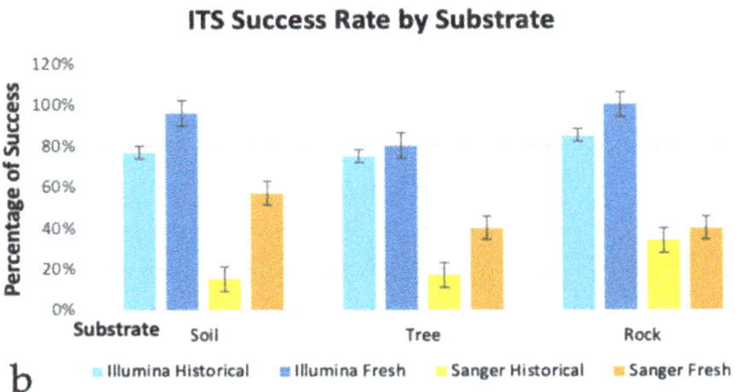

Figure 3. Overall ITS Success Rate in Illumina versus Sanger sequencing with historical and fresh samples (**a**), including a comparison by substrate (**b**). Color pallet proposed by Wong [77].

We did not find a linear pattern in sequencing success dependent on time of historical collections for the Illumina or Sanger sequencing, but overall there was an increase in success for the more recent specimens (Figure 4a,b). There was also no difference in sequencing success when analyzed by habitat in which the specimen was originally collected (Figure 4c), although this information was only available for 17% of the historical samples. Given the high rate of success even for older specimens in the two studied genera, sequencing success may also depend on how the material was collected, dried, and preserved, thus affecting potential initial DNA degradation independent of how long the material has been stored. Unfortunately, data on collection and preservation methods were not available, but we used visual inspection of the condition of the material as a proxy: most successfully sequenced specimens appeared in good condition, as if preserved recently, their colors mostly grey to white-ish, whereas unsuccessfully sequenced specimens often showed discolorations, typically turning yellowish brown. Such discolorations, likely associated with oxidation and DNA degradation, are usually caused when material is collected in the hydrated stage and pressed before fully air-dried, or when a heat source has been used to dry the material, sometimes also in combination with placing the material in alcohol prior to heating, a commonly applied technique in the past.

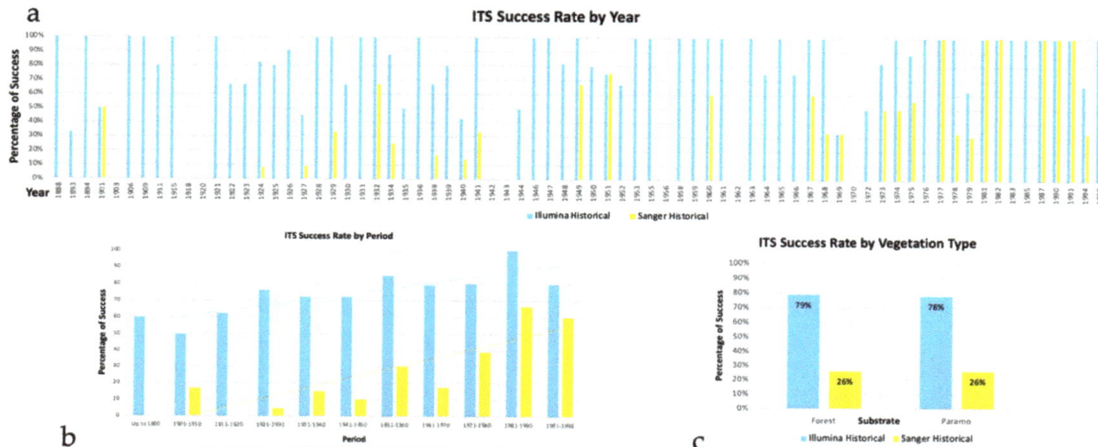

Figure 4. Percentage of success in sequencing historical samples between the two tested sequencing platforms (Illumina, Sanger) by year (**a**), by period of time (**b**), and vegetation type (**c**).

Beyond the target mycobiont sequences, most Illumina-sequenced samples also produced ASVs for the most commonly found contaminants, such as *Penicillium*, *Cladosporium*, *Fusarium*, and *Aspergillus* (all Ascomycota). These fungi may already be present in the living lichen specimen but more commonly originate during the drying process or subsequent preservation. In addition, fungi typically occurring in the substrate where lichens were collected (e.g., soil), as well as human-related contaminants were found, including *Trichobolus* (Ascomycota), *Mortierella* (Mucoromycota), and *Wallemia* (Basidiomycota), despite the washes performed prior to DNA extraction. Beyond that, many mushroom species were detected, most likely by the presence of spores in the samples, as well as several basidiomycetous yeasts in the class Tremellomycetes.

3.2. ITS-Based Phylogeny

Our new ITS tree with 1325 sequences contained 1091 unique samples, collected from 29 different countries (Figures 5 and S1). By utilizing historical specimens, we were able to add records for an additional 11 countries for which the diversity of *Cora* and *Corella* species was previously unknown (Figure 2; a singular fresh collection from Sri Lanka not mapped, see [33]).

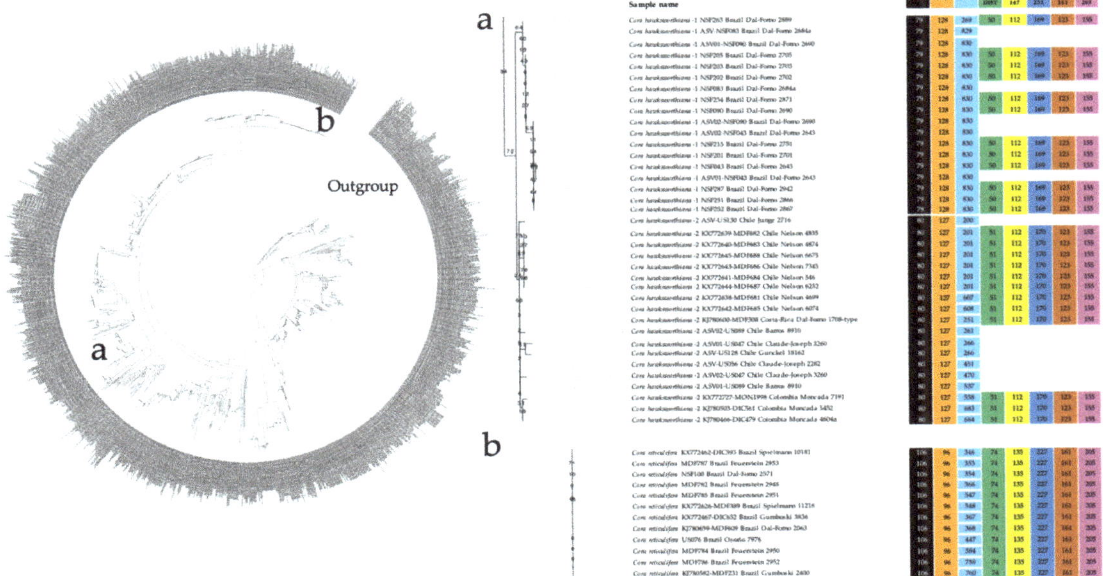

Figure 5. ITS fungal barcoding tree inferred by maximum likelihood for the genus *Cora*, including the *Acantholichen-Corella* clade as outgroup (1325 terminals), and two examples of the results of species delimitation approaches. *Cora hawksworthiana* (**a**) is currently considered a single species, but could potentially be further divided into two—one from Brazil and another for Chile, Colombia, and Costa Rica—based on the ad hoc, GMYC, DNADIST, and ABGD-231 approaches. In contrast, *Cora reticulifera* (**b**), an abundant common species from southeastern Brazil and Uruguay, with a uniform phenotype and ecology, was delimited as a single species by all methods, except in bPTP, which recovered each sample as a separated species. For the full-length tree as well as species delimitations for all specimens, see Figure S1 and Table S4.

Given this information, we were able to extend the distribution of six species in *Corella* and 28 in *Cora*; for example, *Cora davibogotana* previously only known from Colombia, is now also reported from Venezuela; and *Cora* spec-84, before known exclusively from Brazil, is now also known from Uruguay (Figure S1). In all cases, range extensions were to adjacent countries. Countries such as Ecuador, Jamaica, Mexico, Colombia, and Brazil, continue to be well represented by fresh specimens, while for other countries, the addition of new specimens from the herbarium was invaluable. For example, for Panama, we added 18 herbarium collections to compliment the one available fresh specimen, now corresponding to seven or eight species as opposed to just one previously known from the country. Overall, approximately 25–30 additional novel lineages were detected among the herbarium collections alone; however, since these require further studies to be formally described (as either species or at infraspecific level, following the approach proposed by Lücking et al. [34]), they will be treated in a separate publication.

3.3. Comparison between ITS and Astral Six-Marker Tree

Our microfluidics PCR followed by Illumina sequences yielded 239 novel sequences belonging to *EF3*, mtLSU, nuLSU and nuSSU. ITS data were also highly successful with this approach; however, we already had those sequenced with other methods. Our ITS-based ML tree and the ASTRAL six-marker coalescent tree exhibited very similar topologies (Figure 6), with a normalized Robinson–Foulds distance of 0.0278, suggesting that an ITS-based phylogeny reliably recovers a multi-marker phylogeny in this particular genus.

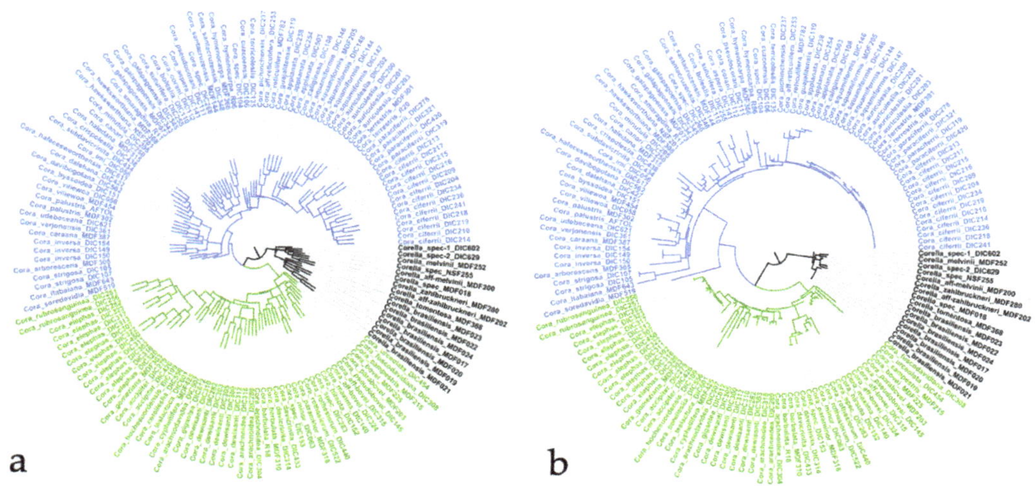

Figure 6. Comparison between inferred ASTRAL six-marker species tree (**a**) and ITS maximum likelihood tree (**b**). The different colors simply denote the two major clades in *Cora*, with *Corella* in black as the outgroup.

3.4. Quantitative Species Delimitation Methods

The bPTP species delimitation method estimated the number of species to be between 708 and 889 (mean: 791, output delimitation with 853 clades), with an acceptance rate of 0.826 (Tables 3 and S4). MCMC generations merged at 249,739 and split at 250,261. The GMYC method, on the other hand, resulted in an estimation of 189 species, with a confidence interval of 145–237 (Tables 3 and S4).

Barcoding gap analysis using DNADIST via BioEdit resulted in an identity threshold value of 99.4% (distance threshold value = 0.006), retaining 94.6% of all within-species and excluding 95.4% of all between-species pairwise identities based on ad hoc delimitations. Most species delimited ad hoc were well delimited using this threshold value, such as the *Cora applanata-reticulifera* clade (Figure 7, File S11, Table S4). However, we detected several cases of ad hoc delimited species that may represent more than one taxon (e.g., *C. arachnoidea*, *C. davidia*, and *C. hawksworthiana*), as well as cases of diffuse species complexes where the threshold value did not fully discriminate between ad hoc-delimited species, including the *C. galapagoensis* clade and the *C. squamiformis-ciferrii* clade (Figure 7, File S11, Table S4). Overall, we found six cases of ad hoc-delimited species complexes to be potentially merged based on the threshold value, resulting in a possible reduction from 15 to six species. On the other hand, four cases would suggest further splitting, resulting in a potential increase from four to nine species. Applying both corrections strictly to the initial number of 175 ad hoc-delimited species (plus one subspecies) would yield 171 species, a minor difference of only 2.3%.

ABGD on the subset of 716 *Cora* terminals returned stable results for relative gap width settings X = 1.0, 0.5, and 0.1. Setting X = 1.0 estimated between 147 and 231 species, corresponding to a barcode gap distance between 0.004 and 0.011 (Table 3). The three best-scoring ASAP scenarios (lowest score) estimated 161 species, with the following two scenarios ranging between 128 and 205 species, corresponding the threshold distances of 0.0086 (n = 3), 0.0130, and 0.0046, respectively (Table 3). The best-scoring scenario (161 species), therefore, appears to be a reasonable approximation.

Table 3. Species delimitation methods utilized and their respective results.

Dataset	Method	Settings	Estimate	Other Metrics:			
Cora 1325 Terminals	bPTP	min	791 (mean)	708 (min)	889 (max)		
Cora 1325 Terminals	GMYC	single	189	145–237 (confidence interval)	36.02878 (likelihood ratio)		1.50E-08 (LR Test)
Cora 1325 Terminals	*ad hoc*		265				
				Extrapolated			Barcode gap/threshold distance
Cora 716 Terminals	*ad hoc*		175	265			N/A
Cora 716 Terminals	DNADIST		171	259			0.0060
					Prior maximal distance		Barcode gap/threshold distance
Cora 716 Terminals	ABGD	Partition 2	231	350	0.001668		0.0040
Cora 716 Terminals	ABGD	Partition 3	231	350	0.002783		0.0040
Cora 716 Terminals	ABGD	Partition 4	147	223	0.004642		0.0110
					ASAP score	*p*-value	Barcode gap/threshold distance
Cora 716 Terminals	ASAP	K80-2	161	244	11.0	0.1620	0.0086
Cora 716 Terminals	ASAP	K80-2	128	194	13.5	0.0025	0.0130
Cora 716 Terminals	ASAP	K80-2	205	310	17.5	0.2020	0.0046
Cora 716 Terminals	ASAP	JC69	161	244	9.5	0.1300	0.0086
Cora 716 Terminals	ASAP	JC69	143	217	18.5	0.0796	0.0102
Cora 716 Terminals	ASAP	JC69	205	310	18.5	0.246	0.0046
Cora 716 Terminals	ASAP	Simple	161	244	8.5	0.1400	0.0086
Cora 716 Terminals	ASAP	Simple	205	310	17.0	0.2260	0.0046

Extrapolation of these distance-based approaches to the entire taxon set (265 ad hoc-delimited species) would result in estimates of 259 species (DNADIST), 223–350 species (ABGD), and (194–)244–310 species (ASAP).

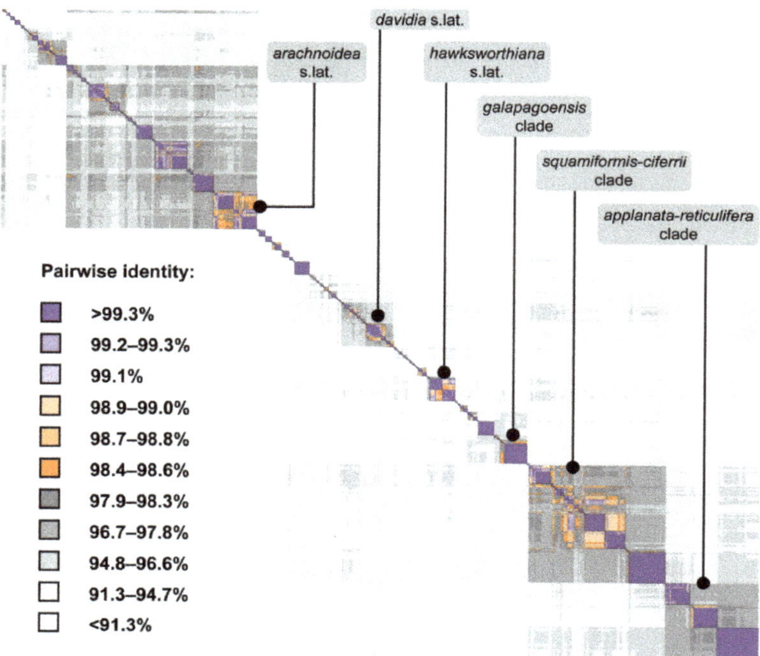

Figure 7. Barcoding gap analysis using DNADIST in BioEdit. The different colors represent pairwise identity percentages (see legend). Highlighted clades show instances where the DNADIST and the ad hoc approaches inferred different species delimitations, potentially over- or underestimating species richness.

3.5. ITS-Based BLAST Performance

BLAST performance was similar overall among the four scenarios (ITS1 region only, ITS1 including ASVs, ITS2 region only, and short subterminal ITS2 region). In all four cases, BLAST hits with other terminals of the same ad hoc-delimited species yielded E scores comparable to those obtained with self hits, although slightly lower on average, reflecting infraspecific variation (Figure 8). E scores also discriminated well between hits and misses. Upon closer examination, however, the four scenarios showed subtle differences. The best level of discrimination between hits and misses, and hence the highest level of accuracy was found with the ITS2 region: 88% of the queries accurately discriminated between hits and misses (min hit score > max miss score); however, in 53% the min hit score was 10% larger than the max miss score and in 23% of the queries it was 25% larger. The second best-performing subset was the ITS1 region, with 74%, 32%, and 13%, respectively, still spanning a large number of accurate queries but overall with a lower BLAST gap. When incorporating all available ASVs (including smaller fragments from Sanger and HTS) the ITS1 region performed slightly worse (66%, 25%, 11%), due to a number of incomplete query sequences missing parts of diagnostic regions. The short subterminal ITS2 string, making up about one fourth of the entire ITS2 region, had a larger BLAST gap than the other three scenarios, with 27% and 41% of the queries having the min hit score more than 25% and 10% above the max miss score, respectively, but, on the other hand, only 54% of the queries had a min hit score larger than the max miss score, resulting in potential inaccuracy.

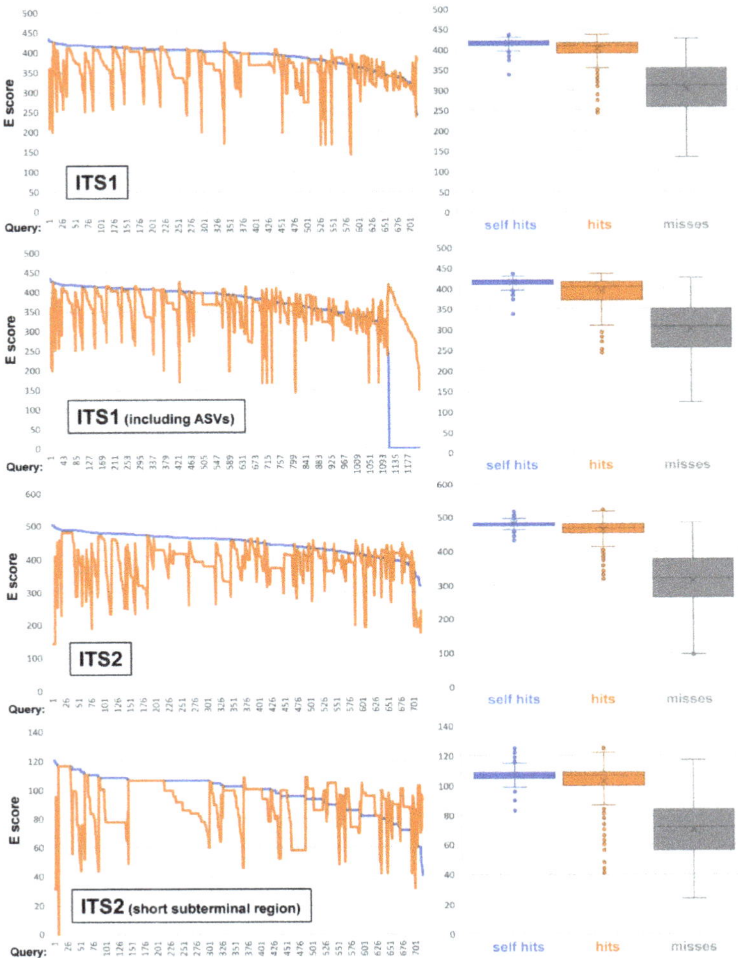

Figure 8. BLAST results for the four different scenarios (ITS1 only, ITS1 including ASVs, ITS2, and short subterminal ITS2 region). The left column of graphs shows the terminals for each subset ranked by the minimum E score for hits (blue line), as opposed by the maximum E score for misses (orange line). The right row of graphs shows the corresponding overall distribution of E scores for self-hits, hits, and misses, for each subset.

3.6. Phenotype Assessment

Total phylogenetic distance between possible pairs of species ranged between 0 and 120 substitutions, and total phenotype distance between 0 and 16 steps or character state differences (Figure 9). There was no discernible correlation between total phylogenetic and phenotype distance, i.e., phenotype distance did not depend on genetic distance. Mean total phenotype distance over all pairs of species was 7.18 (±2.01); if only pairs of species with a phylogenetic distance of 10 or less were considered, mean total phenotype distance was only marginally different (7.29 ± 2.11). Additionally, if only pairs of species representing sister clades were considered, no difference was detected in mean total phenotype distance (7.23 ± 3.19). Thus, on average, more closely related species were not more phenotypically similar to each other than more distantly related species.

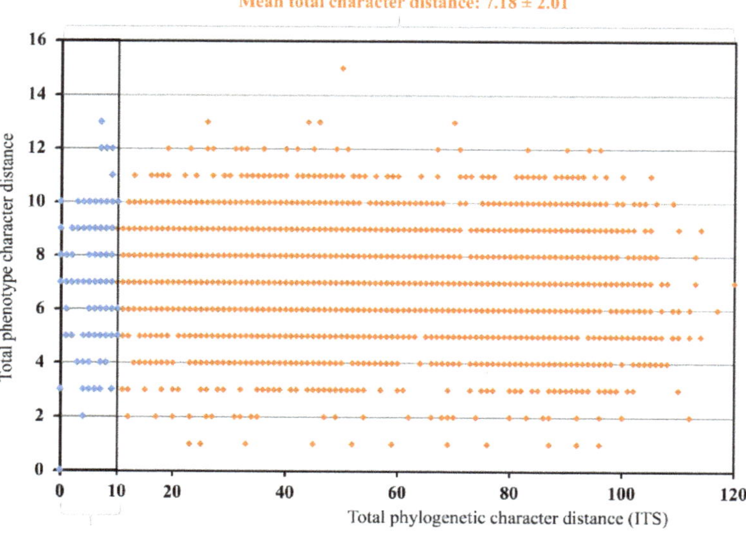

Figure 9. Distance between possible pairs of species for ITS (0–120 bp substitutions), and phenotype (0–16 characters steps). Blue dots on the left side of the graph correspond to closely related species pairs only, orange dots to all other species pair comparisons.

Using only the 11 morphological, anatomical, and chemical characters (columns 2–12 in Table S3), we differentiated 82 distinct phenotypes for the 87 included species of *Cora* (excluding *Corella*). Three phenotypes were shared by two lineages each, and one phenotype was shared between three lineages (Figure 10). There was only one instance of two lineages or species that could be considered eu-(phylo-)cryptic, namely the related *C. squamiformis* and *C. terricoleslia* (Figure 10); both are terrestrial species known from the central Andes (Bolivia). However, the two are not sister species and so it remains unclear whether the congruent phenotype is the result of symplesiomorphy (cryptic speciation) or homoplasy. There were two instances of kapo-(phylo-)cryptic ("near-cryptic") sister species (Figure 10), namely *C. minor* and *C. paraminor* (both known from Costa Rica), and *C. hirsuta* and *C. schizophylloides* (both known from Colombia). In both cases, the two sister species are phylogenetically distinct, with four positional ITS differences in *C. minor* vs. *C. paraminor* and 11 positional ITS differences in *C. hirsuta* vs. *C. schizophylloides* (File S10), and the phylogenetic separation is supported by a single character in each case (i.e., minor differences in the phenotype): short vs. no sutures in *C. minor* vs. *C. paraminor* (a rather subtle feature), and a strigose vs. setose lobe surface in *C. hirsuta* vs. *C. schizophylloides*; the latter two also differing in substrate preference (terrestrial vs. epiphytic). Furthermore, there was one case of two closely related, allo-(phylo-)cryptic (i.e., "semi-cryptic") species, namely *C. applanata* vs. *C. reticulifera*: whereas the first is known from the Andes (documented by 32 sequenced samples), the second appears to be restricted to (south-)eastern Brazil and Uruguay (documented by 66 sequenced samples; Figures 10 and S1, Table S4); both exhibit 16 positional ITS differences (File S10) and are, hence, clearly separated phylogenetically, but do not show any discernable phenotypical differences with respect to the tested characters. Finally, there were two instances of pseudo-(phylo-)cryptic species (i.e., the same phenotype having evolved in homoplasy in distantly related lineages), one encompassing three species (*C. campestris* vs. *C. caliginosa* vs. *C. terrestris* in blue) and the other one two species (*C. cuzcoensis* vs. *C. davibogotana* in green; Figure 10). Since only eu-, kapo-, and allo-(phylo-)cryptic lineages could be considered as potential "false positives" (i.e., taxa interpreted as species that may in fact represent infraspecific lineages), the potential error of overestimation of the species richness estimate in this dataset, by using phylogeny as sole evidence (i.e., potential taxonomic overinflation), would only be four species, out of 87 (i.e., less than 5%).

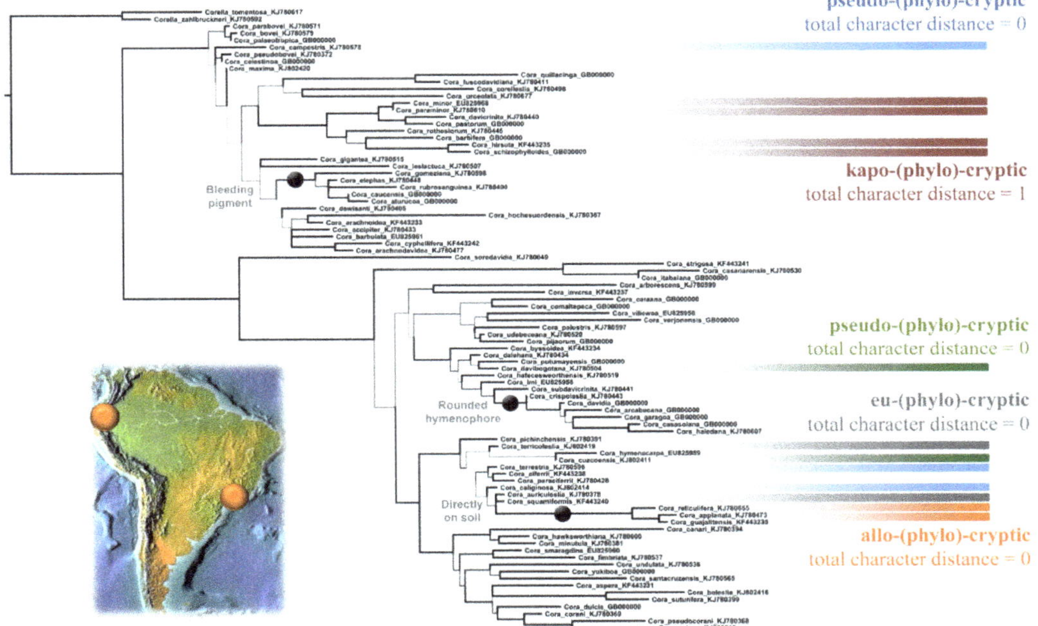

Figure 10. Clade structure (87 *Cora* species plus two *Corella* outgroup) highlighting types of crypticity observed throughout the tree. The inset map highlights the allopatric distribution of *C. applanata* (northern Andes) vs. *C. reticulifera* (southeastern Brazil).

Our analysis of the TL, CI, RI, RC, and HI indices derived from phenotypic characters, comparing random, DNA-based, and phenotype-based trees, showed that trees based on phenotype alone were substantially more structured in terms of their phenotypic characters than DNA-based trees (Figure 11). DNA-based trees were also always closer to random trees than to phenotype-based trees in their phenotypic signal. This suggests a relatively high level of homoplasy for individual character states and a low level of character state intercorrelation. On the other hand, the mean pairwise character distance of over seven steps indicates that species generally have distinct morphologies, supporting their phylogenetic delimitation. Overall, closely related species usually look distinct and have different substrate ecologies, whereas distantly related species are more likely to appear similar.

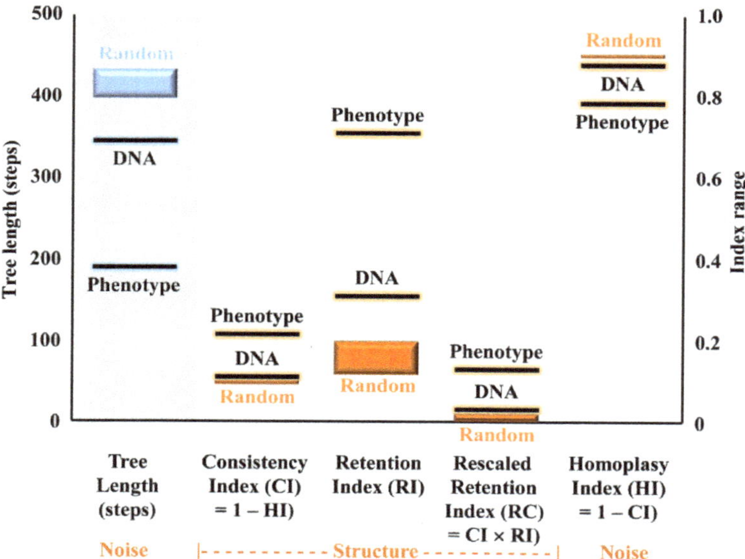

Figure 11. Analyses showing TL, CI, RI, RC, and HI indices derived from phenotypic characters comparing random (Random), DNA-based (DNA), and phenotype-based (Phenotype) trees, assuming that DNA-based trees represent the underlying "true" phylogeny. For DNA-based trees, the corresponding index is always between that of random and phenotype-based trees, but in all cases closer to the random trees, indicating that phenotype structure correlates little with deeper nodes and mostly with terminal lineages.

Although there was a strong correlation between phylogenetically defined clades interpreted as species and their associated phenotypic characters, phenotype structure became generally diffuse at deeper nodes, as evidenced by the lack of correlation between phylogenetic and phenotypic distance. However, limited phylogenetic structure was still found for some characters at higher clade levels. Three small clades uniformly contained species with specific phenotype character states that may represent synapomorphies for these clades: one clade containing all species with a bleeding pigment formed upon rewetting previously dried collections (*C. rubrosanguinea* clade); one clade with species forming adnate, rounded, confluent hymenophores different from the concentrically shaped hymenophores in most other species (*C. garagoa* clade); and one clade with species colonizing naked soil with completely flattened, adnate thalli (*C. reticulifera-applanata* clade; Figure 10).

3.7. Intragenomic ITS Variation

Our side-by-side comparison of the same samples sequenced with Illumina and Sanger sequencing yielded 89 specimens for this analysis. However, for 14 of these samples, the regions sequenced were non-overlapping, with less than 50 bp of overlap, or of low quality and, therefore, not included.

For the rest, or 75 samples, 55 (73%) of the Sanger sequence did not show any ambiguity, 11 (15%) showed one ambiguity (=fixed allele), while nine (12%) showed two or more ambiguities (Figure 12a). As for the ASVs, 290 ASVs were detected for these 75 samples, ranging from 1 to 11 per sample. About 30% of the ASVs matched the Sanger sequence exactly (=0 singletons), while most ASVs (61%) presented only one different base pair in comparison to the Sanger sequence (=1 singleton) (Figure 12b). Most samples produced multiple ASVs, but were usually dominated by one or two haplotypes.

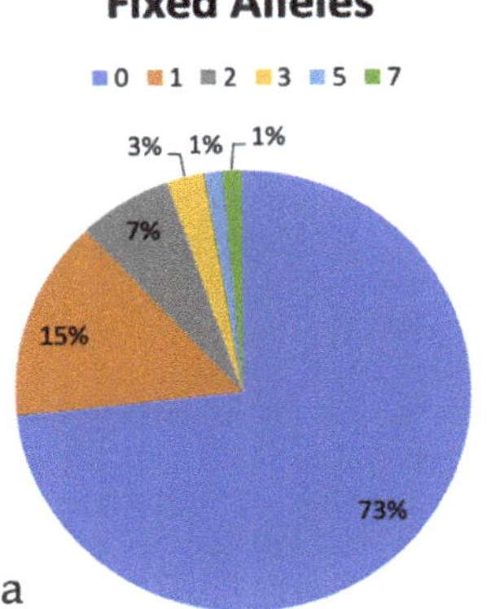

Figure 12. ITS1 variation observed within specimens. The fixed alleles graph (**a**) shows the percentages in which Sanger sequences presented ambiguities (0–7) in the ITS1 (e.g., an "R" whereas the Illumina reads showed either an "A" or "G", the majority showing none). The singleton alleles graph (**b**) shows the percentage of ASVs in which Illumina reads differed from the Sanger sequences by 0–7 base pairs aside from the fixed alleles (e.g., 61% of all ASVs showed 1 bp different then the corresponding Sanger sequence they were being compared to).

4. Discussion

4.1. Use of Archival Specimens

One of the most important results of our study is the finding that over 75% of the historical and nearly 94% of the fresh collected lichen samples yielded sequences with the Illumina platform. Using Sanger sequencing, far lower success rates were achieved for both historical (19%) and fresh (58%) collections. The reasons for the comparatively low success rate using Sanger sequencing on the newly collected material are unknown; in previous attempts to sequence fresh material from these genera, we typically achieved a success rate of 70–90% using Sanger technology. Initial preservation of the material also poses challenges on recent collections, especially if specimens cannot be properly dried when in the field for more than one day and carefully curated shortly afterwards. Nonetheless, the highly successful Illumina sequencing of these added fresh specimens indicates that DNA degradation had not progressed too far and even when these exact samples represented more difficulties with Sanger, Illumina sequencing worked satisfactorily. Regarding historical collections, similar or even higher success rates as in our study have been obtained in other recent studies using HTS to obtain DNA sequences from historical lichen collections [19–21], indicating that archival specimens available in herbaria and fungaria around the world may potentially yield usable sequences if the proper DNA extraction and sequencing approaches are taken. This includes many rare taxa, potentially extinct taxa that are no longer found in nature, and other taxa of unique value for which it is difficult to gather fresh material. The fact that so few published sequences currently exist for historical lichen collections is, thus, likely a consequence of the only recently growing awareness of the potential of advanced molecular sequencing methods to unlock these resources [12,13].

It should not be surprising that Illumina HTS yielded significantly better results than Sanger sequencing for archival specimens. DNA fragmentation in ancient herbarium samples is a well-documented phenomenon [1,78], including in lichen-forming fungi [16,18–20,79,80]. Considering this, and other postmortem damage known to take place in collected specimens [16,80], the success rate for the samples studied here is remarkably high, which may, in part, be explained by the generally higher sequencing success for Basidiomycota among fungal collections [24,25].

The addition of 318 new sequences from historical collections of *Cora* and *Corella*, including those of *Cora timucua* [23], makes this group one of the best represented among lichens with regard to historical sequences. With 92% of the 2988 GBIF occurrences known for this group (as of 31 January 2022) representing preserved samples (i.e., with a voucher deposited somewhere, not based on observations only), it is clear that historical collections are an invaluable resource that can be used in an integrative framework for describing and detecting new species and inferring relationships among them.

Our results demonstrate that the age of a specimen has some effect on the sequencing success rate, especially with regard to Sanger sequencing. Substrate type may also affect the success rate. However, it appears that other factors may play an important role in the degradation of DNA, such as the techniques employed at the time of collection to dry and preserve the specimens. Since these methods are usually not indicated on the specimen label, it is impossible to discern their potential impact from the role of age or substrate type as a determinant of sequencing success. Given the relatively good success rate we achieved, the fact that age is not a main determinant of success is notable. One important factor to consider is poikilohydry, which plays a major role in diurnal metabolism of lichens and directly relates to mechanisms protecting the DNA [81,82]. Perhaps lichens that undergo pronounced and/or prolonged dry periods maintain effective DNA protection mechanisms, whereas in lichen growing under frequently or permanently humid conditions, the DNA may be less protected from desiccation, a hypothesis already considered by Kistenich et al. [19]. One may, for instance, expect that species growing under more extreme water stress conditions, such as in southern South America or in the high Andes above 4000 m, would show better sequencing success even in archival specimens. However, without systematic comparison across different lichen taxa and a variety of habitats, this remains speculation, and how this would translate into sequencing success rates in *Cora* is unclear.

An important challenge to sequencing old samples is contamination, stemming from three potential sources: (1) fungi of the microbiome already present in the sample when collected; (2) fungal contaminants emerging due to specimen handling and preservation (e.g., molds); and (3) laboratory contaminants, which are particularly an issue with the highly sensitive HTS approaches. Although the first two potential sources of contamination in archival specimens are beyond the control of the investigator, laboratory contamination can be avoided or reduced to a minimum by applying recommended best practices, such as: (1) avoiding plate extractions and using individual tubes instead; (2) using extreme care when handling specimens and extracts (e.g., wearing gloves, sterilizing all equipment, especially forceps, etc.); (3) extracting DNA under sterile conditions, such as those found on ancient DNA laboratories; and (4) avoid working simultaneously with fresh and historical materials (e.g., in the same sequencing run), since more recently collected samples will tend to dominate a run, even with careful normalization of PCR input as performed here. In any case, potential contamination can be assessed posteriori by analyzing the taxonomic composition of fungal reads in a given sample.

In addition to the target mycobionts, we detected multiple other fungal taxa in our HTS samples. Although some of these, such as *Aspergillus* or *Penicillium*, may represent post-sampling contaminants, many others are frequent, opportunistic, or stable residents of the lichen mycobiome [34,83], including in *Cora* [84,85]. A great deal of evidence indicates the presence of obligately lichenicolous and endolichenic fungi and/or cortical yeasts in lichens [86–96]. With respect to known fungal groups previously found in lichens, in our material we detected ASVs belonging to members of the orders Cystofilobasidiales, Filobasidiales Tremellales, and Trichosporonales, all within the class Tremellomycetes, in 24% of the fresh samples and 22% of the historical samples (and in none of the negative controls) with varying quantity of reads. More than 21 distinct genera were detected within this class, with the most commonly observed genus being *Hannaella*, a basidiomycetous yeast genus found widely on leaf surfaces of various plants [97]. Even though the presence of fungi may influence humidity and ionic regimen on a thallus surface and subsequently transcriptomic response, without further data and knowledge of the development of these communities of species over time and their role in the symbioses, it is impossible to assign much significance to it at this point.

Presumptive laboratory contaminants were also sometimes observed in some samples. Generally, these were present in very low frequencies (e.g., fewer than 100 reads while the target mycobiont had 30,000 reads) and were not consistent with the inferred ecogeography of the corresponding species, which allowed their recognition and removal. If a sample only showed rare reads (less than 100) and no prevalent ASVs were present, the sample was considered unsuccessfully sequenced and was not included in the downstream analyses.

4.2. Assessment of ITS as a Barcoding Marker and Intragenomic Variation within ITS

Given the reported issues with the use of ITS as a fungal barcoding marker [34–37], and to address the possible argument that the observed phylogenetic diversity in *Cora* and *Corella* may in part be artifactual, we paid special attention to intragenomic variation in the ITS barcoding marker as evidenced by variation in the ASVs and ambiguities in Sanger sequences among our studied samples. Our expectation that potential ambiguities in Sanger sequences, mostly representing double peaks in the sequencing chromatograms, would match dominant and consistent SNPs in the corresponding Illumina ASVs was supported by the data, which allowed us to quantify this phenomenon reliably and in detail.

Since Illumina sequencing did not allow for amplification of the full ITS region, our comparisons were limited to samples for which we successfully sequenced the ITS1 region using both Sanger and Illumina for the same sample. For the 290 ASVs detected in these 75 samples, we detected no variation among the Sanger and the Illumina sequences (30%), one singleton (61%), two singletons (5%), three singletons (2%), and four to seven singletons (equal or less than 1%). Except for two samples for which only Illumina sequences were available, this variation had no effect on the phylogenetic placement of the target reads or the delimitation of phylogenetically defined lineages, supporting our earlier findings based on 454 pyrosequencing data that intragenomic ITS variation in *Cora* is low and does not lead to artifactual lineages [41]. The observed exceptions relate to two issues: either the target sequence was too short to cover lineage-diagnostic variation, then typically clustering at the base of the target clade or nearby; or the variation could be interpreted as potential hybridization and introgression, given that the ASVs detected were unique within the run, we are discarding the option of contamination in these specific cases, since multiple ASVs were available matching distinct alleles. Although this needs to be tested with genomic approaches, it would not be expected to lead to artifactual taxa, at least not in terms of species counts, as a hybrid component of the ITS would correspond to another, closely related species. We also considered mixed thalli (i.e., chimeric thalli between closely related species) as a potential source for this pattern, but with the methodological approach used here, this cannot be resolved.

Following earlier work with 454 pyrosequencing [32,41] and the increasing use of ITS1 for the Earth Microbiome Project and in other lichen studies [98], we adopted the ITS1 region as the default portion of the ITS barcoding marker for the Illumina sequencing employed here. Nonetheless, our analysis of full-length sequences indicates that at least in some clades, ITS2 showed better resolution for accurately detecting species using ITS-based BLAST identifications. This may be due to the more variable subterminal portion of the ITS2, which makes reliable alignments more challenging, but it appears to be highly discriminant, even between closely related species. Therefore, future metabarcoding using short reads should also attempt to sequence the ITS2 region [99–101], or focus on longer amplicon sequencing (PacBio, MINon, etc.) or shotgun sequencing, which might provide ways of overcoming sequence length limitations, although each of these techniques comes with its own disadvantages. However, most of the times, either ITS1 or ITS2 already provide enough resolution for species boundaries, especially within our integrative framework, making an Illumina an ideal method when assessing hundreds of samples [101].

In fungal barcoding approaches, it has been argued that single marker approaches, such as with ITS, may lead to inaccurate results or even cause taxonomic inflation if the data are not properly analyzed and interpreted [34,102]. In the case of *Cora* and *Corella*, ITS appears to work remarkably well, even in portions of the backbone, as evidenced by the high level of congruence between our single-marker ITS-based phylogeny and six-marker ASTRAL coalescent tree. At the level of terminal clades, a threshold ITS-based identify value of 99.4% appears to reliably discriminate between species, although some variation is observed which may depend on how recently a species-level clade evolved. In a few recently emerging species complexes, no absolute threshold value could be established and also the ITS-based BLAST results were partially diffuse, whereas in other cases, our initially delimited species-level clades may represent more than one lineage. Overall, these effects largely balance each other in terms of species counts, but may lead to inaccuracies in delimiting species in certain clades. Beyond single-marker ITS approaches, three other strategies could be used to test species delimitation in these cases: (1) multi-marker coalescent approaches [103]; (2) phylogenomics target capture approaches [104–106]; and (3) population genetics using microsatellites or RADSeq [107–109].

With regard to multi-marker approaches, our data show that ITS performed better for delimiting species than the protein-coding markers *RPB2* or *EF3*, but also the classical markers nuLSU, mtSSU, and mtLSU; in addition, the ITS marker is much easier to generate. Consequently, multi-marker approaches or alternative barcodes do not seem to constitute a promising next step in resolving problematic species complexes or refining the DNA barcoding approach in this group of basidiolichens. Phylogenomic approaches (e.g., target capture), have also shown limitations in resolving recently evolving species [106], and, therefore, we consider the RADseq approach as potentially useful to further assess difficult species complexes in *Dictyonematinae* in addition to our ongoing metagenomic analyses. For a general barcoding approach, however, including metabarcoding with HTS approaches, we recommend the continued use of the ITS marker, due to its high amplification success and the broad molecular framework it provides to establish species hypotheses in this group of lichen-forming Basidiomycota.

4.3. Accurate Assessment of Phylogenetic Diversity in Cora and Corella

The large amount of ITS data now available allowed us to assess phylogenetic diversity in this group of basidiolichens using various quantitative and semi-quantitative approaches. As an initial approach to establish species hypotheses, we used the same ad hoc delimitation employed in our previous studies [32,33], namely a combination of visual inspection of stem branch lengths and support versus within-clade branch length variation versus geographic origin of the samples. In the present case, this led to the distinction of 265 ad hoc species hypotheses for the entire dataset (including all ASVs) and 175 for the subset of near-complete ITS sequences. Distance-based quantitative approaches (DNADIST-based analysis, ABGD, ASAP) all resulted in numbers within the range of 128–231 for the subset tested and 194–350 for the entire dataset (extrapolated). In contrast, the tree-based method bPTP yielded much higher estimates (709–889 species for the entire dataset). GMYC inferred values more similar to those of distance-based methods, with 189 estimated species in an interval of 145–237. To what extent these estimates might be real remains unclear. If the example of *Usnea antarctica* versus *U. aurantiacoatra* is taken as reference, near-identical ITS patterns may indeed hide more than one species [107,108], and lack of ITS-based resolution is also known from other fungi [35–37]. It is, therefore, possible that clades currently delimited as a single species with our ad hoc approach or using distance-based methods represent more than one species, although a three-fold increase seems unlikely based on our current knowledge. Consequently, we consider our ad hoc approach reliable at this point, as it is closer to the middle range of distance-based estimates and far below the bPTP delimitation approach, which, in turn, showed highly contrasting results to all other methods. An integrative approach was also the solution Boluda and colleagues [110] proposed to disentangle the incongruencies of the use of chemistry, morphology, molecular data (including multiple species delimitation methods) or phylogeny alone, for species boundaries in the *Bryoria* sect. *Implexae* complex.

The inclusion of a large number of historical collections extended the geographic range of sequenced samples, but still left many areas with potential occurrence of *Cora* (and *Corella*) unsampled. Thus, compared to the present number of 265 species, our original prediction of 450 species [32] still provides a valid framework and it seems likely that this number will eventually be reached. Undersampled regions notably include the central and southern portion of the Andes (Peru, Bolivia, Chile, Argentina), but also large parts of Central America and western Mexico, as well as the Guyana Highlands.

Overall, the present number of formally described (102), phylogenetically distinguished (by our ad hoc approach, 265), and predicted (>450 [32]) that the species in *Cora* does not differ from the range of accepted species in the 25 largest ascomycete lichen genera, which lies between 170 and 820 [111]. As such, the diversity now recognized in *Cora* aligns well with other megadiverse lichenized genera, showing that certain basidiolichen groups may harbor a species diversity similar to the most speciose ascolichen groups, an idea that would have been dismissed by most lichenologists even just a decade ago. Indeed, the observed diversity in these basidiolichens is striking not because there are so many species but because it has not been recognized before, much less at this magnitude. Prior to Parmasto's monograph [30], six species had been formally described in this group (currently named *Cora bovei*, *C. ciferrii*, *C. glabrata*, *C. gyrolophia*, *C. pavonia*, and *C. reticulifera*) and three more in the genus *Corella* (*C. brasiliensis*, *C. tomentosa*, and *C. zahlbruckneri*); all nine had been synonymized by Parmasto under one taxon (*Dictyonema pavonium*). This historical number is remarkably low compared to other genera of similar size (e.g., *Sticta*), with hundreds of names established in the early literature. The main reason for the comparatively low number of historical epithets in *Cora* is that important field characters, such as color and substrate, are lost in herbarium specimens if not recorded at the time of collection, which were the primary source of access for researchers in the 19th century but also for modern monographers. This led to the lack of perception of size as an important character, as smaller herbarium specimens were sometimes interpreted as immature. Even field experience did not reveal the true nature of this group of basidiolichens, as the differences in ecology and morphology between specimens were interpreted as environmentally induced variation [112], a concept popular in the second half of the 20th century [113].

4.4. Level of Cryptic Speciation and Potential Taxonomic Inflation

The existence of hidden or unrecognized species within presumably well-known taxa is not an isolated phenomenon in fungi. In many presumably well-known taxa, such as the fly agaric (*Amanita muscaria* s.lat. [114]), the chanterelle (*Cantharellus cibarius* s.lat. [115–117]), the Lingzhi mushroom (*Ganoderma lucidum* s.lat. [118]), the true morels (*Morchella esculenta* s.lat. [119]), and the yellow speckleberry lichen (*Pseudocyphellaria crocata* s.lat. [120]), species delimitation studies using ITS and other markers have revealed a large number of previously unrecognized lineages.

Although some of these pose difficulties delimiting species phenotypically, other cases, such as *Pseudocyphellaria crocata* s.lat., often reveal taxonomically useful characters that had not been considered to be diagnostic before. In the genus *Cora*, given previous failures to properly recognize species diversity and the low number of characters useful for taxonomy, one would expect a number of over 250 species hypothesized from molecular data to go along with a high level of evolutionary crypticity, resulting in many species undistinguishable through their phenotype, potentially resulting in taxonomic inflation. Although the number of taxonomically useful characters in *Cora* is indeed limited, lacking for instance the diversity of spore types, vegetative propagules, or secondary compounds found in megadiverse ascolichen genera, the comparatively low number of 11 main phenotypes characters led to no less than 82 distinct phenotypes among the 87 analyzed species, rejecting the notion of largely cryptic speciation or taxonomic inflation in this genus. Instead, even with a low amount of perceived options to reliably distinguish species, we demonstrate that the combination of these characters yields sufficient information allowing to differentiate most species detected by molecular methods. Cases of identical phenotypes were in part found in distantly related lineages only (i.e., pseudophylocryptic), thus representing homoplasies rather than genuine cryptic speciation, whereas closely related species were mostly phenotypically distinct. Indeed, among 87 lineages, we identified only one case where two closely related species could not be distinguished by phenotype or chorology (euphylocryptic). Other cases differed either in one character state (kapophylocryptic) or in distribution (allophylocryptic). This supports the phenotype as useful for species-level taxonomy but renders phenotypic characters as of limited value when inferring phylogenetic relationships within this genus, with the exception of a few characters that correlate with larger clades.

Our results thus suggest that phenotype variation, species delimitation and the level of homoplasy in the basidiolichen genus *Cora* are comparable to large genera of lichenized Ascomycota, in which a limited set of phenotype characters leads to free or partially constrained combinations of character states in individual species. For example, in the crustose genus *Lecanora*, with 550 species [111], species are usually recognized by a combination of thallus morphology, apothecial disc color, epihymenial and excipular crystals, and chemistry [121–130], whereas in *Usnea*, a combination of growth form, branching pattern, thallus sectional structure, branch outgrowths and appendices, and secondary chemistry and pigments define species [131–136]. Other examples can be found in foliose Parmeliaceae review in [137], such as *Bulbothrix* [138], or the crustose genera *Caloplaca* [139,140], *Graphis* and *Allographa* [141,142]. Thus, in both Asco- or Basidiomycota, phenotypical characters may not correspond to molecular phylogenies at all clade levels [31], but they are useful in diagnosing closely related species within clades.

If the remarkable species diversity in *Cora* is largely not cryptic, the question must again be raised: why has it not been recognized before? As mentioned above, reasons can be looked for in the loss of important features in herbarium collections, similar to mushroom taxonomy, but also in the overinterpretation of variation as ecologically induced and not taxonomic. *Cora* is, therefore, not really a case of "hidden" diversity, but one of previously unrecognized or "overlooked" diversity.

The notion that phenotypically similar species of *Cora* are generally only distantly related could be explained by similar selective pressures in ecologically equivalent habitats, but in part also by free variation of a limited set of characters that may not represent functional traits. Once the phylogenetic diversity of *Cora* has been fully assessed phenotypically, this will be an exciting avenue for future studies. Fortunately, given the techniques to assess phenotype characters in herbarium collections [33], archival specimens for which sequence data are now available can be incorporated in such studies, providing a much broader geographical and ecological framework.

5. Conclusions

We conclude that DNA barcoding in the foliose genera *Cora* and *Corella* is indeed a success story, considering that our knowledge of this group of basidiolichens has increased from a single perceived species to one of the largest genera of lichen-forming fungi, detected largely through rigorous application of the fungal ITS barcoding marker within our integrative approach, combining: (1) visual inspection of the molecular data (alignments + phylogeny); (2) phenotypic and anatomical data; and (3) ecology, including habitat and substrate preferences. Almost all species of *Cora* and *Corella* have been documented molecularly from the onset, resulting in a broad, near-complete, and steadily-growing phylogenetic framework where additional sequence data can be quickly added, particularly those obtained from further historical specimens, including types.

Our highly successful HTS results indicate that ITS barcoding of collections collected from a couple decades to over a century ago can be used to extend our knowledge of species-level diversity. In particular, the addition of archival samples not only lead to the discovery of novel lineages but made the phylogenies more robust, helped to better assess species distributions, and thus remarkably improved our understanding of levels of inferred endemism.

The high number of species now recognized in *Cora* is not surprising if compared to large lichenized genera in Ascomycota. What is extraordinary is that the numerous species are not actually cryptic. We emphasize that our broad phylogenetic sampling was only partially responsible for this change, with assessments of the genus in the field being equally important in this case. The key element to uncover this previously unrecognized diversity was our integrative approach, which allowed us to test the reliability of novel taxonomic characters and question classic concepts about environmental variation, in which the addition of historical specimens was invaluable. Traditional concepts largely ignored phenotypically distinct characters, dismissing them as variability. However, it is important to acknowledge that we are at a privileged position in research, where we can employ molecular data to test these phenotypic-based hypotheses. At first glance, phenotypical characters seem less clear in basidiolichens, including its variation and the lack of features usually applicable to lichens in general. In ascolichens, traditional concepts have often focused on both morphological adaptations resulting from the lichen symbiosis as a whole and on characteristic features of the mycobiont. We postulate that the patterns of diversification seen in *Cora*, namely that phenotypic variation previously regarded as environmentally triggered is actually diagnostic of species-level lineages, will also hold true for many ascolichen genera (or for some lineages within) that have not yet been broadly assessed with molecular methods.

To assess evolutionary patterns related to so-called cryptic speciation, we, therefore, found it useful in this study to introduce more precise definitions of our concept of "crypticity", taking into account that species may often appear similar but are nevertheless not actually identical (kapo-(phylo-)cryptic vs. eu-(phylo-)cryptic), that phenotypically similar species might nevertheless not be closely related (pseudo-(phylo-)cryptic), and that species distributions patterns (allo- vs. sympatric) must be taken into account when looking at this phenomenon (allo-(phylo-)cryptic). These definitions might prove useful when evaluating cases of cryptic speciation in other lichens, or even more broadly across fungi, even though, in the case of *Cora*, the majority of the species should no longer be considered cryptic, but instead an example where phenotypic variation had not been yet linked with genetic divergence.

Supplementary Materials: The following supporting information can be downloaded at: https://www.mdpi.com/article/10.3390/d14040284/s1, Figure S1: ITS full-length ML tree; Table S1: List of material used in this study; Table S2: Species utilized on the phenotypic assessment and their GenBank number; Table S3: Scores for the phenotype matrix (20 characters) for the 89 species utilized; Table S4: Full species delimitation results for all specimens; and 11 supplemental files containing data analyzed (File S1: ITS sequence alignment utilized as input in the ML analyses; File S2: Six-marker trees utilized as input in ASTRAL-III, the map file to delimit species in ASTRAL and all six alignments utilized to generate the respective trees; File S3: Only the separated corresponding ITS alignment with the same terminals as ASTRAL six-marker data; File S4: Subset of the ITS sequence data limited to the genus *Cora*, retaining only terminals with less than 30% gaps; File S5: Subset of 758 terminals with (near) complete ITS sequences, similar to the *Cora* subset File S4, but also including *Acantholichen* and *Corella*; File S6: BLAST performance of dataset 1—with the ITS1 region only; File S7: BLAST performance of dataset 2—entire ITS region of all original terminals with less than 10% gaps, including Illumina ASVs; File S8: BLAST performance of dataset 3—with the ITS2 region only; File S9: BLAST performance of dataset 4—with a short terminal region of the ITS2; File S10: Alignment with 89 target sequences for the phenotypic assessment; File S11: Distance matrix utilized for the barcoding gap analyses).

Author Contributions: Conceptualization, M.D.F., J.D.L., M.G. and R.L.; methodology and analyses, M.D.F., R.L.; lab work, M.D.F., B.M.; funding acquisition, M.D.F., J.D.L., F.B., E.S. and R.L.; first manuscript versions, M.D.F., J.D.L. and R.L.; writing, reviewing and editing, all authors. All authors have read and agreed to the published version of the manuscript.

Funding: This research was funded by the National Science Foundation (DEB 0841405 and PRFB 1609022), the Smithsonian Institution (SI) and the National Museum of Natural History (Peter Buck Postdoctoral Fellowship), the Global Genome Initiative and the SI DNA Barcode Network.

Institutional Review Board Statement: Not applicable.

Informed Consent Statement: Not applicable.

Data Availability Statement: The data presented in this study are available in Supplementary Materials and deposited at GenBank.

Acknowledgments: M.D.F., J.D.L. and R.L. would like to thank the National Science Foundation for financial support and all collaborators who have helped in field expeditions, as well as those who sent material throughout the years. M.D.F. also wishes to thank the Smithsonian Institution and the National Museum of Natural History for a Peter Buck Postdoctoral Fellowship, a grant from the SI DNA Barcode Network (FY18 Award Cycle, PI MDF) and an award from the Global Genome Initiative (GGI-Exp_Sci-2016-069, PIs ES and MDF). M.D.F. would also like to thank Katherine (Katie) Murphy, Niamh Redmond, José Zuniga, Jeffrey Hunt, Jonathan (Jon) Coddington, Sergei Drovetski, Kenan Matterson, Gabriel (Gabe) Johnson, and Julia Steier, for their help in the LAB and project support at the SI-NMNH. Nancy McInerney is thanked for her help and permission to utilize equipment at the CCG. Morgan Gostel and Jerrod Stone from the Botanical Research Institute of Texas are acknowledged for their help with the microfluidics PCR and assessment of herbarium Sanger sequences, respectively. Galapagos specimens examined as part of this study were collected and examined from 2009 to 2013 as part of an NSF sponsored project (DEB 0841405), a project included in the Annual Operative Plan as part of the Galapagos Lichen Inventory by the Charles Darwin Foundation (CDF), approved at the time by the Directorate of the Galapagos National Park (DPNG). We are particularly grateful to successive scientific directors of the research station and technical directors of the Galapagos National Park for their continued support. The Galapagos Lichen Inventory is now part of the national biodiversity assessment "Biodiversidad Genética del Ecuador", led by the Instituto Nacional de Biodiversidad del Ecuador (INABIO). For research and specimen permits we are particularly indebted to Danny Rueda, Daniel Lara Solís, Galo Quedaza and Victor Carrión from DPNG, and Diego Inclán, Francisco Prieto and Rosa Batallas (INABIO). This publication is contribution number 2447 of the Charles Darwin Foundation for the Galapagos Islands. Lastly, we thank the reviewers for their contributions to the improvement of this manuscript.

Conflicts of Interest: The authors declare no conflict of interest.

References

1. Funk, V.A. Collections-based science in the 21st Century. *J. Syst. Evol.* **2018**, *56*, 175–193. [CrossRef]
2. Holmes, M.W.; Hammond, T.T.; Wogan, G.; Walsh, R.E.; LaBarbera, K.; Wommack, E.A.; Martins, F.M.; Crawford, J.C.; Mack, K.L.; Bloch, L.M.; et al. Natural history collections as windows on evolutionary processes. *Mol. Ecol.* **2016**, *25*, 864–881. [CrossRef] [PubMed]
3. Salvador, R.B.; Cunha, C.M. Natural history collections and the future legacy of ecological research. *Oecologia* **2020**, *192*, 641–646. [CrossRef] [PubMed]
4. MacLean, H.J.; Nielsen, M.E.; Kingsolver, J.G.; Buckley, L.B. Using museum specimens to track morphological shifts through climate change. *Philos. Trans. R. Soc. B Biol. Sci.* **2019**, *374*, 20170404. [CrossRef]
5. Hedrick, B.P.; Heberling, J.M.; Meineke, E.K.; Turner, K.G.; Grassa, C.J.; Park, D.S.; Kennedy, J.; A Clarke, J.; A Cook, J.; Blackburn, D.C.; et al. Digitization and the Future of Natural History Collections. *BioScience* **2020**, *70*, 243–251. [CrossRef]
6. Schindel, D.E.; Cook, J.A. The next generation of natural history collections. *PLOS Biol.* **2018**, *16*, e2006125. [CrossRef]
7. Kauserud, H.; Stige, L.C.; Vik, J.O.; Økland, R.H.; Høiland, K.; Stenseth, N.C. Mushroom fruiting and climate change. *Proc. Natl. Acad. Sci. USA* **2008**, *105*, 3811–3814. [CrossRef]
8. Ristaino, J.B. The Importance of Mycological and Plant Herbaria in Tracking Plant Killers. *Front. Ecol. Evol.* **2020**, *7*. [CrossRef]
9. Purvis, O.; Dubbin, W.; Chimonides, P.; Jones, G.; Read, H. The multi-element content of the lichen *Parmelia sulcata*, soil, and oak bark in relation to acidification and climate. *Sci. Total Environ.* **2008**, *390*, 558–568. [CrossRef]
10. Agnan, Y.; Séjalon-Delmas, N.; Probst, A. Comparing early twentieth century and present-day atmospheric pollution in SW France: A story of lichens. *Environ. Pollut.* **2013**, *172*, 139–148. [CrossRef]
11. Minganti, V.; Drava, G.; de Pellegrini, R.; Modenesi, P.; Malaspina, P.; Giordani, P. Temporal trends (1981–2007) of trace and rare earth elements in the lichen *Cetraria islandica* (L.) Ach. from Italian herbaria. *Chemosphere* **2014**, *99*, 180–185. [CrossRef] [PubMed]
12. Stiller, M.; Knapp, M.; Stenzel, U.; Hofreiter, M.; Meyer, M. Direct multiplex sequencing (DMPS)—A novel method for targeted high-throughput sequencing of ancient and highly degraded DNA. *Genome Res.* **2009**, *19*, 1843–1848. [CrossRef] [PubMed]

13. Bieker, V.; Martin, M.D. Implications and future prospects for evolutionary analyses of DNA in historical herbarium collections. *Bot. Lett.* **2018**, *165*, 409–418. [CrossRef]

14. Staats, M.; Erkens, R.; van de Vossenberg, B.T.; Wieringa, J.; Kraaijeveld, K.; Stielow, B.; Geml, J.; Richardson, J.-E.; Bakker, F.T. Genomic Treasure Troves: Complete Genome Sequencing of Herbarium and Insect Museum Specimens. *PLoS ONE* **2013**, *8*, e69189. [CrossRef]

15. Cong, Q.; Shen, J.; Zhang, J.; Li, W.; Kinch, L.N.; Calhoun, J.V.; Warren, A.D.; Grishin, N.V. Genomics Reveals the Origins of Historical Specimens. *Mol. Biol. Evol.* **2021**, *38*, 2166–2176. [CrossRef]

16. Sohrabi, M.; Myllys, L.; Stenroos, S. Successful DNA sequencing of a 75 year-old herbarium specimen of *Aspicilia aschabadensis* (J. Steiner) Mereschk. *Lichenol.* **2010**, *42*, 626–628. [CrossRef]

17. Redchenko, O.; Vondrák, J.; KOŠNAR, J. The oldest sequenced fungal herbarium sample. *Lichenol.* **2012**, *44*, 715–718. [CrossRef]

18. Bendiksby, M.; Mazzoni, S.; Jørgensen, M.H.; Halvorsen, R.; Holien, H. Combining genetic analyses of archived specimens with distribution modelling to explain the anomalous distribution of the rare lichen *Staurolemma omphalarioides*: Long-distance dispersal or vicariance? *J. Biogeogr.* **2014**, *41*, 2020–2031. [CrossRef]

19. Kistenich, S.; Halvorsen, R.; Schrøder-Nielsen, A.; Thorbek, L.; Timdal, E.; Bendiksby, M. DNA Sequencing Historical Lichen Specimens. *Front. Ecol. Evol.* **2019**, *7*. [CrossRef]

20. Gueidan, C.; Elix, J.A.; McCarthy, P.M.; Roux, C.; Mallen-Cooper, M.; Kantvilas, G. PacBio amplicon sequencing for metabarcoding of mixed DNA samples from lichen herbarium specimens. *MycoKeys* **2019**, *53*, 73–91. [CrossRef]

21. Gueidan, C.; Li, L. A long-read amplicon approach to scaling up the metabarcoding of lichen herbarium specimens. *MycoKeys* **2022**, *86*, 195–212. [CrossRef] [PubMed]

22. Schmull, M.; Dal-Forno, M.; Lücking, R.; Cao, S.; Clardy, J.; Lawrey, J.D. *Dictyonema huaorani* (Agaricales: Hygrophoraceae), a new lichenized basidiomycete from Amazonian Ecuador with presumed hallucinogenic properties. *Bryologist* **2014**, *117*, 386–394. [CrossRef]

23. Lücking, R.; Kaminsky, L.; Perlmutter, G.B.; Lawrey, J.D.; Forno, M.D. *Cora timucua* (Hygrophoraceae), a new and potentially extinct, previously misidentified basidiolichen of Florida inland scrub documented from historical collections. *Bryologist* **2020**, *123*, 657–673. [CrossRef]

24. Osmundson, T.W.; Robert, V.A.; Schoch, C.L.; Baker, L.J.; Smith, A.; Robich, G.; Mizzan, L.; Garbelotto, M.M. Filling Gaps in Biodiversity Knowledge for Macrofungi: Contributions and Assessment of an Herbarium Collection DNA Barcode Sequencing Project. *PLoS ONE* **2013**, *8*, e62419. [CrossRef] [PubMed]

25. Voitk, A.; Saar, I.; Lücking, R.; Moreau, P.-A.; Corriol, G.; Krisai-Greilhuber, I.; Thorn, R.G.; Hay, C.R.; Moncada, B.; Gulden, G. Surprising Morphological, Ecological and ITS Sequence Diversity in the *Arrhenia acerosa* Complex (Basidiomycota: Agaricales: Hygrophoraceae). *Sydowia* **2021**, *73*, 133. [CrossRef]

26. Schoch, C.L.; Seifert, K.A.; Huhndorf, S.; Robert, V.; Spouge, J.L.; Levesque, C.A.; Chen, W.; Fungal Barcoding Consortium; Fungal Barcoding Consortium Author List; Bolchacova, E.; et al. Nuclear ribosomal internal transcribed spacer (ITS) region as a universal DNA barcode marker for Fungi. *Proc. Natl. Acad. Sci. USA* **2012**, *109*, 6241–6246. [CrossRef]

27. Lawrey, J.D.; Lücking, R.; Sipman, H.J.; Chaves, J.L.; Redhead, S.A.; Bungartz, F.; Sikaroodi, M.; Gillevet, P.M. High concentration of basidiolichens in a single family of agaricoid mushrooms (Basidiomycota: Agaricales: Hygrophoraceae). *Mycol. Res.* **2009**, *113*, 1154–1171. [CrossRef]

28. Dal-Forno, M.; Lawrey, J.D.; Sikaroodi, M.; Bhattarai, S.; Gillevet, P.M.; Sulzbacher, M.; Lücking, R. Starting from scratch: Evolution of the lichen thallus in the basidiolichen *Dictyonema* (Agaricales: Hygrophoraceae). *Fungal Biol.* **2013**, *117*, 584–598. [CrossRef]

29. Lodge, D.J.; Padamsee, M.; Matheny, P.B.; Aime, M.C.; Cantrell, S.A.; Boertmann, D.; Kovalenko, A.; Vizzini, A.; Dentinger, B.T.M.; Kirk, P.M.; et al. Molecular phylogeny, morphology, pigment chemistry and ecology in Hygrophoraceae (Agaricales). *Fungal Divers.* **2014**, *64*, 1–99. [CrossRef]

30. Parmasto, E. The Genus *Dictyonema* ('Thelephorolichenes'). *Nova Hedwig.* **1978**, *29*, 99–144.

31. Hawksworth, D.L. A New Name for *Dictyonema pavonium* (Swartz) Parmasto. *Lichenologist* **1988**, *20*, 101. [CrossRef]

32. Lücking, R.; Dal-Forno, M.; Sikaroodi, M.; Gillevet, P.M.; Bungartz, F.; Moncada, B.; Yánez-Ayabaca, A.; Chaves, J.L.; Coca, L.F.; Lawrey, J.D. A single macrolichen constitutes hundreds of unrecognized species. *Proc. Natl. Acad. Sci. USA* **2014**, *111*, 11091–11096. [CrossRef] [PubMed]

33. Lücking, R.; Forno, M.D.; Moncada, B.; Coca, L.F.; Vargas-Mendoza, L.Y.; Aptroot, A.; Arias, L.J.; Besal, B.; Bungartz, F.; Cabrera-Amaya, D.M.; et al. Turbo-taxonomy to assemble a megadiverse lichen genus: Seventy new species of *Cora* (Basidiomycota: Agaricales: Hygrophoraceae), honouring David Leslie Hawksworth's seventieth birthday. *Fungal Divers.* **2016**, *84*, 139–207. [CrossRef]

34. Lücking, R.; Leavitt, S.D.; Hawksworth, D.L. Species in lichen-forming fungi: Balancing between conceptual and practical considerations, and between phenotype and phylogenomics. *Fungal Divers.* **2021**, *109*, 99–154. [CrossRef]

35. Lücking, R.; Aime, M.C.; Robbertse, B.; Miller, A.N.; Ariyawansa, H.A.; Aoki, T.; Cardinali, G.; Crous, P.W.; Druzhinina, I.S.; Geiser, D.M.; et al. Unambiguous identification of fungi: Where do we stand and how accurate and precise is fungal DNA barcoding? *IMA Fungus* **2020**, *11*, 1–32. [CrossRef]

36. Stielow, J.; Lévesque, C.; Seifert, K.; Meyer, W.; Irinyi, L.; Smits, D.; Renfurm, R.; Verkley, G.; Groenewald, M.; Chaduli, D.; et al. One fungus, which genes? Development and assessment of universal primers for potential secondary fungal DNA barcodes. *Persoonia-Mol. Phylogeny Evol. Fungi* **2015**, *35*, 242–263. [CrossRef]

37. Al-Hatmi, A.M.; Ende, A.G.V.D.; Stielow, J.B.; van Diepeningen, A.D.; Seifert, K.A.; McCormick, W.; Assabgui, R.; Gräfenhan, T.; De Hoog, G.S.; Levesque, C.A. Evaluation of two novel barcodes for species recognition of opportunistic pathogens in *Fusarium. Fungal Biol.* **2016**, *120*, 231–245. [CrossRef]

38. Kelly, L.J.; Hollingsworth, P.M.; Coppins, B.J.; Ellis, C.J.; Harrold, P.; Tosh, J.; Yahr, R. DNA barcoding of lichenized fungi demonstrates high identification success in a floristic context. *N. Phytol.* **2011**, *191*, 288–300. [CrossRef]

39. Pino-Bodas, R.; Martín, M.P.; Burgaz, A.R.; Lumbsch, H.T. Species delimitation in *Cladonia* (Ascomycota): A challenge to the DNA barcoding philosophy. *Mol. Ecol. Resour.* **2013**, *13*, 1058–1068. [CrossRef]

40. Dal Forno, M.; Bungartz, F.; Yánez-Ayabaca, A.; Lücking, R.; Lawrey, J.D. High levels of endemism among Galapagos basidiolichens. *Fungal Divers.* **2017**, *85*, 45–73. [CrossRef]

41. Lücking, R.; Lawrey, J.D.; Gillevet, P.M.; Sikaroodi, M.; Dal-Forno, M.; Berger, S.A. Multiple ITS Haplotypes in the Genome of the Lichenized Basidiomycete *Cora inversa* (Hygrophoraceae): Fact or Artifact? *J. Mol. Evol.* **2014**, *78*, 148–162. [CrossRef] [PubMed]

42. Bradshaw, M.; Grewe, F.; Thomas, A.; Harrison, C.H.; Lindgren, H.; Muggia, L.; Clair, L.L.S.; Lumbsch, H.T.; Leavitt, S.D. Characterizing the ribosomal tandem repeat and its utility as a DNA barcode in lichen-forming fungi. *BMC Evol. Biol.* **2020**, *20*, 2. [CrossRef] [PubMed]

43. Dal Forno, M.; Lawrey, J.D.; Sikaroodi, M.; Gillevet, P.M.; Schuettpelz, E.; Lücking, R. Extensive photobiont sharing in a rapidly radiating cyanolichen clade. *Mol. Ecol.* **2021**, *30*, 1755–1776. [CrossRef] [PubMed]

44. Dal Forno, M.; Kaminsky, L.; Lücking, R. IUCN Red List of Threatened Species: *Cora timucua*. 2020. Available online: https://www.iucnredlist.org/ (accessed on 23 January 2022).

45. Marotz, C.; Amir, A.; Humphrey, G.; Gaffney, J.; Gogul, G.; Knight, R. DNA extraction for streamlined metagenomics of diverse environmental samples. *BioTechniques* **2017**, *62*, 290–293. [CrossRef]

46. Gardes, M.; Bruns, T.D. ITS primers with enhanced specificity for basidiomycetes—Application to the identification of mycorrhizae and rusts. *Mol. Ecol.* **1993**, *2*, 113–118. [CrossRef]

47. White, T.J.; Bruns, T.; Lee, S.; Taylor, J. Amplification and direct sequencing of fungal ribosomal RNA genes for phylogenetics. In *PCR Protocols: A Guide to Methods and Applications*; Innis, M.A., Gelfand, D.H., Sninsky, J.J., White, T.J., Eds.; Academic Press: San Diego, CA, USA, 1990; pp. 315–322.

48. Caporaso, J.G.; Lauber, C.L.; Walters, W.A.; Berg-Lyons, D.; Huntley, J.; Fierer, N.; Owens, S.M.; Betley, J.; Fraser, L.; Bauer, M.; et al. Ultra-high-throughput microbial community analysis on the Illumina HiSeq and MiSeq platforms. *ISME J.* **2012**, *6*, 1621–1624. [CrossRef]

49. ITS Illumina Amplicon Protocol. Earth Microbiome Project. Available online: https://earthmicrobiome.org/protocols-and-standards/its/ (accessed on 5 July 2017).

50. Bolyen, E.; Rideout, J.R.; Dillon, M.R.; Bokulich, N.A.; Abnet, C.; Al-Ghalith, G.A.; Alexander, H.; Alm, E.J.; Arumugam, M.; Asnicar, F. QIIME 2: Reproducible, Interactive, Scalable, and Extensible Microbiome Data Science. *Peer J. Prepr.* **2018**. [CrossRef]

51. Callahan, B.J.; Mcmurdie, P.J.; Rosen, M.J.; Han, A.W.; Johnson, A.J.A.; Holmes, S.P. DADA2: High-resolution sample inference from Illumina amplicon data. *Nat. Methods* **2016**, *13*, 581–583. [CrossRef]

52. Abarenkov, K.; Zirk, A.; Piirmann, T.; Pöhönen, R.; Ivanov, F.; Nilsson, R.H.; Kõljalg, U. *UNITE QIIME Release Fungi 2*; UNITE Community: Tuckingmill, UK, 2020. [CrossRef]

53. Hydra Smithsonian Institution. *High Performance Computing Cluster*; Smithsonian Institution: Washington, DC, USA, 2022.

54. Gostel, M.R.; Coy, K.A.; Weeks, A. Microfluidic PCR-based target enrichment: A case study in two rapid radiations of *Commiphora* (Burseraceae) from Madagascar. *J. Syst. Evol.* **2015**, *53*, 411–431. [CrossRef]

55. Gostel, M.R.; Zúñiga, J.D.; Kress, W.J.; Funk, V.A.; Puente-Lelievre, C. Microfluidic Enrichment Barcoding (MEBarcoding): A new method for high throughput plant DNA barcoding. *Sci. Rep.* **2020**, *10*, 1–13. [CrossRef]

56. Katoh, K.; Rozewicki, J.; Yamada, K.D. MAFFT online service: Multiple sequence alignment, interactive sequence choice and visualization. *Briefings Bioinform.* **2019**, *20*, 1160–1166. [CrossRef] [PubMed]

57. Sela, I.; Ashkenazy, H.; Katoh, K.; Pupko, T. GUIDANCE2: Accurate detection of unreliable alignment regions accounting for the uncertainty of multiple parameters. *Nucleic Acids Res.* **2015**, *43*, W7–W14. [CrossRef] [PubMed]

58. Stamatakis, A. RAxML version 8: A tool for phylogenetic analysis and post-analysis of large phylogenies. *Bioinformatics* **2014**, *30*, 1312–1313. [CrossRef] [PubMed]

59. Miller, M.A.; Pfeiffer, W.; Schwartz, T. The CIPRES science gateway: Enabling high-impact science for phylogenetics researchers with limited resources. In Proceedings of the 1st Conference of the Extreme Science and Engineering Discovery Environment: Bridging from the Extreme to the Campus and Beyond, New York, NY, USA, 16–19 July 2012; pp. 1–8.

60. Zhang, C.; Rabiee, M.; Sayyari, E.; Mirarab, S. ASTRAL-III: Polynomial time species tree reconstruction from partially resolved gene trees. *BMC Bioinform.* **2018**, *19*, 15–30. [CrossRef] [PubMed]

61. Smith, M.R. TreeDist: Distances between Phylogenetic Trees. Comprehensive R Archive Network. *Bioinformatics* **2020**, *36*, 5007–5013. [CrossRef]

62. Smith, M.R. Information theoretic generalized Robinson–Foulds metrics for comparing phylogenetic trees. *Bioinformatics* **2021**, *37*, 2077–2078. [CrossRef]

63. Fujisawa, T.; Barraclough, T.G. Delimiting Species Using Single-Locus Data and the Generalized Mixed Yule Coalescent Approach: A Revised Method and Evaluation on Simulated Data Sets. *Syst. Biol.* **2013**, *62*, 707–724. [CrossRef]

64. Suchard, M.A.; Lemey, P.; Baele, G.; Ayres, D.L.; Drummond, A.J.; Rambaut, A. Bayesian phylogenetic and phylodynamic data integration using BEAST 1.10. *Virus Evol.* **2018**, *4*, vey016. [CrossRef]

65. Drummond, A.J.; Rambaut, A. BEAST: Bayesian evolutionary analysis by sampling trees. *BMC Evol. Biol.* **2007**, *7*, 1–8. [CrossRef]

66. Drummond, A.J.; Suchard, M.A.; Xie, D.; Rambaut, A. Bayesian Phylogenetics with BEAUti and the BEAST 1.7. *Mol. Biol. Evol.* **2012**, *29*, 1969–1973. [CrossRef]

67. Zhang, J.; Kapli, P.; Pavlidis, P.; Stamatakis, A. A general species delimitation method with applications to phylogenetic placements. *Bioinformatics* **2013**, *29*, 2869–2876. [CrossRef] [PubMed]

68. Felsenstein, J. *PHYLIP—Phylogeny Inference Package (Version 3.57c)*; University of Washington: Seattle, WA, USA, 1995.

69. Hall, T.A. BioEdit: A User-Friendly Biological Sequence Alignment Editor and Analysis Program for Windows 95/98/NT. *Nucleic Acids Symp. Ser.* **1999**, *41*, 95–98. [CrossRef]

70. Puillandre, N.; Lambert, A.; Brouillet, S.; Achaz, G. ABGD, Automatic Barcode Gap Discovery for primary species delimitation. *Mol. Ecol.* **2012**, *21*, 1864–1877. [CrossRef] [PubMed]

71. Puillandre, N.; Brouillet, S.; Achaz, G. ASAP: Assemble species by automatic partitioning. *Mol. Ecol. Resour.* **2021**, *21*, 609–620. [CrossRef] [PubMed]

72. Hall, T.; Biosciences, I.; Carlsbad, C. BioEdit: An Important Software for Molecular Biology. *GERF Bull. Biosci.* **2011**, *2*, 60–61.

73. Altschul, S.F.; Gish, W.; Miller, W.; Myers, E.W.; Lipman, D.J. Basic local alignment search tool. *J. Mol. Biol.* **1990**, *215*, 403–410. [CrossRef]

74. Madden, T. The BLAST Sequence Analysis Tool. In *The NCBI Handbook*; National Center for Biotechnology Information (US): Bethesda, MD, USA, 2003.

75. Swofford, D. *PAUP*: Phylogenetic Analysis Using Parsimony (*and Other Methods)*; Sinauer Associates: Sunderland, MA, USA, 2003. [CrossRef]

76. Swofford, D.L.; Sullivan, J. *Phylogeny Inference Based on Parsimony and Other Methods Using PAUP*. The Phylogenetic Handbook: A Practical Approach to DNA and Protein Phylogeny*; Sinauer Associates: Sunderland, MA, USA, 2003; Volume 7, pp. 160–206.

77. Wong, B. Points of view: Color blindness. *Nat. Methods* **2011**, *8*, 441. [CrossRef]

78. Bakker, F.T.; Bieker, V.C.; Martin, M.D. Editorial: Herbarium Collection-Based Plant Evolutionary Genetics and Genomics. *Front. Ecol. Evol.* **2020**, *8*, 603948. [CrossRef]

79. Ohmura, Y.; Uno, K.; Hosaka, K.; Hosoya, T. DNA Fragmentation of herbarium specimens of lichens, and significance of epitypification for threatened species of Japan. In Proceedings of the 10th International Mycological Congress, Bangkok, Thailand, 3–9 August 2014; p. 151.

80. Gueidan, C.; Aptroot, A.; Cáceres, M.E.D.S.; Binh, N.Q. Molecular phylogeny of the tropical lichen family Pyrenulaceae: Contribution from dried herbarium specimens and FTA card samples. *Mycol. Prog.* **2016**, *15*. [CrossRef]

81. Kranner, I.; Zorn, M.; Turk, B.; Wornik, S.; Beckett, R.P.; Batič, F. Biochemical traits of lichens differing in relative desiccation tolerance. *N. Phytol.* **2003**, *160*, 167–176. [CrossRef] [PubMed]

82. Kranner, I.; Beckett, R.; Hochman, A.; Nash, T.H. Desiccation-Tolerance in Lichens: A Review. *Bryologist* **2008**, *111*, 576–593. [CrossRef]

83. Hawksworth, D.L.; Grube, M. Lichens redefined as complex ecosystems. *N. Phytol.* **2020**, *227*, 1281–1283. [CrossRef]

84. Sierra, M.A.; Danko, D.C.; Sandoval, T.A.; Pishchany, G.; Moncada, B.; Kolter, R.; Mason, C.E.; Zambrano, M.M. The Microbiomes of Seven Lichen Genera Reveal Host Specificity, a Reduced Core Community and Potential as Source of Antimicrobials. *Front. Microbiol.* **2020**, *11*, 398. [CrossRef]

85. Hodkinson, B.P.; Gottel, N.R.; Schadt, C.W.; Lutzoni, F. Photoautotrophic symbiont and geography are major factors affecting highly structured and diverse bacterial communities in the lichen microbiome. *Environ. Microbiol.* **2012**, *14*, 147–161. [CrossRef]

86. Prillinger, H.; Kraepelin, G.; Lopandic, K.; Schweigkofler, W.; Molnár, O.; Weigang, F.; Dreyfuss, M.M. New species of Fellomyces Isolated from Epiphytic Lichen Species. *Syst. Appl. Microbiol.* **1997**, *20*, 572–584. [CrossRef]

87. Mark, K.; Laanisto, L.; Bueno, C.G.; Niinemets, Ü.; Keller, C.; Scheidegger, C. Contrasting co-occurrence patterns of photobiont and cystobasidiomycete yeast associated with common epiphytic lichen species. *N. Phytol.* **2020**, *227*, 1362–1375. [CrossRef]

88. Arnold, A.E.; Miadlikowska, J.; Higgins, K.L.; Sarvate, S.D.; Gugger, P.; Way, A.; Hofstetter, V.; Kauff, F.; Lutzoni, F. A Phylogenetic Estimation of Trophic Transition Networks for Ascomycetous Fungi: Are Lichens Cradles of Symbiotrophic Fungal Diversification? *Syst. Biol.* **2009**, *58*, 283–297. [CrossRef]

89. Duarte, A.; Passarini, M.R.Z.; Delforno, T.; Pellizzari, F.; Cipro, C.V.Z.; Montone, R.C.; Petry, M.; Putzke, J.; Rosa, L.H.; Sette, L.D. Yeasts from macroalgae and lichens that inhabit the South Shetland Islands, Antarctica. *Environ. Microbiol. Rep.* **2016**, *8*, 874–885. [CrossRef]

90. Lindgren, H.; Diederich, P.; Goward, T.; Myllys, L. The phylogenetic analysis of fungi associated with lichenized ascomycete genus *Bryoria* reveals new lineages in the Tremellales including a new species *Tremella huuskonenii* hyperparasitic on *Phacopsis huuskonenii*. *Fungal Biol.* **2015**, *119*, 844–856. [CrossRef] [PubMed]

91. Millanes, A.M.; Diederich, P.; Wedin, M. Cyphobasidium gen. nov., a new lichen-inhabiting lineage in the Cystobasidiomycetes (Pucciniomycotina, Basidiomycota, Fungi). *Fungal Biol.* **2015**, *120*, 1468–1477. [CrossRef] [PubMed]

92. DePriest, P.T.; Ivanova, N.V.; Fahselt, D.; Alstrup, V.; Gargas, A. Sequences of Psychrophilic Fungi Amplified from Glacier-Preserved Ascolichens. *Can. J. Bot.* **2000**, *78*, 1450–1459. [CrossRef]
93. Spribille, T.; Tuovinen, V.; Resl, P.; Vanderpool, D.; Wolinski, H.; Aime, M.C.; Schneider, K.; Stabentheiner, E.; Toome-Heller, M.; Thor, G.; et al. Basidiomycete yeasts in the cortex of ascomycete macrolichens. *Science* **2016**, *353*, 488–492. [CrossRef] [PubMed]
94. Lawrey, J.D.; Diederich, P. Lichenicolous Fungi: Interactions, Evolution, and Biodiversity. *Bryologist* **2003**, *106*, 80–120. [CrossRef]
95. Diederich, P.; Lawrey, J.D.; Ertz, D. The 2018 classification and checklist of lichenicolous fungi, with 2000 non-lichenized, obligately lichenicolous taxa. *Bryologist* **2018**, *121*. [CrossRef]
96. Tuovinen, V.; Ekman, S.; Thor, G.; Vanderpool, D.; Spribille, T.; Johannesson, H. Two Basidiomycete Fungi in the Cortex of Wolf Lichens. *Curr. Biol.* **2019**, *29*, 476–483.e5. [CrossRef]
97. Li, Q.; Li, L.; Feng, H.; Tu, W.; Bao, Z.; Xiong, C.; Wang, X.; Qing, Y.; Huang, W. Characterization of the Complete Mitochondrial Genome of Basidiomycete Yeast *Hannaella oryzae*: Intron Evolution, Gene Rearrangement, and Its Phylogeny. *Front. Microbiol.* **2021**, *12*. [CrossRef]
98. Fernández-Mendoza, F.; Fleischhacker, A.; Kopun, T.; Grube, M.; Muggia, L. ITS1 metabarcoding highlights low specificity of lichen mycobiomes at a local scale. *Mol. Ecol.* **2017**, *26*, 4811–4830. [CrossRef]
99. Banchi, E.; Ametrano, C.G.; Stanković, D.; Verardo, P.; Moretti, O.; Gabrielli, F.; Lazzarin, S.; Borney, M.; Tassan, F.; Tretiach, M.; et al. DNA metabarcoding uncovers fungal diversity of mixed airborne samples in Italy. *PLoS ONE* **2018**, *13*, e0194489. [CrossRef]
100. Wright, B.; Clair, L.L.S.; Leavitt, S.D. Is targeted community DNA metabarcoding suitable for biodiversity inventories of lichen-forming fungi? *Ecol. Indic.* **2019**, *98*, 812–820. [CrossRef]
101. Olds, C.G.; Berta-Thompson, J.W.; Loucks, J.J.; Levy, R.A.; Wilson, A.W. Applying a Modified Metabarcoding Approach for the Sequencing of Macrofungal Specimens from Fungarium Collections. *bioRxiv* **2021**. [CrossRef]
102. Leavitt, S.D.; Johnson, L.; St. Clair, L.L. Species Delimitation and Evolution in Morphologically and Chemically Diverse Communities of the Lichen-forming Genus *Xanthoparmelia* (Parmeliaceae, Ascomycota) in Western North America. *Am. J. Bot.* **2011**, *98*, 175–188. [CrossRef] [PubMed]
103. Jiang, S.-H.; Lücking, R.; Liu, H.-J.; Wei, X.-L.; Xavier-Leite, A.B.; Portilla, C.V.; Ren, Q.; Wei, J.-C. Twelve New Species Reveal Cryptic Diversification in Foliicolous Lichens of *Strigula* s.lat. (*Strigulales, Ascomycota*). *J. Fungi* **2022**, *8*, 2. [CrossRef]
104. Leavitt, S.D.; Grewe, F.; Widhelm, T.; Muggia, L.; Wray, B.; Lumbsch, H.T. Resolving evolutionary relationships in lichen-forming fungi using diverse phylogenomic datasets and analytical approaches. *Sci. Rep.* **2016**, *6*, 22262. [CrossRef]
105. Pizarro, D.; Divakar, P.K.; Grewe, F.; Leavitt, S.D.; Huang, J.-P.; Grande, F.D.; Schmitt, I.; Wedin, M.; Crespo, A.; Lumbsch, T. Phylogenomic analysis of 2556 single-copy protein-coding genes resolves most evolutionary relationships for the major clades in the most diverse group of lichen-forming fungi. *Fungal Divers.* **2018**, *92*, 31–41. [CrossRef]
106. Lücking, R.; Moncada, B.; Widhelm, T.J.; Lumbsch, H.T.; Blanchon, D.J.; de Lange, P.J. The Sticta Filix—Sticta Lacera Conundrum (Lichenized Ascomycota: Peltigeraceae Subfamily Lobarioideae): Unresolved Lineage Sorting or Developmental Switch? *Bot. J. Linn. Soc.* **2021**, boab083. [CrossRef]
107. Grewe, F.; Lagostina, E.; Wu, H.; Printzen, C.; Lumbsch, T. Population genomic analyses of RAD sequences resolves the phylogenetic relationship of the lichen-forming fungal species *Usnea antarctica* and *Usnea aurantiacoatra*. *MycoKeys* **2018**, *43*, 91–113. [CrossRef]
108. Lagostina, E.; Grande, F.D.; Andreev, M.; Printzen, C. The use of microsatellite markers for species delimitation in Antarctic *Usnea* subgenus *Neuropogon*. *Mycologia* **2018**, *110*, 1047–1057. [CrossRef]
109. Widhelm, T.J.; Grewe, F.; Huang, J.; Ramanauskas, K.; Mason-Gamer, R.; Lumbsch, H.T. Using RADseq to understand the circum-Antarctic distribution of a lichenized fungus, *Pseudocyphellaria glabra*. *J. Biogeogr.* **2021**, *48*, 78–90. [CrossRef]
110. Boluda, C.; Rico, V.; Divakar, P.; Nadyeina, O.; Myllys, L.; McMullin, T.; Zamora, J.C.; Scheidegger, C.; Hawksworth, D. Evaluating methodologies for species delimitation: The mismatch between phenotypes and genotypes in lichenized fungi (*Bryoria* sect. *Implexae*, Parmeliaceae). *Persoonia-Mol. Phylogeny Evol. Fungi* **2019**, *42*, 75–100. [CrossRef] [PubMed]
111. Lücking, R.; Hodkinson, B.P.; Leavitt, S.D. The 2016 classification of lichenized fungi in the Ascomycota and Basidiomycota—Approaching one thousand genera. *Bryologist* **2017**, *119*, 361. [CrossRef]
112. Larcher, W.; Vareschi, V. Variation in Morphology and Functional Traits of *Dictyonema glabratum* from Contrasting Habitats in the Venezuelan Andes. *Lichenologist* **1988**, *20*, 269–277. [CrossRef]
113. Hawksworth, D.L. Some advances in the study of lichens since the time of E. M. Holmes. *Bot. J. Linn. Soc.* **1973**, *67*, 3–31. [CrossRef]
114. Geml, J.; Laursen, G.A.; O'Neill, K.; Nusbaum, H.C.; Taylor, D.L. Beringian origins and cryptic speciation events in the fly agaric (Amanita muscaria). *Mol. Ecol.* **2006**, *15*, 225–239. [CrossRef] [PubMed]
115. Ogawa, W.; Endo, N.; Fukuda, M.; Yamada, A. Phylogenetic analyses of Japanese golden chanterelles and a new species description, *Cantharellus anzutake* sp. nov. *Mycoscience* **2018**, *59*, 153–165. [CrossRef]
116. Leacock, P.R.; Riddell, J.; Wilson, A.W.; Zhang, R.; Ning, C.; Mueller, G.M. *Cantharellus chicagoensis* sp. nov. is supported by molecular and morphological analysis as a new yellow chanterelle in midwestern United States. *Mycologia* **2016**, *108*, 765–772. [CrossRef]
117. Buyck, B.; Hofstetter, V. The contribution of tef-1 sequences to species delimitation in the *Cantharellus cibarius* complex in the southeastern USA. *Fungal Divers.* **2011**, *49*, 35–46. [CrossRef]

118. Hapuarachchi, K.K.; Wen, T.C.; Deng, C.Y.; Kang, J.C.; Hyde, K.D. Mycosphere Essays 1: Taxonomic Confusion in the *Ganoderma lucidum* Species Complex. *Mycosphere* **2015**, *6*, 542–559. [CrossRef]

119. Du, X.-H.; Zhao, Q.; O'Donnell, K.; Rooney, A.P.; Yang, Z.L. Multigene molecular phylogenetics reveals true morels (*Morchella*) are especially species-rich in China. *Fungal Genet. Biol.* **2012**, *49*, 455–469. [CrossRef]

120. Moncada, B.; Reidy, B.; Lücking, R. A phylogenetic revision of Hawaiian *Pseudocyphellaria* sensu lato (lichenized Ascomycota: Lobariaceae) reveals eight new species and a high degree of inferred endemism. *Bryologist* **2014**, *117*, 119–160. [CrossRef]

121. Lumbsch, H.T.; Elix, J. Lecanora. In *Flora of Australia Volume 56A, Lichens 4*; CSIRO Publishing: Clayton, VIC, Australia, 2004; ISBN 0-643-09056-8.

122. Guderley, R. Die *Lecanora subfusca*-Gruppe in Süd-Und Mittelamerika. *J. Hattori Bot. Lab.* **1999**, *87*, 131–257. [CrossRef]

123. Guderley, R.; Lumbsch, H.T. Notes on Multispored Species of *Lecanora* Sensu Stricto. *Lichenologist* **1999**, *31*, 197–203. [CrossRef]

124. Lumbsch, H.T.; Plümper, M.; Guderley, R.; Feige, G.B. The Corticolous Species of *Lecanora* Sensu Stricto with Pruinose Apothecial Discs. *Symb. Bot. Ups.* **1997**, *32*, 131–161.

125. Lumbsch, H.T.; Feige, G.B.; Elix, J.A. A Revision of the Usnic Acid Containing Taxa Belonging to *Lecanora* sensu stricto (Lecanorales: Lichenized Ascomycotina). *Bryologist* **1995**, *98*, 561–577. [CrossRef]

126. Lumbsch, H.T.; Guderley, R.; Elix, J.A. A Revision of Some Species in *Lecanora* sensu stricto with a Dark Hypothecium (Lecanorales, Ascomycotina). *Bryologist* **1996**, *99*, 269–291. [CrossRef]

127. Zhao, X.; Leavitt, S.D.; Zhao, Z.T.; Zhang, L.L.; Arup, U.; Grube, M.; Pérez-Ortega, S.; Printzen, C.; Śliwa, L.; Kraichak, E.; et al. Towards a revised generic classification of lecanoroid lichens (Lecanoraceae, Ascomycota) based on molecular, morphological and chemical evidence. *Fungal Divers.* **2016**, *78*, 293–304. [CrossRef]

128. Bungartz, F.; Elix, J.A.; Printzen, C. Lecanoroid lichens in the Galapagos Islands: The genera *Lecanora, Protoparmeliopsis*, and *Vainionora* (Lecanoraceae, Lecanoromycetes). *Phytotaxa* **2020**, *431*, 1–85. [CrossRef]

129. Guzow-Krzemińska, B.; Łubek, A.; Malíček, J.; Tønsberg, T.; Oset, M.; Kukwa, M. *Lecanora stanislai*, a new, sterile, usnic acid containing lichen species from Eurasia and North America. *Phytotaxa* **2017**, *329*, 201–211. [CrossRef]

130. Tripp, E.A.; Morse, C.A.; Keepers, K.G.; Stewart, C.A.; Pogoda, C.S.; White, K.H.; Hoffman, J.R.; Kane, N.C.; McCain, C.M. Evidence of substrate endemism of lichens on Fox Hills Sandstone: Discovery and description of *Lecanora lendemeri* as new to science. *Bryologist* **2019**, *122*, 246–259. [CrossRef]

131. Gerlach, A.D.C.L.; Clerc, P.; da Silveira, R.M.B. Taxonomy of the corticolous, shrubby, esorediate, neotropical species of *Usnea* Adans. (Parmeliaceae) with an emphasis on southern Brazil. *Lichenologist* **2017**, *49*, 199–238. [CrossRef]

132. Gerlach, A.D.C.L.; Toprak, Z.; Naciri, Y.; Caviró, E.A.; da Silveira, R.M.B.; Clerc, P. New insights into the *Usnea cornuta* aggregate (Parmeliaceae, lichenized Ascomycota): Molecular analysis reveals high genetic diversity correlated with chemistry. *Mol. Phylogenet. Evol.* **2019**, *131*, 125–137. [CrossRef]

133. Truong, C.; Clerc, P. The lichen genus *Usnea* (Parmeliaceae) in tropical South America: Species with a pigmented medulla, reacting C+ yellow. *Lichenologist* **2012**, *44*, 625–637. [CrossRef]

134. Truong, C.; Bungartz, F.; Clerc, P. The lichen genus *Usnea* (Parmeliaceae) in the tropical Andes and the Galapagos: Species with a red-orange cortical or subcortical pigmentation. *Bryologist* **2011**, *114*, 477–503. [CrossRef]

135. Truong, C.; Rodriguez, J.M.; Clerc, P. Pendulous *Usnea* species (Parmeliaceae, lichenized Ascomycota) in tropical South America and the Galapagos. *Lichenologist* **2013**, *45*, 505–543. [CrossRef]

136. Herrera-Campos, M.A.; Clerc, P.; Nash, T.H. Pendulous Species of *Usnea* from the Temperate Forests in Mexico. *Bryologist* **1998**, *101*, 303–329. [CrossRef]

137. Thell, A.; Crespo, A.; Divakar, P.K.; Kärnefelt, I.; Leavitt, S.D.; Lumbsch, H.T.; Seaward, M.R.D. A review of the lichen family Parmeliaceae—History, phylogeny and current taxonomy. *Nord. J. Bot.* **2012**, *30*, 641–664. [CrossRef]

138. Kirika, P.M.; Divakar, P.K.; Buaruang, K.; Leavitt, S.D.; Crespo, A.; Gatheri, G.W.; Mugambi, G.; Benatti, M.N.; Lumbsch, H.T. Molecular phylogenetic studies unmask overlooked diversity in the tropical lichenized fungal genus *Bulbothrix* s.l. (Parmeliaceae, Ascomycota). *Bot. J. Linn. Soc.* **2017**, *184*, 387–399. [CrossRef]

139. Vondrák, J.; Frolov, I.; Arup, U.; Khodosovtsev, A. Methods for phenotypic evaluation of crustose lichens with emphasis on Teloschistaceae. *Chornomorski Bot. J.* **2013**, *9*, 382–405. [CrossRef]

140. Frolov, I.; Vondrák, J.; Fernández-Mendoza, F.; Wilk, K.; Khodosovtsev, A.; Halıcı, M.G. Three New, Seemingly-Cryptic Species in the Lichen Genus *Caloplaca* (Teloschistaceae) Distinguished in Two-Phase Phenotype Evaluation. *Ann. Bot. Fenn.* **2016**, *53*, 243–262. [CrossRef]

141. Lücking, R.; Archer, A.W.; Aptroot, A. A world-wide key to the genus *Graphis* (Ostropales: Graphidaceae). *Lichenologist* **2009**, *41*, 363–452. [CrossRef]

142. Lücking, R.; Kalb, K. Formal Instatement of *Allographa* (Graphidaceae): How to Deal with a Hyperdiverse Genus Complex with Cryptic Differentiation and Paucity of Molecular Data. *Herzogia* **2018**, *31*, 535–561. [CrossRef]

Article

DNA Barcodes for Accurate Identification of Selected Medicinal Plants (Caryophyllales): Toward Barcoding Flowering Plants of the United Arab Emirates

Rahul Jamdade [1,2], Kareem A. Mosa [1,3,*], Ali El-Keblawy [1], Khawla Al Shaer [2], Eman Al Harthi [2], Mariam Al Sallani [2], Mariam Al Jasmi [2], Sanjay Gairola [2], Hatem Shabana [2,4] and Tamer Mahmoud [2,4]

[1] Department of Applied Biology, College of Sciences, University of Sharjah, Sharjah P.O. Box 27272, United Arab Emirates; rahul.arvind@epaa.shj.ae (R.J.); akeblawy@sharjah.ac.ae (A.E.-K.)

[2] Sharjah Seed Bank and Herbarium, Environment and Protected Areas Authority, Sharjah P.O. Box 2926, United Arab Emirates; khawla.alali@epaa.shj.ae (K.A.S.); eman.khalid@epaa.shj.ae (E.A.H.); mariam.alsallani@epaa.shj.ae (M.A.S.); mariam.aljasmi@epaa.shj.ae (M.A.J.); sanjay.gairola@epaa.shj.ae (S.G.); hatem.ahmed@epaa.shj.ae (H.S.); tamer.mahmoud@epaa.shj.ae (T.M.)

[3] Department of Biotechnology, Faculty of Agriculture, Al-Azhar University, Cairo 11751, Egypt

[4] Nature Conservation Sector, Egyptian Environmental Affairs Agency, Cairo 11728, Egypt

* Correspondence: kmosa@sharjah.ac.ae

Citation: Jamdade, R.; Mosa, K.A.; El-Keblawy, A.; Al Shaer, K.; Al Harthi, E.; Al Sallani, M.; Al Jasmi, M.; Gairola, S.; Shabana, H.; Mahmoud, T. DNA Barcodes for Accurate Identification of Selected Medicinal Plants (Caryophyllales): Toward Barcoding Flowering Plants of the United Arab Emirates. *Diversity* **2022**, *14*, 262. https://doi.org/10.3390/d14040262

Academic Editors: Morgan Gostel and W. John Kress

Received: 27 January 2022
Accepted: 17 March 2022
Published: 30 March 2022

Publisher's Note: MDPI stays neutral with regard to jurisdictional claims in published maps and institutional affiliations.

Abstract: The need for herbal medicinal plants is steadily increasing. Hence, the accurate identification of plant material has become vital for safe usage, avoiding adulteration, and medicinal plant trading. DNA barcoding has shown to be a valuable molecular identification tool for medicinal plants, ensuring the safety and efficacy of plant materials of therapeutic significance. Using morphological characters in genera with closely related species, species delimitation is often difficult. Here, we evaluated the capability of the nuclear barcode ITS2 and plastid DNA barcodes rbcL and matK to identify 20 medicinally important plant species of Caryophyllales. In our analysis, we applied an integrative approach for species discrimination using pairwise distance-based unsupervised operational taxonomic unit "OTU picking" methods, viz., ABGD (Automated Barcode Gap Analysis) and ASAP (Assemble Species by Automatic Partitioning). Along with the unsupervised OTU picking methods, Supervised Machine Learning methods (SML) were also implemented to recognize divergent taxa. Our results indicated that ITS2 was more successful in distinguishing between examined species, implying that it could be used to detect the contamination and adulteration of these medicinally important plants. Moreover, this study suggests that the combination of more than one method could assist in the resolution of morphologically similar or closely related taxa.

Keywords: medicinal plants; DNA barcoding; nuclear barcode; plastid barcodes; unsupervised learning; supervised learning

1. Introduction

Large numbers of people in developing countries rely on wild plant species for their medicinal needs. Over thousands of plant species are used in traditional medicine in different parts of the world. During ancient and modern culture, the healing properties of certain plants have been identified, and these plants currently play a significant role in the treatment of various diseases [1]. Due to their medicinal properties, there is a continuous and perpetual interest in researching and utilizing these valuable natural resources, as demonstrated by a plethora of literature (e.g., [2–10]). Different plant species have been used in ethnomedicine in the Arabian Peninsula since ancient times [5,10]. Sakkir [11] provided an overview of the medicinal plants in the United Arab Emirates (UAE) flora and indicated that roughly 18% of the total plant species identified have medicinal values.

These authors attributed such a low representation of medicinal plants in UAE flora to be most likely due to the unknown medicinal properties of the remaining taxa or lack of documentation of their traditional uses. Cybulska et al. [12] reviewed the available information on the medicinal uses of the halophytes in the UAE flora and highlighted the presence of valuable medicinal plants. These medicinally important plants display a specific tolerance to environmental stresses such as high temperature, drought, and salinity. It is expected that these medicinal halophytes might represent a valuable source of phytochemicals in salt marshes, where harsh conditions induce the production of both enzymes and phytochemicals in response to Reactive Oxygen Species (ROS) [12].

A recent survey on the ethnomedicinal knowledge of commonly used medicinal plants in a part of the UAE highlighted the importance of traditional medicinal plants and the need for knowledge documentation [13]. Further, due to various threats to medicinal plants, such as habitat loss and alteration, and overgrazing, there is a need for proper identification and conservation. Phondani et al. [8] documented the ethnobotanical uses of 58 medicinally important plants of the Arabian deserts. They emphasized the need to document ethnobotanical knowledge for sustainability and scientific validation to conserve these valuable medicinal plants native to the Arabian deserts. The above proves the importance of using many UAE plants in folk medicine.

There is a continued increase in the demand for herbal medicinal plants. Therefore, there are some accidental or intentional contaminations and adulteration with non-medicinal plants or other undesirable plant tissues [14,15]. Such contamination could reduce the effectiveness of the active ingredients, which might lead to detrimental health consequences [16,17]. The authentication of plant material has become necessary for safe use, avoiding adulteration, and trade in medicinal plants. Therefore, there is a need for fast authentication methods to authenticate dried herbal medicinal plants from the other components [18]. The detection of adulteration requires accurate, fast molecular techniques for plant identification, especially if it is difficult to discriminate between closely related plants morphologically [19]. In addition, the proper identification and documentation of medicinal plants in the region could add to their conservation and sustainable utilization.

The classical taxonomic techniques based on morphology and anatomy complement molecular techniques for accurately identifying morphologically similar closely related plant taxa. Currently, there are initiatives for generating DNA barcode libraries of vascular plant flora. Completing such libraries and making them available will help fast, accurate identification, which would lead to the better conservation and utilization of native plants, particularly those used in herbal medicine [20–22]. In this context, a tool such as DNA barcoding could help resolve these issues and lead to the rapid and accurate identification of medicinal species. Moreover, DNA barcoding could be helpful in the identification of medicinal plants in trade, as most herbal material is traded as dried leaves, roots, and bark or in powdered form, thus contributing to their safe use and avoiding adulteration [23].

DNA barcoding has become a useful complementary tool in diverse disciplines of biological sciences. The application of plant DNA barcodes, especially in floristic investigations, ecology, evolution, and conservation fields, has gained momentum over the last decade [24,25]. Several studies have highlighted the potential applications of DNA barcoding in the accurate identification of taxa, discovery of cryptic species, and as an crucial component in phylogenetics investigations [24,26–28]. However, this technique also has some potential limitations, especially in plants where the selected barcode region might lack enough information to provide DNA level species-specific differences and the concurrent observation of such differences at the secondary metabolite level, similar to that observed by Celiński et al. [29].

In medicinal plant research, DNA barcoding is emerging as a valuable molecular identification technique that has greatly ensured the safety and effectiveness of plant materials of medicinal value [30]. The reviews by Techen et al. [31] and Nazar et al. [32] have discussed the selection of the genomic regions as possible barcodes for medicinal plants, including new achievements in the field of DNA barcoding. Those reviews provide a comprehensive overview of DNA-based methods, technologies, and a combination of three or more genomic regions that were investigated for medicinal plant identification by various researchers worldwide.

Over the years, researchers have suggested different coding and noncoding genes in the nuclear and plastid genomes as potential barcodes for plants [33–37]. The types of DNA barcode markers used for plant identification range from a single chloroplast region to a combination of different regions (see [30,31,38]). Significant progress has been made in the identification of medicinal plants using DNA barcoding (e.g., [25,31,39–47]. For medicinal plant identification, some researchers have used a combination of markers between matK, rbcL, trnH-psbA, and ITS2 sequences. For example, Schori et al. [48] analyzed the rbcL, matK, and psbA-trnH loci of fourteen species of medicinal plants and found that depending on the plant to be identified, one region was preferred over the other, as a single barcode region is not enough to ensure the species identification.

Moreover, along with the selection of efficient DNA barcode markers for species identification, it is necessary to utilize competent methods for effective species discrimination. The results produced by one or more methods sometimes differ, which could require the implementation of more than one method that must be applied and compared jointly [49,50]. The most conventional DNA barcode analysis method is the pairwise distance-based unsupervised Operational Taxonomic Unit (OTU) picking method, where Automated Barcode Gap Discovery (ABGD) is the most widely used tool, followed by the recently developed Assemble Species by Automatic Partitioning (ASAP) [50,51]. Comparatively, some studies have shown a higher rate of species discrimination using the supervised learning approach [21,52–55]. Thus, in this study, we used a comparative approach of implementing both unsupervised and supervised methods for species delimitation.

The UAE has not received much attention to digitally record flora in the form of DNA barcodes [22], as there are only three studies cataloging flora of the UAE to date. Moreover, there are scarce studies on the DNA barcodes of medicinal plants [56], and existing studies have amplified three barcode loci for the coding genes matK, rbcL, and rpoC1 in 10 flowering plants from the UAE. Maloukh et al. [57] focused on authenticating the morphological identification of 51 plant species using rbcL and matK regions. Further, Mosa et al. [15] provided evidence that DNA barcoding was efficient in the detection of adulteration in plant-based herbal products in the UAE. Based on the results obtained, these authors also suggested rbcL as a promising barcode locus for resolving their studied species.

Since 2018, the Sharjah Seed Bank and Herbarium have engaged in the process of DNA barcoding the entire UAE flora [20–22]. Here, we assessed the capability of plastid DNA barcode markers rbcL and matK, and a nuclear marker, ITS2, for the identification of 20 medicinally important plant species belonging to the order Caryophyllales. The core Caryophyllales represent one of the largest eudicot orders with about 12,000 species and 30 families worldwide [58], and some species are used medicinally [59,60]. Various molecular systematic studies on Caryophyllales are available that have substantially increased our knowledge of their phylogeny [61]. The Caryophyllales is represented in the UAE's flora by 11 families. These are Aizoaceae, Amaranthaceae, Caryophyllaceae, Frankeniaceae, Gisekiaceae, Molluginaceae, Nyctaginaceae, Plumbaginaceae, Polygonaceae, Portulacaceae, and Tamaricaceae [62,63]. Among the families of Caryophyllales that have difficulties in morphological discrimination, especially during vegetative stages of the life cycles, are Amaranthaceae (e.g., *Haloxylon persicum*, *H. salicornicum*, *Salicornia persica*), Polygonaceae (e.g., *Calligonum comosum* and *C. crinitum*), and Tamaricaceae (*Tamarix aucheriana* and *T. nilotica*).

The objective of the present study was to barcode the medicinal plant species, compare the discriminatory power of the standard barcode regions, and explore the taxonomic implications in the studied taxa. Establishing a DNA barcoding system could facilitate the conservation of the UAE's medicinal taxa, help overcome the limitations of morphological characters, and contribute to species identification for their efficient utilization. The study results could help generate DNA barcode libraries of the UAE vascular plant flora, which could be a step toward completing the UAE and global DNA barcoding libraries.

2. Materials and Methods

2.1. Specimen Collection and Data Acquisition

Twenty species from the order Caryophyllales were included in this study. Of these, 13 species (36 samples) were collected from the field between 2019 and 2020, and 7 species were retrieved from the NCBI GenBank. All the studied species are from the United Arab Emirates (Figure 1), and their information is provided in Table S1. Overall, we collected between one and eight specimens per species. Our collection did not involve protected areas or endangered species.

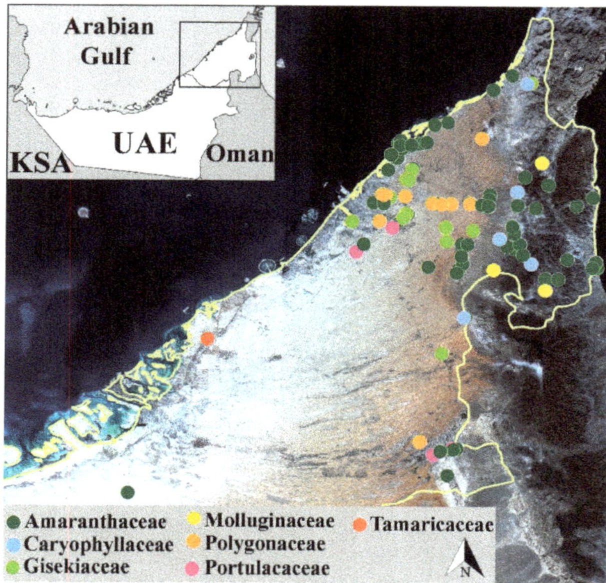

Figure 1. Map showing collection sites of plant samples for DNA barcoding and seed bank collections.

The collected species included herbaceous plants, shrubs, and trees, which occur in a range of habitats, including sandy, coastal plains, mountain ridges, and wadi beds. Species such as *Haloxylon salicornicum*, *Gisekia pharnacioides*, and *Aerva javanica* had the highest number of accessions, while *Tamarix nilotica* and *Amaranthus hybridus* had the lowest number. Morphological identification of all plant species was based on reliable diagnostic characters and available literature on local flora [62–65], produced by researchers who have worked in the field exploring UAE's flora for about one decade. The voucher specimens for collected species were curated by the Sharjah Seedbank and Herbarium (SSBH), Environment and Protected Areas Authority (EPAA), for record and references. According to the literature survey, the plant species included in the present study have medicinal values and are used in the treatment of various ailments (Table S2).

2.2. Tissue Sampling and DNA Extraction

Plant tissues were sampled from 36 individuals and dried immediately with silica gel at room temperature for DNA extraction. Unique identifiers were provided to the specimens and the tissue samples. Further, the tissue samples were ground to a fine powder using BeadBlaster 24 microcentrifuge homogenizer. Genomic DNA extraction was then performed using the DNeasy Plant Mini Kit (Qiagen, Hilden, Germany) according to the manufacturer's protocol with minor modifications, where proteinase K was added, followed by the AP1 buffer and RNase A, and the incubation time was increased to 1–3 h. Samples were eluted in Nuclease-Free Water. The isolated DNA was tested for quality by gel electrophoresis (BioRad, Hercules, CA, USA) on a 1% agarose gel and quantity using spectrophotometric analysis (Denovix, Wilmington, DE, USA).

2.3. PCR Amplification and Purification

Two plastid barcode regions, rbcL and matK, and one nuclear ribosomal barcode regions, Internal Transcribed Spacer (ITS2), were amplified via Polymerase Chain Reaction (PCR) (Biorad, Biorad Laboratories, Singapore and Applied Biosystems Veriti Thermal Cycler, Life Technologies Holdings Pte. Ltd., Singapore) using forward and reverse primers of rbcL [35,66], matK (proposed by Ki-Joong Kim, see [67]), and ITS2 [68,69] (Table S3). The 25 µL PCR reaction using a 5× FIREPol master mix was prepared to amplify the respective barcode region. The standard thermal profile of all primers used is shown in Table S3. Modification in the annealing temperature was performed wherever required. PCR products were then verified through gel electrophoresis on a 2% agarose gel. Amplified products were purified using the MEGAquick-spinTM plus total fragment DNA Purification Kit (Intron Biotechnology) and then sequenced commercially.

2.4. Sequence Analysis

Bidirectional sequencing was performed for rbcL, matK, and ITS2 barcode markers. The obtained sequences were assembled and aligned in Geneious Prime v2021 (geneious.com (accessed on 27 December 2021).) and MEGA X. [70] using the Muscle algorithm. The sequences were then submitted to NCBI GenBank through a web-based sequence submission tool 'BankIt,' and accessions numbers were obtained for all the studied barcode markers (Table S1).

Further, the sequences were subjected to the taxonomic evaluation using the NCBI GenBank BLASTn to obtain homologies between the fragments [71]. In addition, unsupervised OTU picking methods were employed, where the phylogenetic analysis was performed using MEGA, and the assessment of OTUs was performed using ABGD and ASAP.

Along with the unsupervised OTU picking methods, Supervised Machine Learning methods (SML) were also implemented to recognize divergent taxa. The aligned datasets were formatted to the WEKA's required file format using the FASTA to WEKA converter [54]. Further, in WEKA machine learning, the random forest and sequential minimal optimization classifiers were used through the 10-folds of cross-validation [72].

Phylogenetic tree construction was carried out in MEGAX. Initially, the phylogenetic model test was performed to determine the best-fit nucleotide substitution model with the lowest BIC scores (Bayesian Information Criterion). Accordingly, Maximum Likelihood (ML) phylogeny was inferred using the Kimura-2-Parameter (K2P) model with discrete gamma distribution was selected for the rbcL dataset. For the matK dataset, a ML phylogenetic tree was constructed using the General Time Reversible model (GTR) with discrete Gamma distribution (G). For the ITS2 dataset, ML phylogeny was achieved using the K2P model with discrete gamma distribution and invariant sites (G + I). All phylogenetic trees were given a bootstrap support of 1000. Moreover, for the phylogenetic tree annotation, the Interactive Tree of Life webserver was used. In addition, the ASAP webserver was used to build the partitions for species delimitation.

3. Results

3.1. Barcode Amplification and Sequencing Success

The core plant barcode markers rbcL, matK, and ITS2 were amplified at various temperature gradients and sequenced successfully (Figure 2).

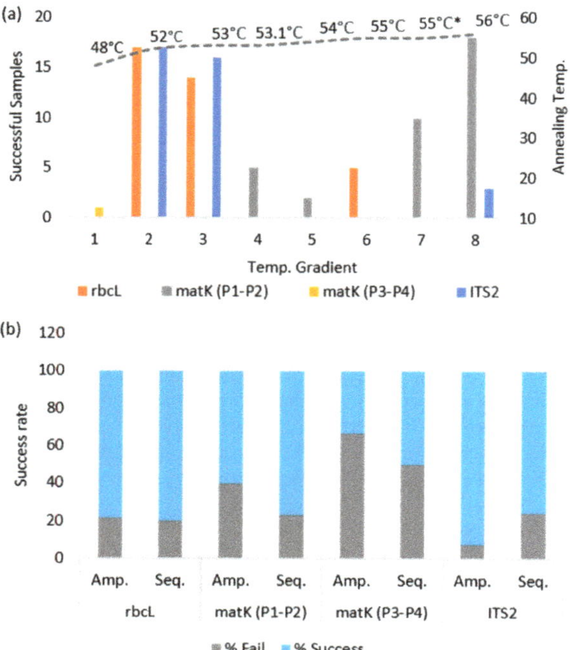

Figure 2. PCR amplification and sequencing success rate for the DNA barcode markers. (**a**) Annealing temperature gradients used for the barcode amplification, (* represents the dual annealing temperature 50 | 55 °C), (**b**) PCR amplification and sequencing success of the attempted samples.

The rbcL and ITS2 markers showed significant success rates in PCR amplification ranging from 70% to 95%, while matK exhibited a lower success rate of about 60% (Figure 2). In sequencing, the rbcL marker showed the highest (80%) success rate, followed by the matK (76%) and ITS2 (75%) (Figure 2). In addition, another pair of matK markers was used for the amplicon recovery, where only one sample was successfully amplified. Overall, 99 sequences were obtained belonging to the rbcL (35), matK (34), and ITS2 (30) markers, representing about 13 species.

3.2. Taxonomic Assignment

The taxonomic identification of the collected specimens was initially made based on their key morphological characters. Further, the taxonomic evaluation was performed at the NCBI GenBank's nucleotide database using the NCBI-BLASTn tool.

Overall, we obtained 99 barcodes belonging to 36 specimens and 13 species in the present study. In addition, we retrieved 18 more barcodes belonging to 18 specimens and 11 species from the NCBI GenBank based on the records from previous studies performed on the flora of the UAE. Altogether, the dataset comprised about 117 barcodes, 54 specimens, and 20 species in common, viz., rbcL ($n = 49$), matK ($n = 38$), and ITS2 ($n = 30$).

Those barcode datasets were further analyzed using the unsupervised OTU picking methods, viz., ABGD and ASAP. The ABGD recognized groups of about 10 to 16 species only using J69 and K80 metrics. In addition, the initial partition exhibited lower accuracy in the species resolution than the recursive partition. Thus, the recursive partitions were further taken into consideration. The rbcL showed 6 partitions of which the fifth recursive partition resolved about 28 specimens and 7 species correctly (a priori intraspecific divergence of (P) = 0.0077, relative gap width (X) = 1.0) (Figure 3a and Table 1). In the case of matK, about 9 partitions were recognized, of which the eighth recursive partition was able to successfully resolve 29 specimens and 9 species (at $p = 0.035$ and X = 1) (Figure 3b and Table 1). In the ITS2 dataset, about 10 partitions were recognized, of which the seventh recursive partition was found to resolve 29 specimens and 10 species (at $p = 0.0215$ and X = 1) (Figure 3c and Table 1). The simple distance metric showed the lowest accuracy compared to JC69 and K80. Thus, it was not considered.

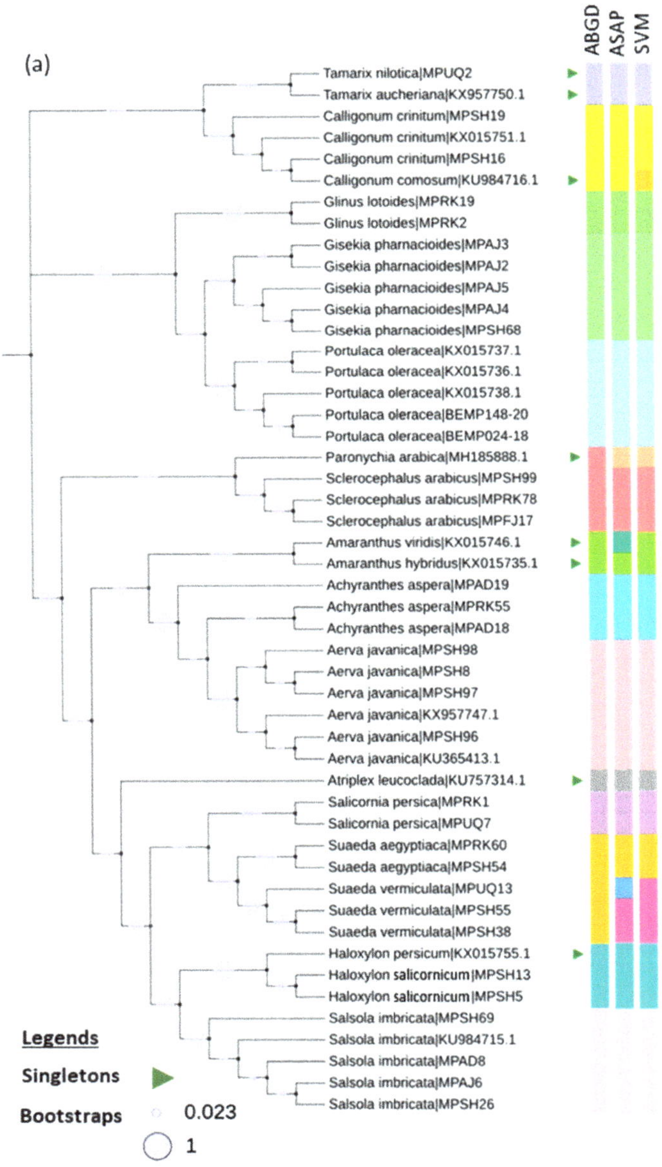

Figure 3. *Cont.*

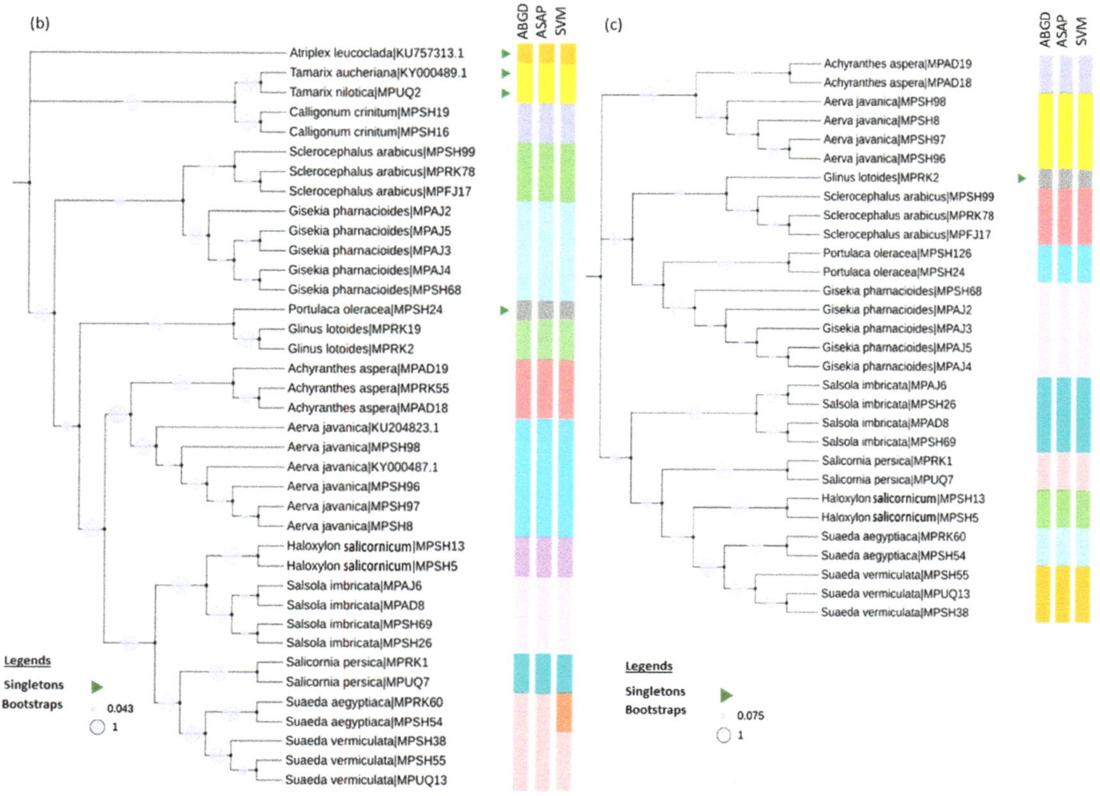

Figure 3. Taxonomic evaluation using unsupervised (ABGD and ASAP) and supervised learning (SVM) methods. (**a**) RbcL maximum likelihood (ML) phylogeny inferred using Kimura-2-parameter (K2P) model and discrete gamma distribution with 100 bootstrap support. (**b**) MatK ML phylogeny inferred using the General Time Reversible model and discrete gamma distribution with 1000 bootstrap support. (**c**) ITS2 ML phylogeny inferred by K2P model along using discrete gamma distribution and invariant sites with 1000 bootstrap support.

Table 1. Summary of species identification using unsupervised and supervised learning methods.

Barcode Marker	Method		Correct (%)	Incorrect (%)	Ambiguous (%)	Singleton (%)
rbcL (Spm. = 49, Sp. = 20)	ABGD	SPECIMENS	57.14	40.82	0.00	2.04
		SPECIES	35.00	60.00	0.00	5.00
	ASAP	SPECIMENS	67.35	18.37	6.12	8.16
		SPECIES	45.00	30.00	5.00	20.00
	SVM	SPECIMENS	79.59	4.08	0.00	16.33
		SPECIES	55.00	5.00	0.00	40.00
matK (Spm. = 38, Sp. = 15)	ABGD/ASAP	SPECIMENS	76.32	18.42	0.00	5.26
		SPECIES	60.00	26.67	0.00	13.33
	SVM	SPECIMENS	89.47	0.00	0.00	10.53
		SPECIES	73.33	0.00	0.00	26.67
ITS2 (Spm. = 29, Sp. = 12)	ABGD/ASAP/SVM	SPECIMENS	96.55	0.00	0.00	3.45
		SPECIES	90.91	0.00	0.00	9.09

As seen in the ABGD analysis, the algorithm identified several species partitions for each *p*-value (priori), which might derive uncertainty from the data [50]. Therefore, it is recommended to implement an integrative taxonomic approach to evaluate the relevance of the ABGD partitions [50]. Thus, the species assignment was further validated using the ASAP, followed by the supervised machine learning approach. In the case of ASAP, it appeared to provide a gap-width score, *p*-value, threshold distance dT, and the number of species corresponding to each defined partition, and thus overcame the challenge of a priori defined by ABGD. The partition could then be prioritized by considering the smallest ASAP score and the asterisk marks that represent the overall best scores.

Accordingly, the partitions with the highest species resolution were discovered for the matK and ITS2 datasets at the threshold distance of 0.029 and 0.0134, respectively (Figure 4a,b). In the matK dataset, about 29 specimens and 9 species were resolved, while in the ITS2, about 29 specimens and 10 species were resolved successfully (Figure 4a,b and Table 1). However, for the rbcL dataset, the second successive partition at the threshold distance of 0.0045 with lower ASAP scores was found to be the best partition showing a higher resolution (Figure 4c), further accurately discriminating 33 specimens and 9 species, and was thus taken into consideration (Figure 3a, and Table 1).

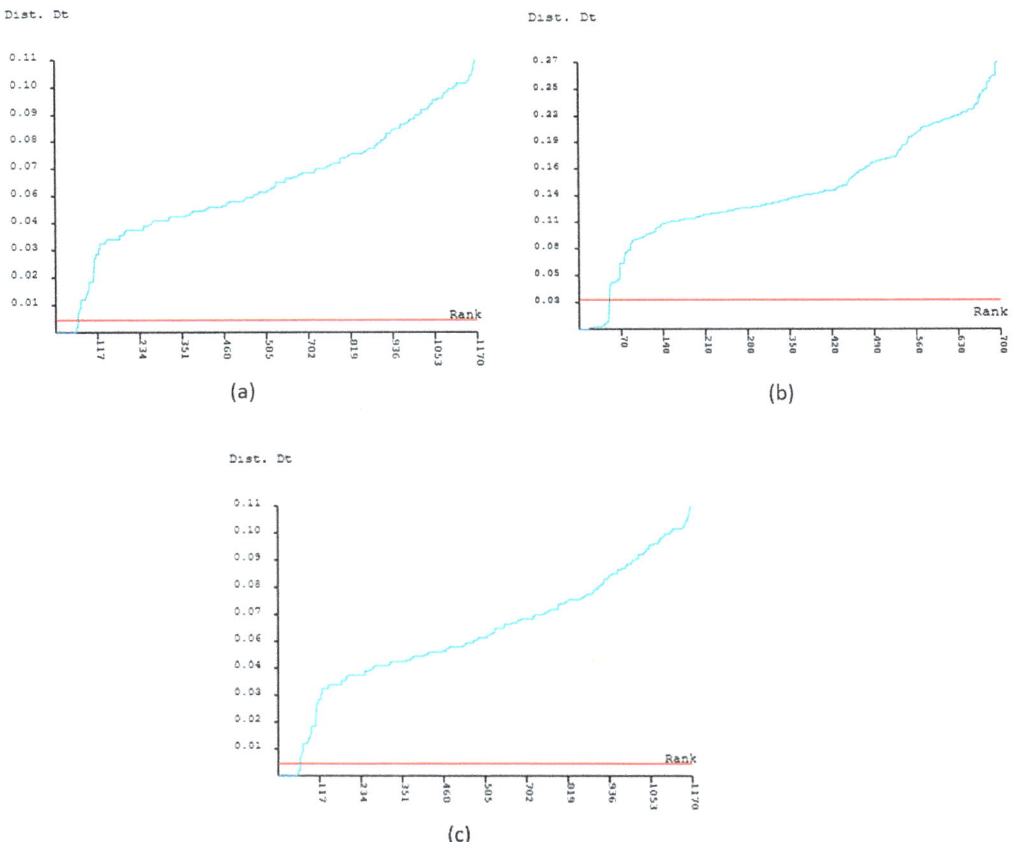

Figure 4. Threshold distance ranking the best partition for species delimitation. (**a**) matK 2nd partition (ASAP score = 2.50, Nb = 13, *p* = 0.0027); (**b**) ITS2 2nd partition (ASAP score= 2.50, Nb = 11, *p* = 0.042); (**c**) rbcL 2nd partition (ASAP score = 4.50, number of species (Nb) = 18, *p* = 0.0052).

Following the unsupervised approach, the analysis through the SML approach exhibited the highest species resolution in all three markers, rbcL, matK, and ITS2. SML appeared to resolve about 39 specimens and 10 species in rbcL, 34 specimens and 10 species in matK, and about 29 specimens and 10 species in ITS2 (Figure 3a–c and Table 1).

Overall, in the rbcL dataset, the ASAP and SVM methods successfully differentiated *Paronychia arabica* from *Scelerocephalus arabicus*, and resolved the *Suaeda* genus (*Suaeda aegyptiaca* from *Suaeda vermiculata*, whereas *Haloxylon persicum* and *Haloxylon salicornicum* could not be discriminated using all three methods (ABGD, ASAP, and SVM) (Figure 3a). Moreover, in the rbcL dataset, ASAP alone was able to differentiate the *Amaranthus* genus (*Amaranthus viridis* and *Amaranthus hybridus*), while SVM alone was able to delimit the Calligonum genus (*Calligonum crinitum* and *Calligonum comosum*) (Figure 3a).

In the matK dataset, *Suaeda aegyptiaca* and *Suaeda vermiculata* were only resolved by SVM (Figure 3b), while in the ITS2 dataset, both the species seemed to be accurately differentiated using all three methods (ABGD, ASAP, and SVM) (Figure 3c). Altogether, 17 species were successfully resolved from the 20 barcoded species using the rbcL, matK, and ITS2 markers, though the matK and ITS2 datasets lacked enough species memberships for all 20 species.

3.3. Genetic Divergence

The genetic divergence analysis was conducted for the rbcL, matK and ITS2 datasets using the TaxonDNA (Table 2). The highest intraspecific distance of 2.45% was observed in the ITS2 dataset among the individuals of *Salsola imbricata*, while the highest interspecific divergence of 2.58% was observed between the species of the genus *Suaeda*. Similarly, the *Suaeda* genus seemed to exhibit maximum interspecific distances for the rbcL dataset, wherein the matK *Tamarix* genus showed higher interspecific divergence, followed by *Suaeda*. In the case of the species from the genus *Suaeda*, the genetic divergence in the rbcL (1.55%) and matK (1.21%) was not enough for the ABGD (rbcL and matK datasets) and ASAP (matK datasets) to discriminate the *S. vermiculata* from the *S. aegyptiaca*. However, all methods employed successfully resolved species from the *Suaeda* genus using the ITS2 dataset at the genetic divergence of 2.58%.

Table 2. Intra- and interspecific genetic divergence.

Barcode Marker	Max Intra-Sp. Dist.	Avg. Intra-Sp. Dist.	Min. Intra-Sp. Dist.	Max. Inter-Sp. Dist.	Avg. Inter-Sp. Dist.	Min. Inter-Sp. Dist.
rbcL	0.77	0.06	0	1.55	0.45	0
matK	0.69	0.18	0	1.21	0.74	0
ITS2	2.45	0.06	0	2.58	2.32	2.19

4. Discussion

The use of herbal medicine traditionally for disease treatment and as a precursor for developing several important drugs [2,73] necessitates the accurate identification of medicinal plants. The results of our study suggest that the applications of DNA barcoding techniques can enhance the accurate identification of medicinally important species. Our study is among the first to utilize different DNA barcode markers and confirms the potential of the barcoding approach for the accurate identification of medicinally useful plants from the UAE that will help generate a reference dataset for research and other applications.

We investigated the efficacy of the three DNA barcode regions (rbcL, matK, and ITS2) for discriminating selected medicinal plant species belonging to the order Caryophyllales. The first step in assessing the potential candidate barcodes was to estimate the universality of the amplification and sequencing success rate across the studied taxa. The matK region showed a lower amplification rate (60%) than rbcL and ITS2, although two matK primers pairs were used with several attempts under different conditions (Figure 2a). The MatK (P1 and P2) pair was highly effective in the amplification success. However, the matK

(P3 and P4) pair resulted in low recovery (only one sample was amplified successfully). The inconsistent success rate has been reported for matK. Several studies have indicated that the matK region was less amplified than other regions in different angiosperms and gymnosperms, including some arid desert plants [74–76]. The universality issues of the matK primer could be attributed to the nucleotide variations in the respective binding site that could inhibit the PCR amplification [74], or to the large amplified product size ($\approx$900 bps) that could be susceptible to the degradation [77]. Cräutlein et al. [78] suggested the need for further efforts to improve primer design in matK to achieve higher efficiency. For sequencing, a higher number of good-quality sequences (80%) were obtained for rbcL than the other two regions. This result is aligned with previous studies that compared the three barcode loci for the coding genes (matK, rbcL, and rpoC1) for the discrimination of different plants of the UAE and concluded that the rbcL was more effective in discrimination between species [15,56,57].

Several different approaches based on the DNA barcoding technique have been advised for assigning species to their relevant taxa [52,54,79,80]. Our analysis applied an integrative approach for the delimitation of species using unsupervised "OTU picking" methods, viz., ABGD and ASAP that use only pairwise genetic distances, along with supervised methods for more data reliability. The ABGD method automatically identifies where the barcode gap is located in their distribution. This gap marks the limit between minimum interspecific and maximum intraspecific divergence. Thus, it is crucial to ensure the distance-based method's effectiveness [51,81]. Our results showed that the recursive partitions in ABGD recognized more OTUs than primary ones, exhibiting a higher accuracy in species resolution under the analysis, which corroborates with previous observations [51,82,83]. Further, ASAP was performed to evaluate the relevance of the ABGD partitions, as any species partition must be subsequently tested against other evidence as recommended in an integrative taxonomy approach [50].

Our results indicated that the unsupervised ABGD method showed taxonomic conflicts in rbcL between *Amaranthus* species (*A. hybridus* and *A. viridis*), and between *Paronychia arabica* and *Sclerocephalus arabicus*. Interestingly, these species differed morphologically and could be discriminated easily (Figure 3a). Moreover, merged taxa were observed for the genus *Suaeda* (*S. aegyptiaca* and *S. vermiculata*) in the rbcL dataset using ABGD, as well as in the matK datasets using both the ABGD and ASAP methods (Figure 3a,b). Moreover, a low pairwise interspecific divergence of rbcL (=1.55%) and matK (=1.21%) was observed between the species of *Suaeda*, thereby exhibiting a monophyletic relationship. A similar result was observed by Kapralov et al. [80], who provided strong statistical support for the monophyly. The taxonomic relationships might be confusing due to the absence of a barcode gap, which can result from a limited number of sequences per species (i.e., <3–5) [51].

Following the ABGD and ASAP methods, species delimitation through character-based supervised machine learning methods was utilized to understand better the confirmation of the initial identification [84]. So far, several studies have performed the character-based barcoding approach, which has proved its usefulness in identifying plant species better than the conventional unsupervised methods [52–54,85]. In our analysis, the unsupervised ASAP method tended to provide a better resolution potential for the rbcL dataset than its neighboring ABGD method (Table 1). In addition, ASAP was able to resolve two singleton species in the rbcL dataset that were not even recognized using the ABGD method (Figure 3a). Moreover, when compared with the supervised learning approach, the SVM method stood out as the more efficient method to provide an accurate identification than the unsupervised approach with the higher number of species, as observed in the rbcL and matK datasets (Table 1 and Figure 3a,b). In addition, *S. aegyptiaca* and *S. vermiculata* were also recovered as separate clades, which indicates that the intraspecific diversity could be hidden [34,86].

It has been reported that OTUs proposed by one or more methods could be inconsistent in distinguishing between the members of closely related genera [49]. In our study, we observed that the members of genus *Amaranthus* (*A. viridis* and *A. hybridus*) were only discriminated through ASAP, but members of *Calligonum* (*C. crinitum* and *C. comosum*) were distinguished only by SVM. This supports the importance of using more than one method, especially for closely related species that are difficult to discriminate morphologically, such as *C. crinitum* and *C. comosum*. The use of more than one method can maximize the probability of identifying morphologically similar species and overcoming the limitation associated with each method [50,87].

Overall, the taxonomic performance of SVM was stronger than that of ABGD and ASAP in the rbcL dataset. The SVM delivered the highest incidence of correct matches (55.0%) across the 20 species compared to 35.0% and 45.0% for ABGD and ASAP, respectively (Table 1). In the matK dataset, the performance of ABGD was similar to ASAP (60.0%) and was improved to 73.33% using supervised learning methods. However, all the methods delivered a similar percentage of correct matches in the ITS2 dataset (Table 1). Considerably, it is now a well-known fact that the combination of the two plastid markers, ribulose 1,5-bisphosphate carboxylase gene (rbcL) and maturase K (matK), that were accepted as the core barcoding regions [33], do not grant a suitable coverage of plant species. Thus, they must often be implemented along with the other hypervariable sequences, such as nuclear ITS or the plastid interspacer region trnH-psbA [88].

Moreover, the efficiency of the utilized markers and methods depends on the sample size, as the singleton species or small sample size could lead to skewed results [21]. In our study, we had about 10 singleton species, which were considered as singletons and not independent OTUs to reduce the probability of biased identification. Thus, an adequate sample size and proper implementation of the DNA barcoding technique can provide a scientific basis for the molecular identification and conservation of valuable medicinal species. Our study is among the first to utilize different DNA barcode markers and to confirm the potential of DNA barcoding in the accurate identification of medicinally important plants from the UAE. The dataset generated through this study will assist in developing the reference library, and allows others to contribute and explore the genetic potential of the available germplasm for various applications.

5. Conclusions

The results support the potential use of DNA barcoding in discriminating closely related taxa of Caryophyllales. The ITS2 was more effective in the discrimination between studied species, indicating its potential for distinguishing between Caryophyllales medicinally important plants and non-medicinal plants or other undesirable plant tissues. However, due to the inability of one DNA barcoding analysis method to discriminate between members of closely related genera, we propose combining two or more methods. The results of this study could fill a small gap of generating DNA barcodes for local (i.e., the UAE), regional (i.e., the Arab Gulf region), and global libraries of vascular plant flora.

Supplementary Materials: The following supporting information can be downloaded at: https://www.mdpi.com/article/10.3390/d14040262/s1, Table S1: List of specimen samples used in this study. Table S2: Information on ecological characteristics and medicinal benefits of studied plant species, Table S3: List of Primers with their thermal profile used in this study.

Author Contributions: K.A.M., A.E.-K. and R.J. conceived the ideas and designed the experiment; R.J., S.G., H.S. and T.M. collected the samples; R.J., K.A.S., E.A.H., M.A.S. and M.A.J. performed the Laboratory analysis; R.J. conducted the bioinformatics analysis; R.J., K.A.S., E.A.H., M.A.S., M.A.J., S.G., H.S. and T.M. were involved in data validation and curation; R.J., K.A.S., E.A.H., M.A.S., M.A.J., S.G., H.S. and T.M. prepared the original draft; K.A.M. and A.E.-K. reviewed and edited the manuscript; K.A.M. and A.E-K were involved in project supervision, administration, and funding acquisition. All authors have read and agreed to the published version of the manuscript.

Funding: This research was funded by a collaborative research grant from the University of Sharjah and Sharjah Research Academy, project # 1702145054-P.

Institutional Review Board Statement: Not applicable.

Informed Consent Statement: Not applicable.

Data Availability Statement: The data presented in this study are openly available in the NCBI GenBank (Table S1).

Conflicts of Interest: The authors declare no conflict of interest.

References

1. Singh, A.; Singh, S.; Mohan Prasad, S. Role of Medicinal Plants for Health Perspective: Special Reference to Antioxidant Potential. *J. Chem. Biol. Ther.* **2016**, *1*, 106. [CrossRef]
2. Hostettmann, K.; Marston, A.; Ndjoko, K.; Wolfender, J.-L. The Potential of African Plants as a Source of Drugs. *Curr. Org. Chem.* **2005**, *4*, 973–1010. [CrossRef]
3. Ljubuncic, P.; Azaizeh, H.; Portnaya, I.; Cogan, U.; Said, O.; Saleh, K.A.; Bomzon, A. Antioxidant activity and cytotoxicity of eight plants used in traditional Arab medicine in Israel. *J. Ethnopharmacol.* **2005**, *99*, 43–47. [CrossRef] [PubMed]
4. Cope, T.A.; Miller, A.G.; Morris, M. Plants of Dhofar: The Southern Region of Oman. Traditional, Economic and Medicinal Uses. *Geogr. J.* **1990**, *156*, 89. [CrossRef]
5. Ghazanfar, S.A. *Handbook of Arabian Medicinal Plants*; CRC Press: Boca Raton, FL, USA, 1994.
6. Bussmann, R.W.; Malca, G.; Glenn, A.; Sharon, D.; Nilsen, B.; Parris, B.; Dubose, D.; Ruiz, D.; Saleda, J.; Martinez, M.; et al. Toxicity of medicinal plants used in traditional medicine in Northern Peru. *J. Ethnopharmacol.* **2011**, *137*, 121–140. [CrossRef]
7. Chandra, S.; Rawat, D.S. Medicinal plants of the family Caryophyllaceae: A review of ethno-medicinal uses and pharmacological properties. *Integr. Med. Res.* **2015**, *4*, 123–131. [CrossRef]
8. Phondani, P.C.; Bhatt, A.; Elsarrag, E.; Horr, Y.A. Ethnobotanical magnitude towards sustainable utilization of wild foliage in Arabian Desert. *J. Tradit. Complement. Med.* **2016**, *6*, 209–218. [CrossRef]
9. Ullah, R.; Alqahtani, A.S.; Noman, O.M.A.; Alqahtani, A.M.; Ibenmoussa, S.; Bourhia, M. A review on ethno-medicinal plants used in traditional medicine in the Kingdom of Saudi Arabia. *Saudi J. Biol. Sci.* **2020**, *27*, 2706–2718. [CrossRef]
10. Aati, H.; El-Gamal, A.; Shaheen, H.; Kayser, O. Traditional use of ethnomedicinal native plants in the Kingdom of Saudi Arabia. *J. Ethnobiol. Ethnomed.* **2019**, *15*, 2. [CrossRef]
11. Sakkir, S. Medicinal plants diversity and their conservation status in the United Arab Emirates (UAE). *J. Med. Plants Res.* **2012**, *6*, 1304–1322. [CrossRef]
12. Cybulska, I.; Brudecki, G.; Alassali, A.; Thomsen, M.; Jed Brown, J. Phytochemical composition of some common coastal halophytes of the United Arab Emirates. *Emirates J. Food Agric.* **2014**, *26*, 1046–1056. [CrossRef]
13. Sajjad, A.; Syed, A.; Hasnain, A.; Mohamed, E. Ethno Botanical Study of Traditional Native Plants in Al Ain UAE. *Int. J. Adv. Res. Biol. Sci.* **2017**, *4*, 1–10. [CrossRef]
14. Parvathy, V.A.; Swetha, V.P.; Sheeja, T.E.; Sasikumar, B. Detection of plant-based adulterants in turmeric powder using DNA barcoding. *Pharm. Biol.* **2015**, *53*, 1774–1779. [CrossRef]
15. Mosa, K.A.; Soliman, S.; El-Keblawy, A.; Ali, M.A.; Hassan, H.A.; Tamim, A.A.B.; Al-Ali, M.M. Using DNA Barcoding to Detect Adulteration in Different Herbal Plant-Based Products in the United Arab Emirates: Proof of Concept and Validation. *Recent Pat. Food. Nutr. Agric.* **2018**, *9*, 55–64. [CrossRef]
16. Abdel-Aziz, S.M.; Aeron, A.; Kahil, T.A.; Abdel-Aziz, S.M.; Kahil, T.A.; Aeron, A. Health Benefits and Possible Risks of Herbal Medicine. In *Microbes in Food and Health*; Garg, N., Abdel-Aziz, S., Aeron, A., Eds.; Springer: Cham, Switzerland, 2016. [CrossRef]
17. Pratiwi, R.; Dipadharma, R.H.F.; Prayugo, I.J.; Layandro, O.A. Recent Analytical Method for Detection of Chemical Adulterants in Herbal Medicine. *Molecules* **2021**, *26*, 6606. [CrossRef]
18. Tungmunnithum, D.; Renouard, S.; Drouet, S.; Blondeau, J.P.; Hano, C. A Critical Cross-Species Comparison of Pollen from Nelumbo nucifera Gaertn. vs. Nymphaea lotus L. for Authentication of Thai Medicinal Herbal Tea. *Plants* **2020**, *9*, 921. [CrossRef]
19. Wu, L.; Wu, M.; Cui, N.; Xiang, L.; Li, Y.; Li, X.; Chen, S. Plant super-barcode: A case study on genome-based identification for closely related species of Fritillaria. *Chinese Med.* **2021**, *16*, 1–11. [CrossRef]
20. Mosa, K.A.; Gairola, S.; Jamdade, R.; El-Keblawy, A.; Al Shaer, K.I.; Al Harthi, E.K.; Shabana, H.A.; Mahmoud, T. The Promise of Molecular and Genomic Techniques for Biodiversity Research and DNA Barcoding of the Arabian Peninsula Flora. *Front. Plant. Sci.* **2019**, *9*, 1929. [CrossRef]
21. Jamdade, R.; Upadhyay, M.; Al Shaer, K.; Al Harthi, E.; Al Sallani, M.; Al Jasmi, M.; Ketbi, A. Al Evaluation of Arabian Vascular Plant Barcodes (rbcL and matK): Precision of Unsupervised and Supervised Learning Methods towards Accurate Identification. *Plants* **2021**, *10*, 2741. [CrossRef]
22. Jamdade, R.A.; Mahmoud, T.; Gairola, S. Prospects of genomic resources available at the global databases for the flora of United Arab Emirates. *3 Biotech.* **2019**, *9*, 1–9. [CrossRef]
23. Urumarudappa, S.K.J.; Tungphatthong, C.; Prombutara, P.; Sukrong, S. DNA metabarcoding to unravel plant species composition in selected herbal medicines on the National List of Essential Medicines (NLEM) of Thailand. *Sci. Rep.* **2020**, *10*, 18259. [CrossRef]
24. Kress, W.J. Plant DNA barcodes: Applications today and in the future. *J. Syst. Evol.* **2017**, *55*, 291–307. [CrossRef]
25. Aghayeva, P.; Cozzolino, S.; Cafasso, D.; Ali-zade, V.; Fineschi, S.; Aghayeva, D. DNA barcoding of native Caucasus herbal plants: Potentials and limitations in complex groups and implications for phylogeographic patterns. *Biodivers. Data J.* **2021**, *9*, 1–28. [CrossRef]

26. Hebert, P.D.N.; Penton, E.H.; Burns, J.M.; Janzen, D.H.; Hallwachs, W. Ten species in one: DNA barcoding reveals cryptic species in the neotropical skipper butterfly Astraptes fulgerator. *Proc. Natl. Acad. Sci. USA* **2004**, *101*, 14812–14817. [CrossRef]

27. Burns, J.M.; Janzen, D.H.; Hajibabaei, M.; Hallwachs, W.; Hebert, P.D.N. DNA barcodes and cryptic species of skipper butterflies in the genus Perichares in Area de Conservación Guanacaste, Costa Rica. *Proc. Natl. Acad. Sci. USA* **2008**, *105*, 6350–6355. [CrossRef]

28. Dick, C.W.; Webb, C.O. Plant DNA Barcodes, Taxonomic Management, and Species Discovery in Tropical Forests. *Methods Mol. Biol.* **2012**, *858*, 379–393. [CrossRef]

29. Celiński, K.; Kijak, H.; Wojnicka-Półtorak, A.; Buczkowska-Chmielewska, K.; Sokołowska, J.; Chudzińska, E. Effectiveness of the DNA barcoding approach for closely related conifers discrimination: A case study of the Pinus mugo complex. *Comptes Rendus Biol.* **2017**, *340*, 339–348. [CrossRef]

30. Yu, J.; Wu, X.; Liu, C.; Newmaster, S.; Ragupathy, S.; Kress, W.J. Progress in the use of DNA barcodes in the identification and classification of medicinal plants. *Ecotoxicol. Environ. Saf.* **2021**, *208*, 111691. [CrossRef]

31. Techen, N.; Parveen, I.; Pan, Z.; Khan, I.A. DNA barcoding of medicinal plant material for identification. *Curr. Opin. Biotechnol.* **2014**, *25*, 103–110. [CrossRef]

32. Nazar, N.; Howard, C.; Slater, A.; Sgamma, T. Challenges in Medicinal and Aromatic Plants DNA Barcoding—Lessons from the Lamiaceae. *Plants* **2022**, *11*, 137. [CrossRef]

33. CBOL Plant Working Group; Hollingsworth, P.M.; Forrest, L.L.; Spouge, J.L.; Hajibabaei, M.; Ratnasingham, S.; van der Bank, M.; Chase, M.W.; Cowan, R.S.; Erickson, D.L.; et al. A DNA barcode for land plants. *Proc. Natl. Acad. Sci. USA* **2009**, *106*, 12794–12797. [CrossRef]

34. Kress, W.J.; Wurdack, K.J.; Zimmer, E.A.; Weigt, L.A.; Janzen, D.H. Use of DNA barcodes to identify flowering plants. *Proc. Natl. Acad. Sci. USA* **2005**, *102*, 8369–8374. [CrossRef] [PubMed]

35. Kress, W.J.; Erickson, D.L. A Two-Locus Global DNA Barcode for Land Plants: The Coding rbcL Gene Complements the Non-Coding trnH-psbA Spacer Region. *PLoS ONE* **2007**, *2*, e508. [CrossRef] [PubMed]

36. Chase, M.W.; Salamin, N.; Wilkinson, M.; Dunwell, J.M.; Kesanakurthi, R.P.; Haidar, N.; Savolainen, V. Land plants and DNA barcodes: Short-term and long-term goals. *Philos. Trans. R. Soc. B Biol. Sci.* **2005**, *360*, 1889–1895. [CrossRef]

37. Taberlet, P.; Coissac, E.; Pompanon, F.; Gielly, L.; Miquel, C.; Valentini, A.; Vermat, T.; Corthier, G.; Brochmann, C.; Willerslev, E. Power and limitations of the chloroplast trnL (UAA) intron for plant DNA barcoding. *Nucleic Acids Res.* **2007**, *35*. [CrossRef]

38. Mishra, P.; Kumar, A.; Nagireddy, A.; Mani, D.N.; Shukla, A.K.; Tiwari, R.; Sundaresan, V. DNA barcoding: An efficient tool to overcome authentication challenges in the herbal market. *Plant. Biotechnol. J.* **2016**, *14*, 8–21. [CrossRef]

39. Hashim, A.M.; Alatawi, A.; Altaf, F.M.; Qari, S.H.; Elhady, M.E.; Osman, G.H.; Abouseadaa, H.H. Phylogenetic relationships and DNA barcoding of nine endangered medicinal plant species endemic to Saint Katherine protectorate. *Saudi J. Biol. Sci.* **2021**, *28*, 1919–1930. [CrossRef]

40. Gao, T.; Yao, H.; Song, J.; Liu, C.; Zhu, Y.; Ma, X.; Pang, X.; Xu, H.; Chen, S. Identification of medicinal plants in the family Fabaceae using a potential DNA barcode ITS2. *J. Ethnopharmacol.* **2010**, *130*, 116–121. [CrossRef]

41. Jiao, J.; Huang, W.; Bai, Z.; Liu, F.; Ma, C.; Liang, Z. DNA barcoding for the efficient and accurate identification of medicinal polygonati rhizoma in China. *PLoS ONE* **2018**, *13*, e0201015. [CrossRef]

42. Pathak, M.R.; Mohamed, A.A.M.; Farooq, M. DNA Barcoding and Identification of Medicinal Plants in the Kingdom of Bahrain. *Am. J. Plant. Sci.* **2018**, *9*, 2757–2774. [CrossRef]

43. Tahir, A.; Hussain, F.; Ahmed, N.; Ghorbani, A.; Jamil, A. Assessing universality of DNA barcoding in geographically isolated selected desert medicinal species of Fabaceae and Poaceae. *Peer J.* **2018**, *6*, e4499. [CrossRef]

44. Zhang, J.; Chen, M.; Dong, X.; Lin, R.; Fan, J.; Chen, Z. Evaluation of four commonly used DNA barcoding loci for Chinese medicinal plants of the family Schisandraceae. *PLoS ONE* **2015**, *10*, e0125574. [CrossRef]

45. Moon, B.C.; Kim, W.J.; Ji, Y.; Lee, Y.M.; Kang, Y.M.; Choi, G. Molecular identification of the traditional herbal medicines, Arisaematis Rhizoma and Pinelliae Tuber, and common adulterants via universal DNA barcode sequences. *Genet. Mol. Res.* **2016**, *15*, gmr7064. [CrossRef]

46. Kim, W.J.; Ji, Y.; Choi, G.; Kang, Y.M.; Yang, S.; Moon, B.C. Molecular identification and phylogenetic analysis of important medicinal plant species in genus Paeonia based on rDNA-ITS, matK, and rbcL DNA barcode sequences. *Genet. Mol. Res.* **2016**, *15*, gmr.15038472. [CrossRef]

47. Theodoridis, S.; Stefanaki, A.; Tezcan, M.; Aki, C.; Kokkini, S.; Vlachonasios, K.E. DNA barcoding in native plants of the Labiatae (Lamiaceae) family from Chios Island (Greece) and the adjacent Çeşme-Karaburun Peninsula (Turkey). *Mol. Ecol. Resour.* **2012**, *12*, 620–633. [CrossRef]

48. Schori, M.; Showalter, A.M. DNA barcoding as a means for identifying medicinal plants of Pakistan. *Pakistan J. Bot.* **2011**, *43*, 1–4.

49. Ducasse, J.; Ung, V.; Lecointre, G.; Miralles, A. LIMES: A tool for comparing species partition. *Bioinformatics* **2020**, *36*, 2282–2283. [CrossRef]

50. Puillandre, N.; Brouillet, S.; Achaz, G. ASAP: Assemble species by automatic partitioning. *Mol. Ecol. Resour.* **2021**, *21*, 609–620. [CrossRef]

51. Puillandre, N.; Lambert, A.; Brouillet, S.; Achaz, G. ABGD, Automatic Barcode Gap Discovery for primary species delimitation. *Mol. Ecol.* **2012**, *21*, 1864–1877. [CrossRef]

52. He, T.; Jiao, L.; Wiedenhoeft, A.C.; Yin, Y. Machine learning approaches outperform distance- and tree-based methods for DNA barcoding of Pterocarpus wood. *Planta* **2019**, *249*, 1617–1625. [CrossRef]

53. Emu, M.; Sakib, S. Species Identification using DNA Barcode Sequences through Supervised Learning Methods. In Proceedings of the 2019 International Conference on Electrical, Computer and Communication Engineering (ECCE), Cox'sBazar, Bangladesh, 7–9 February 2019. [CrossRef]

54. Weitschek, E.; Fiscon, G.; Felici, G. Supervised DNA Barcodes species classification: Analysis, comparisons and results. *BioData Min.* **2014**, *7*, 1–18. [CrossRef]

55. Patil, T.S.; Jamdade, R.A.; Patil, S.M.; Govindwar, S.P.; Muley, D.V. DNA barcode based delineation of freshwater fishes from northern Western Ghats of India, one of the world's biodiversity hotspots. *Biodivers. Conserv.* **2018**, *27*, 3349–3371. [CrossRef]

56. Enan, M.R.; Palakkott, A.R.; Ksiksi, T.S. DNA barcoding of selected UAE medicinal plant species: A comparative assessment of herbarium and fresh samples. *Physiol. Mol. Biol. Plants* **2017**, *23*, 221–227. [CrossRef]

57. Maloukh, L.; Kumarappan, A.; Jarrar, M.; Salehi, J.; El-wakil, H.; Rajya Lakshmi, T.V. Discriminatory power of rbcL barcode locus for authentication of some of United Arab Emirates (UAE) native plants. *3 Biotech.* **2017**, *7*, 144. [CrossRef]

58. Sukhorukov, A.P.; Mavrodiev, E.V.; Struwig, M.; Nilova, M.V.; Dzhalilova, K.K.; Balandin, S.A.; Erst, A.; Krinitsyna, A.A. One-seeded fruits in the core caryophyllales: Their origin and structural diversity. *PLoS ONE* **2015**, *10*, e0130783. [CrossRef]

59. Kool, A.; de Boer, H.J.; Krüger, Å.; Rydberg, A.; Abbad, A.; Björk, L.; Martin, G. Molecular identification of commercialized medicinal plants in Southern Morocco. *PLoS ONE* **2012**, *7*, e39459. [CrossRef]

60. Hernández-Ledesma, P.; Berendsohn, W.G.; Borsch, T.; Von Mering, S.; Akhani, H.; Arias, S.; Castañeda-Noa, I.; Eggli, U.; Eriksson, R.; Flores-Olvera, H.; et al. A taxonomic backbone for the global synthesis of species diversity in the angiosperm order caryophyllales. *Willdenowia* **2015**, *45*, 281–383. [CrossRef]

61. Cuénoud, P.; Savolainen, V.; Chatrou, L.W.; Powell, M.; Grayer, R.J.; Chase, M.W. Molecular phylogenetics of Caryophyllales based on nuclear 18S rDNA and plastid rbcL, atpB, and matK DNA sequences. *Am. J. Bot.* **2002**, *89*, 132–144. [CrossRef]

62. Jongbloed, M.V.D.; Feulner, G.R.; Böer, B.B.; Western, A.R. *The Comprehensive Guide to the Wild Flowers of the United Arab Emirates*; Environmental Research and Wildlife Development Agency: Abu Dhabi, United Arab Emirates, 2003; ISBN 978-9948408246.

63. Karim, F.M.; Fawzi, N.M. *Flora of the United Arab Emirates*; Publications Department; United Arab Emirates University: Al Ain, United Arab Emirates, 2007; ISBN 9789948021407.

64. Feulner, G.R. The Flora of the Ru'us al-Jibal—the Mountains of the Musandam Peninsula: An Annotated Checklist and Selected Observations. Available online: http://www.enhg.org/Portals/1/trib/V19/TribulusV19.pdf (accessed on 6 July 2021).

65. Feulner, G.R. The Olive Highlands: A Unique "Island" of Biodiversity within the Hajar Mountains of the United Arab Emirates. Available online: http://www.enhg.org/Portals/1/trib/V22/TribulusV22.pdf (accessed on 6 July 2021).

66. Levin, R.A.; Wagner, W.L.; Hoch, P.C.; Nepokroeff, M.; Pires, J.C.; Zimmer, E.A.; Sytsma, K.J. Family-level relationships of Onagraceae based on chloroplast *rbc L* and *ndh F* data. *Am. J. Bot.* **2003**, *90*, 107–115. [CrossRef]

67. Lee, H.L.; Yi, D.K.; Kim, J.S. Development of plant DNA barcoding markers from the variable noncoding regions of chloroplast genome. In Proceedings of the Abstract Presented at the Second International Barcode of Life Conference, Academia Sinica, Taipei, Taiwan, 18–20 September 2007.

68. Chen, S.; Yao, H.; Han, J.; Liu, C.; Song, J.; Shi, L.; Zhu, Y.; Ma, X.; Gao, T.; Pang, X.; et al. Validation of the ITS2 Region as a Novel DNA Barcode for Identifying Medicinal Plant Species. *PLoS ONE* **2010**, *5*, e8613. [CrossRef]

69. White, T.; Bruns, T.; Lee, S.; Taylor, J. No Amplification and direct sequencing of fungal ribosomal RNA genes for phylogenetics. In *PCR Protocols: A Guide to Methods and Applications*; MA, W.T., Gelfand, D., Sninsky, J., Eds.; Academic Press: New York, NY, USA, 1990; pp. 315–322.

70. Kumar, S.; Stecher, G.; Li, M.; Knyaz, C.; Tamura, K. MEGA X: Molecular Evolutionary Genetics Analysis across Computing Platforms. *Mol. Biol. Evol.* **2018**, *35*, 1547–1549. [CrossRef] [PubMed]

71. Altschul, S.F.; Gish, W.; Miller, W.; Myers, E.W.; Lipman, D.J. Basic local alignment search tool. *J. Mol. Biol.* **1990**, *215*, 403–410. [CrossRef]

72. Hall, M.; Frank, E.; Holmes, G.; Pfahringer, B.; Reutemann, P.; Witten, I.H. The WEKA data mining software. *ACM SIGKDD Explor. Newsl.* **2009**, *11*, 10–18. [CrossRef]

73. Balunas, M.J.; Kinghorn, A.D. Drug discovery from medicinal plants. *Life Sci.* **2005**, *78*, 431–441. [CrossRef] [PubMed]

74. Bafeel, S.O.; Arif, I.A.; Bakir, M.A.; Khan, H.A.; Al Farhan, A.H.; Al Homaidan, A.A.; Ahamed, A.; Thomas, J. Comparative Evaluation of PCR Success with Universal Primers of Maturase K (matK) and Ribulose-1, 5-Bisphosphate Carboxylase Oxygenase Large Subunit (rbcL) for Barcoding of Some Arid Plants. *Plant. Omics* **2011**, *4*, 195–198.

75. Costion, C.; Ford, A.; Cross, H.; Crayn, D.; Harrington, M.; Lowe, A. Plant DNA Barcodes Can Accurately Estimate Species Richness in Poorly Known Floras. *PLoS ONE* **2011**, *6*, e26841. [CrossRef] [PubMed]

76. YU, J.; XUE, J.-H.; ZHOU, S.-L. New universal matK primers for DNA barcoding angiosperms. *J. Syst. Evol.* **2011**, *49*, 176–181. [CrossRef]

77. Fazekas, A.J.; Kuzmina, M.L.; Newmaster, S.G.; Hollingsworth, P.M. DNA barcoding methods for land plants. *Methods Mol. Biol.* **2012**, *858*, 223–252. [CrossRef]

78. Von Cräutlein, M.; Korpelainen, H.; Pietiläinen, M.; Rikkinen, J. DNA barcoding: A tool for improved taxon identification and detection of species diversity. *Biodivers. Conserv.* **2011**, *20*, 373–389. [CrossRef]

79. Casiraghi, M.; Labra, M.; Ferri, E.; Galimberti, A.; de Mattia, F. DNA barcoding: A six-question tour to improve users' awareness about the method. *Brief. Bioinform.* **2010**, *11*, 440–453. [CrossRef]

80. Kapralov, M.V.; Akhani, H.; Voznesenskaya, E.V.; Edwards, G.; Franceschi, V.; Roalson, E.H. Phylogenetic Relationships in the Salicornioideae / Suaedoideae / Salsoloideae s.l. (Chenopodiaceae) Clade and a Clarification of the Phylogenetic Position of Bienertia and Alexandra Using Multiple DNA Sequence Datasets. *Syst. Bot.* **2006**, *31*, 571–585. [CrossRef]

81. Wyler, S.C.; Naciri, Y. Evolutionary histories determine DNA barcoding success in vascular plants: Seven case studies using intraspecific broad sampling of closely related species. *BMC Evol. Biol.* **2016**, *16*, 103. [CrossRef]

82. Liu, X.F.; Yang, C.H.; Han, H.L.; Ward, R.D.; Zhang, A. Identifying species of moths (Lepidoptera) from Baihua Mountain, Beijing, China, using DNA barcodes. *Ecol. Evol.* **2014**, *4*, 2472–2487. [CrossRef]

83. Yang, Z.; Landry, J.-F.; Hebert, P.D.N. A DNA Barcode Library for North American Pyraustinae (Lepidoptera: Pyraloidea: Crambidae). *PLoS ONE* **2016**, *11*, e0161449. [CrossRef]

84. Zou, S.; Fei, C.; Song, J.; Bao, Y.; He, M.; Wang, C. Combining and Comparing Coalescent, Distance and Character-Based Approaches for Barcoding Microalgaes: A Test with Chlorella-Like Species (Chlorophyta). *PLoS ONE* **2016**, *11*, e0153833. [CrossRef]

85. Jaén-Molina, R.; Marrero-Rodríguez, Á.; Reyes-Betancort, J.A.; Santos-Guerra, A.; Naranjo-Suárez, J.; Caujapé-Castells, J. Molecular taxonomic identification in the absence of a 'barcoding gap': A test with the endemic flora of the Canarian oceanic hotspot. *Mol. Ecol. Resour.* **2015**, *15*, 42–56. [CrossRef]

86. Jiang, K.W.; Zhang, R.; Zhang, Z.F.; Pan, B.; Tian, B. DNA barcoding and molecular phylogeny of Dumasia (Fabaceae: Phaseoleae) reveals a cryptic lineage. *Plant. Divers.* **2020**, *42*, 376–385. [CrossRef]

87. Barley, A.J.; Thomson, R.C. Assessing the performance of DNA barcoding using posterior predictive simulations. *Mol. Ecol.* **2016**, *25*, 1944–1957. [CrossRef]

88. Morello, L.; Braglia, L.; Gavazzi, F.; Gianì, S.; Breviario, D. Tubulin-Based DNA Barcode: Principle and Applications to Complex Food Matrices. *Genes* **2019**, *10*, 229. [CrossRef]

Article

Data Release: DNA Barcodes of Plant Species Collected for the Global Genome Initiative for Gardens (GGI-Gardens) II

Morgan R. Gostel [1,2,*], Mónica M. Carlsen [3], Amanda Devine [2], Katharine B. Barker [2], Jonathan A. Coddington [2] and Julia Steier [2]

1. Botanical Research Institute of Texas, Fort Worth, TX 76132, USA
2. National Museum of Natural History, Smithsonian Institution, P.O. Box 37012, Washington, DC 20013, USA; devinea@si.edu (A.D.); barkerk@si.edu (K.B.B.); coddington@si.edu (J.A.C.); steierj@si.edu (J.S.)
3. Missouri Botanical Garden, St. Louis, MO 63110, USA; monica.carlsen@mobot.org
* Correspondence: mgostel@brit.org; Tel.: +1-(817)-463-4199

Abstract: The Global Genome Initiative for Gardens (GGI-Gardens) is an international partnership of botanic gardens and arboreta that aims to preserve and understand the genomic diversity of plants on Earth. GGI-Gardens has organized a collection program focused on the living collections that partner institutions and supports the preservation of herbarium and genomic vouchers. Collections made through GGI-Gardens are deposited in recognized herbaria and Global Genome Biodiversity Network-partnered biorepositories worldwide, meaning that they are made available to the public. With support from its parent organization, the Global Genome Initiative (GGI), plant DNA barcode sequencing is performed using tissues collected through this partnership that represent taxa without barcode sequences in GenBank. This is the second data release published by GGI-Gardens and constitutes 2722 barcode sequences from 174 families and 702 genera of land plants. All DNA barcodes generated in this study are now available through the Barcode of Life Data Systems (BOLD) and GenBank.

Keywords: biobanking; DNA barcoding; GenBank; ITS2; *matK*; *psbA-trnH*; *rbcL*; viridiplantae

Citation: Gostel, M.R.; Carlsen, M.M.; Devine, A.; Barker, K.B.; Coddington, J.A.; Steier, J. Data Release: DNA Barcodes of Plant Species Collected for the Global Genome Initiative for Gardens (GGI-Gardens) II. *Diversity* **2022**, *14*, 234. https://doi.org/10.3390/d14040234

Academic Editor: Mario A. Pagnotta

Received: 11 February 2022
Accepted: 18 March 2022
Published: 23 March 2022

Publisher's Note: MDPI stays neutral with regard to jurisdictional claims in published maps and institutional affiliations.

1. Introduction

Founded in 2015, the Global Genome Initiative for Gardens (GGI-Gardens, [1]) is an international partnership of botanic gardens and arboreta that aims to preserve and understand Earth's genomic diversity of plants. GGI-Gardens supports the collection of both herbarium and genomic voucher material from the living collections in these partner gardens following best practices for herbarium and genomics research [2]. Collections made through this program are stored in Global Genome Biodiversity Network (GGBN)-partnered DNA banks [3], meaning that they can be utilized for applications ranging from whole genome sequencing [4] to DNA barcoding [5], as well as other genomic research.

Since their conception in 2003 [6], DNA barcode sequences have been used as powerful tools that enable the large-scale and rapid taxonomic identification of species for myriad purposes, including conservation [7], forensics [8], and the quantification of species diversity [9], among others. Emerging techniques, such as metabarcoding [10,11], leverage high-throughput sequencing technology and are capable of sequencing a mixed or pooled sample of species and identify them from their barcode sequence.

An important limiting factor for these and other studies that utilize DNA barcode sequences, however, is the representation of species diversity in reference databases [12]. DNA barcode reference databases are growing in both their taxonomic and geographic scope thanks to a number of large initiatives, which often focus on a particular branch of the tree of life or geographic area. For example, since 2005, the African Centre for DNA Barcoding has been contributing DNA barcode reference sequences from Africa to facilitate improved DNA barcoding applications from this continent [13]. The basic concept of DNA barcoding has also expanded during recent years, thanks to high-throughput sequencing (HTS) technology and methods to "extend" the traditional barcode concept include the use of so-called "genome-skim" data [14] or whole organelle genome sequences as "super-" [15] or "ultra-barcodes" [16]. Clade-based approaches are contributing large-scale DNA barcode reference sequences for entire groups of organisms that are often regionally focused [17] or even hyper locally focused (e.g., sequencing living collections from botanic gardens, [18]), and in this paper we provide a large contribution from collections made by the GGI-Gardens program.

Facilitated by the Global Genome Initiative based at Smithsonian Institution (https://naturalhistory.si.edu/research/global-genome-initiative, accessed on 30 January 2022), new families and genera collected by GGI-Gardens partners to date have been extracted and sequenced using four plant DNA barcode loci (*rbcL*, *matK*, ITS2, and *psbA-trnH*). Past collections made through the GGI-Gardens program have been published as part of large DNA barcode "data releases", the first of which included the publication of nearly 2000 barcode sequences [5]. This manuscript represents the second data release for samples collected through the GGI-Gardens program and will serve as a significant contribution to available plant DNA barcode sequences in public repositories. These barcode sequences will facilitate future plant biodiversity research by improving the ability of researchers to use DNA barcode sequences to accurately identify species from DNA barcode reference databases through general plant inventories, ecological studies, and metabarcoding studies.

2. Methods and Materials

2.1. Tissue Collection

DNA barcode sequences published as part of this data release comprise collections conducted from two GGI-Gardens partners—the Botanical Research Institute of Texas (BRIT) and the Missouri Botanical Garden (MOBOT) between 2017 and 2020. A total of 817 collections (from 788 species) are represented in this data release, and these include 174 families and 702 genera (Supplementary Table S1). All collections were conducted following published best practices [2] and include herbarium vouchers, as well as genomic vouchers that include tissue preserved in silica gel and flash frozen in liquid nitrogen. Collections were prioritized using the GGI Gap Analysis Tool (https://globalgeno.me, accessed on 30 January 2022) following the scheme proposed in Linsky and Gostel [19] and whether DNA barcode quality sequences were available in GenBank. Silica-dried leaf tissues were sampled to generate DNA barcode sequences published in this study.

2.2. DNA Extraction

Silica-preserved leaf tissues (~10 µg of each sample) were sampled with 25 µL ETOH (to mitigate static) into a 96-well plate preloaded with glass and ceramic beads and then disrupted using a FastPrep 96 instrument (MP Biomedicals, Santa Ana, CA, USA). Whole genomic DNA was isolated using an AutoGenprep 965 (Autogen, Holliston, MA, USA) automatic extractor following the manufacturer's protocol for plant tissue.

2.3. PCR Amplification and Sequencing

Four standard plant DNA barcode loci (Table 1) were amplified, including the two core barcoding regions for land plants, *rbcL* and *matK* [20] and two additional loci that have been proposed as additional plant DNA barcoding loci, ITS2 and *psbA-trnH* [21–23]. PCR was performed using Bioline Taq polymerase (New England Biolabs, Ipswich, MA, USA) and a standard thermal cycling profile including an initial denaturation for 5 min at 95 °C, 35 cycles each including 95 °C denaturation for 30 s, a locus-specific annealing temperature (see Table 1) for 30 s, and an extension cycle at 72 °C for 40 s, followed by a final extension of 72 °C for 10 min. Amplified PCR products were cleaned using ExoSAP-IT (ThermoFisher Scientific, Waltham, MA, USA) following the manufacturer's protocols. Cycle sequencing was performed in 96-well plates using the same PCR primers, the BigDye® Terminator v3.1 Cycle Sequencing Kit (Applied Biosystems®, Norwalk, CT, USA), and sequenced on the Automated ABI3730 Sequencer (Life Technologies, Carlsbad, CA, USA). Raw chromatograms were edited in Geneious Prime (2021, https://www.geneious.com, accessed on 15 January 2022) and annotated before uploading to BOLD (https://www.boldsystems.org, accessed on 30 January 2022) and GenBank (See Supplementary Table S1 for accession numbers).

Table 1. Locus information for each plant DNA barcoding marker used in this study, including forward and reverse primer names, primer sequences, annealing temperature, and citation.

Locus	Primer Name	Forward Primer Sequence	Annealing Temperature	Citation
rbcL	rbcLa-F	ATGTCACCACAAACAGAGACTAAAGC	55 °C	[24]
	rbcLa-R	GTAAAATCAAGTCCACCRCG		[25]
matK	matK-xf	TAATTTACGATCAATTCATTC	54 °C	[26]
	matK-MALP	ACAAGAAAGTCGAAGTAT		[27]
ITS2	ITS_S2F	ATGCGATACTTGGTGTGAAT	56 °C	[28]
	ITS4	TCCTCCGCTTATTGATATGC		[29]
psbA-trnH	psbA3_f	GTTATGCATGAACGTAATGCTC	64 °C	[30]
	trnHf_05	CGCGCATGGTGGATTCACAATCC		[31]

3. Results and Data Resources
Sequence Characteristics and Upload to BOLD and GenBank

A total of 2722 DNA barcode sequences were generated and uploaded to the BOLD and GenBank reference databases (both publicly available, Supplementary Table S1). Most DNA barcode sequences represented the *rbcL* locus (789), and the fewest sequences were generated for the *matK* locus (597 sequences). Overall, 650 and 686 sequences were generated for ITS2 and *psbA-trnH*, respectively. Among the DNA barcode sequences presented in this data release are 12 families, 292 genera, and 604 species that previously did not have barcode sequence data available in GenBank. All sequences uploaded to BOLD are contained within the BOLD projects GRDTX and GRDMO. All sequences uploaded to GenBank are part of the GGI-Gardens Bio-Projects (ID: PRJNA791936 & PRJNA485943).

These sequences represent important resources for biodiversity studies and will facilitate rapid species identification and ecological studies that seek to understand plant community composition [9] and species interactions [32], as well as conservation assessments that depend upon DNA barcoding to identify and control invasive species, e.g., [33] and enforce policies regarding the trade in endangered plants, e.g., [34,35]. Expanding the plant DNA barcode sequence reference library can also help botanic gardens to accurately identify species in their living collections, i.e., [36]. We hope this work encourages others who work with plant DNA barcodes to contribute to growing DNA barcode reference databases to improve these public resources.

Supplementary Materials: The following supporting information can be downloaded at: https://www.mdpi.com/article/10.3390/d14040234/s1. Table S1. List of samples collected for the Global Genome Initiative for Gardens projects selected for DNA barcoding in this study, with GenBank accession numbers and genomic voucher identification numbers. All the sequences are included in the GGI-Gardens BioProjects PRJNA791936 and PRJNA485943 in GenBank and BOLD projects GRDTX and GRDMO. Each Collector Number includes a link to the digital genomic voucher record stored in the Smithsonian Biorepository.

Author Contributions: Conceptualization, M.R.G. and J.S.; methodology, M.R.G., M.M.C. & J.S.; software, A.D.; formal analysis, J.S.; resources, K.B.B. and J.A.C.; data curation, J.S.; writing—original draft preparation, M.R.G.; writing—review and editing, J.S., M.M.C., M.R.G., A.D., K.B.B., J.A.C.; project administration, M.R.G., K.B.B., J.A.C. All authors have read and agreed to the published version of the manuscript.

Funding: This research was funded by the Global Genome Initiative under grants GGI-Gardens-2019-206, GGI-Gardens-2020-246, and GGI-Partnerships-2017-167.

Institutional Review Board Statement: Not applicable.

Data Availability Statement: The data presented in this study are available in Supplementary Table S1.

Acknowledgments: All laboratory work was conducted in the Laboratories for Analytical Biology (LAB) at the National Museum of Natural History in Washington, DC and at the Museum Support Center in Suitland, MD. Genomic vouchers are deposited in the Smithsonian Institution Biorepository as well as the DNA banks at the Botanical Research Institute of Texas (BRIT) and the Missouri Botanical Garden (MO). Herbarium vouchers for samples associated with Genbank Projects PRJNA791936 and PRJNA485943 have been deposited in the BRIT and MO herbaria, respectively. M.R.G. thanks Faranhoz Khojayori, Seth Hamby, and Jerrod Stone for their assistance with the collection of samples used in this work from BRIT. M.M.C. thanks Danielle Hopkins, Stephanie Keil, Ella Ludwig, and Gabrielle McAuley for their assistance with collections made at the Missouri Botanical Garden. The sequences published as part of thisfFigur work were supported by the Global Genome Initiative, with assistance from Jose Zúñiga. Any paper(s) resulting directly from this specimen-processing project should reference support from the Global Genome Initiative and the Laboratories of Analytical Biology, National Museum of Natural History, Smithsonian Institution.

Conflicts of Interest: The authors declare no conflict of interest.

References

1. Gostel, M.R.; Kelloff, C.L.; Wallick, K.; Funk, V.A. A workflow to preserve genome-quality tissue samples from plants in botanical gardens and arboreta. *Appl. Plant. Sci.* **2016**, *4*, 1600039. [CrossRef] [PubMed]
2. Funk, V.A.; Gostel, M.R.; Devine, A.; Kelloff, C.L.; Wurdack, K.; Tuccinardi, C.; Radosavljevic, A.; Peters, M.; Coddington, J.A. Guidelines for collecting vouchers and tissues intended for genomic work (Smithsonian Institution): Botany Best Practices. *Biodivers. Data J.* **2017**, *5*, e11625. [CrossRef] [PubMed]
3. Seberg, O.; Droege, G.; Barker, K.B.; Coddington, J.A.; Funk, V.A.; Gostel, M.R.; Peterson, G.; Smith, P.P. Global Genome Biodiversity Network: Saving a blueprint of the Tree of Life—A botanical perspective. *Ann. Bot.* **2016**, *118*, 393–399. [CrossRef] [PubMed]
4. Kress, W.J.; Soltis, D.E.; Kersey, P.J.; Wegrzyn, J.L.; Leebens-Mack, J.H.; Gostel, M.R.; Liu, X.; Soltis, P.S. Green plant genomes: What we know in an era of rapidly expanding opportunities. *Proc. Natl. Acad. Sci. USA* **2022**, *119*, e2115640118. [CrossRef] [PubMed]
5. Zúñiga, J.D.; Gostel, M.R.; Mulcahy, D.G.; Barker, K.B.; Hill, A.; Sedaghatpour, M.; Vo, S.Q.; Funk, V.A.; Coddington, J.A. Data release: DNA barcodes of plant species collected for the Global Genome Initiative for Gardens program, National Museum of Natural History, Smithsonian Institution. *PhytoKeys* **2017**, *88*, 119–122. [CrossRef]
6. Hebert, P.D.N.; Cywinska, A.; Ball, S.L.; deWaard, J.R. Biological identifications through DNA barcodes. *Proc. R. Soc. Lond. B.* **2003**, *270*, 313–321. [CrossRef] [PubMed]
7. Costion, C.M.; Kress, W.J.; Crayn, D.M. DNA barcodes confirm the taxonomic and conservation status of a species of tree on the brink of extinction in the Pacific. *PLoS ONE* **2016**, *11*, e0155118. [CrossRef]
8. Staats, M.; Arulandhu, A.J.; Gravendeel, B.; Holst-Jensen, A.; Scholtens, I.; Peelen, T.; Prins, T.W.; Kok, E. Advances in DNA metabarcoding for food and wildlife forensic species identification. *Anal. Bioanal. Chem.* **2016**, *408*, 4615–4630. [CrossRef]
9. Costion, C.; Ford, A.; Cross, H.; Crayn, D.; Harrington, M.; Lowe, A. Plant DNA barcodes can accurately estimate species richness in poorly known floras. *PLoS ONE* **2011**, *6*, e26841. [CrossRef]
10. Taberlet, P.; Coissac, E.; Pompanon, F.; Brochmann, C.; Willerslev, E. Towards next-generation biodiversity assessment using DNA metabarcoding. *Mol. Ecol.* **2012**, *21*, 2045–2050. [CrossRef]
11. Gostel, M.R.; Zúñiga, J.D.; Kress, W.J.; Funk, V.A.; Puente-Lelievre, C. Microfluidic enrichment barcoding (MEBarcoding): A new method for high throughput plant DNA barcoding. *Sci. Rep.* **2020**, *10*, 8701. [CrossRef] [PubMed]
12. Cowart, D.A.; Pinheiro, M.; Mouchel, O.; Maguer, M.; Grall, J.; Miné, J.; Arnaud-Haond, S. Metabarcoding is powerful yet still blind: A comparative analysis of morphological and molecular surveys of seagrass communities. *PLoS ONE* **2015**, *10*, e0117562. [CrossRef] [PubMed]
13. Bezeng, B.S.; Davies, T.J.; Daru, B.H.; Kabongo, R.M.; Maurin, O.; Yessoufou, K.; van der Bank, H.; van der Bank, M. Ten years of barcoding at the African Centre for DNA Barcoding. *Genome* **2017**, *60*, 629–638. [CrossRef] [PubMed]
14. Coissac, E.; Hollingworth, P.M.; Lavergne, S.; Taberlet, P. From barcodes to genomes: Extending the concept of DNA barcod-ing. *Mol. Ecol.* **2016**, *25*, 1423–1428. [CrossRef]
15. Kane, N.C.; Cronk, Q. Botany without borders: Barcoding in focus. *Mol. Ecol.* **2008**, *17*, 5175–5176. [CrossRef]
16. Li, X.; Yang, Y.; Henry, R.J.; Rossetto, M.; Wang, Y.; Chen, S. Plant DNA barcoding: From gene to genome. *Biol. Rev.* **2015**, *90*, 157–166. [CrossRef]
17. Chua, P.Y.S.; Leerhøi, F.; Langkjaer, E.M.R.; Margaryan, A.; Noer, C.L.; Richter, S.R.; Restrup, M.E.; Bruun, H.H.; Hartvig, I.; Coissac, E.; et al. Towards the extended barcode concept: Generating DNA reference data through genome skimming of danish plants. *bioRxiv* **2021**. [CrossRef]

18. Liu, H.; Wei, J.; Yang, T.; Mu, W.; Song, B.; Yang, T.; Fu, Y.; Wang, X.; Hu, G.; Li, W.; et al. Molecular digitization of a botanical garden: High-depth whole-genome sequencing of 689 vascular plant species from the Ruili Botanical Garden. *GigaScience* **2019**, *8*, giz007. [CrossRef]
19. Linsky, J.; Gostel, M.R. The Global Genome Iniatitive for Gardens: Conservation priorities at the interface of botanic gardens and biodiversity genomics. *BGJournal* **2021**, *18*, 21–23.
20. CBOL Plant Working Group. A DNA barcode for land plants. *Proc. Natl. Acad. Sci. USA* **2009**, *106*, 12794–12797. [CrossRef]
21. Kress, W.J.; Wurdack, K.J.; Zimmer, E.A.; Weigt, L.A.; Janzen, D.H. Use of DNA barcodes to identify flowering plants. *Proc. Natl. Acad. Sci. USA* **2005**, *102*, 8369–8374. [CrossRef] [PubMed]
22. Kress, W.J.; Erickson, D.L. (Eds.) *DNA Barcodes: Methods and Protocols*; Humana Press: Totowa, NJ, USA; Springer Science+Publishing Media, LLC.: New York, NY, USA, 2012; Volume 858, pp. 3–8.
23. Hollingsworth, P.M.; Graham, S.W.; Little, D.P. Choosing and using a plant DNA barcode. *PLoS ONE* **2011**, *6*, e19254. [CrossRef] [PubMed]
24. Levin, R.A.; Wagner, W.L.; Hoch, P.C.; Nepokroeff, M.; Pires, J.C.; Zimmer, E.A.; Sytsma, K.J. Family-level relationships of Onagraceae based on chloroplast *rbcL* and *ndhF* data. *Am. J. Bot.* **2003**, *90*, 107–115. [CrossRef] [PubMed]
25. Kress, W.J.; Erickson, D.L.; Jones, F.A.; Swenson, N.G.; Perez, R.; Sanjur, O.; Bermingham, E. Plant DNA barcodes and a community phylogeny of a tropical forest dynamics plot in Panama. *Proc. Natl. Acad. Sci. USA* **2009**, *106*, 18621–18626. [CrossRef]
26. Ford, C.S.; Ayres, K.L.; Toomey, N.; Haider, N.; Van Alphen Stahl, J.; Kelly, L.J.; Wikström, N.; Hollingsworth, P.M.; Duff, R.J.; Hoot, S.B. Selection of candidate coding DNA barcoding regions for use on land plants. *Bot. J. Linn. Soc.* **2009**, *159*, 1–11. [CrossRef]
27. Dunning, L.T.; Savolainen, V. Broad-scale amplification of *matK* for DNA barcoding plants, a technical note. *Bot. J. Linn. Soc.* **2010**, *164*, 1–9. [CrossRef]
28. Chen, S.; Yao, H.; Han, J.; Liu, C.; Song, J.; Shi, L.; Zhu, Y.; Ma, X.; Gao, T.; Pang, X. Validation of the ITS2 region as a novel DNA barcode for identifying medicinal plant species. *PLoS ONE* **2010**, *5*, e8613. [CrossRef]
29. White, T.J.; Bruns, T.; Lee, S.; Taylor, J. *PCR Protocols: A Guide to Methods and Applications*; Innis, M., Gelfand, D., Sninsky, J., White, T., Eds.; Academic Press: New York, NY, USA, 1990; pp. 315–322.
30. Sang, T.; Crawford, D.; Stuessy, T. Chloroplast DNA phylogeny, reticulate evolution, and biogeography of *Paeonia* (Paeoniaceae). *Am. J. Bot.* **1997**, *84*, 1120–1136. [CrossRef]
31. Tate, J.; Simpson, B. Paraphyly of *Tarasa* (Malvaceae) and diverse origins of the polyploid species. *Syst. Bot.* **2003**, *28*, 723.
32. Lowe, A.; Jones, L.; Witter, L.; Creer, S.; de Vere, N. Using DNA Metabarcoding to Identify Floral Visitation by Pollinators. *Diversity* **2022**, *14*, 2022020018. [CrossRef]
33. Wang, A.; Wu, H.; Zhu, X.; Lin, J. Species Identification of Conyza Bonariensis Assisted by Chloroplast Genome Sequencing. *Front. Genet.* **2018**, *9*, 374. [CrossRef] [PubMed]
34. Nithaniyal, S.; Newmaster, S.G.; Ragupathy, S.; Krishnamoorthy, D.; Vassou, S.L.; Parani, M. DNA Barcode Authentication of Wood Samples of Threatened and Commercial Timber Trees within the Tropical Dry Evergreen Forest of India. *PLoS ONE* **2014**, *9*, e107669. [CrossRef] [PubMed]
35. Hassold, S.; Lowry, P.P.; Bauert, M.R.; Razafintsalama, A.; Ramamonjisoa, L.; Widmer, A. DNA Barcoding of Malagasy Rosewoods: Towards a Molecular Identification of CITES-Listed Dalbergia Species. *PLoS ONE* **2016**, *11*, e0157881. [CrossRef] [PubMed]
36. Le, D.-T.; Zhang, Y.-Q.; Xu, Y.; Guo, L.-X.; Ruan, Z.-P.; Burgess, K.S.; Ge, X.-J. The utility of DNA barcodes to confirm the identification of palm collections in botanical gardens. *PLoS ONE* **2020**, *15*, e0235569. [CrossRef]

Article

The Efficiency of DNA Barcoding in the Identification of Afromontane Forest Tree Species

David Kenfack [1,*], Iveren Abiem [2,3] and Hazel Chapman [3]

1   Smithsonian Tropical Research Institute, 10th & Constitution Avenue NW, Washington, DC 20013-7012, USA
2   Department of Plant Science and Biotechnology, University of Jos, P.M.B. 2084, Jos 930001, Nigeria; abiemiveren@gmail.com
3   School of Biological Sciences, University of Canterbury, Private Bag 4800, Christchurch 8140, New Zealand; hazel.chapman@canterbury.ac.nz
*   Correspondence: kenfackd@si.edu; Tel.: +1-202-633-4706

Abstract: The identification of flowering plants using DNA barcoding proposed in last decades has slowly gained ground in Africa, where it has been successfully used to elucidate the systematics and ecology of several plant groups, and to understand their evolutionary history. Existing inferences on the effectiveness of DNA barcoding to identify African trees are mostly based on lowland forests, whereas adjacent montane forests significantly differ from the latter floristically and structurally. Here, we tested the efficiency of chloroplast DNA barcodes (*rbcLa*, *matK*, and *trnH-psbA*) to identify Afromontane Forest tree species in a 20.28 ha permanent plot in Ngel Nyaki, Taraba state, Nigeria. We collected, identified, and vouchered 274 individuals with diameter at breast height $\geq$ 1 cm belonging to 101 morphospecies, 92 genera, and 48 families. *rbcLa* and *matK* used alone or in combination performed better than in lowland forests, with the best species discrimination obtained with the two-locus combination of *matK* + *rbcLa*. The intragenic spacer *trnH-psbA* was too variable to align and could not be tested using the genetic distance method employed. Classic DNA barcode can be a powerful tool to identify Afromontane tree species, mainly due to the non-prevalence in these communities of species—rich genera (low species-to-genus ratio) that constitute the biggest challenge of DNA barcoding of flowering plants.

Keywords: DNA barcoding; ForestGEO; montane forest; Ngel Nyaki; species identification

Citation: Kenfack, D.; Abiem, I.; Chapman, H. The Efficiency of DNA Barcoding in the Identification of Afromontane Forest Tree Species. *Diversity* **2022**, *14*, 233. https://doi.org/10.3390/d14040233

Academic Editors: W. John Kress, Morgan Gostel and Michael Wink

Received: 16 February 2022
Accepted: 21 March 2022
Published: 23 March 2022

Publisher's Note: MDPI stays neutral with regard to jurisdictional claims in published maps and institutional affiliations.

1. Introduction

Africa includes the second largest tropical forest block in the world, considered as one of the most important pool of biological diversity [1]. Yet, African forests are threatened by expanding human activities such as industrial logging, mining, agriculture, and road networks [2,3], but are also highly susceptible to the impact of climate change [4]. Despite the growing international concern about the future of these forests, the diversity, the ecology and the evolutionary processes that have shaped African forests remain relatively poorly understood, compared to the Amazon forest block [5]. In this regard, there is an urgent need to increase our efforts in documenting and describing the diversity of these forests as many of the species might go extinct before they are discovered. Therefore, large-scale biodiversity inventories of African forests will be critical to develop sound conservation strategies for these forests [6]. During the past decades, significant progress has been made in the study of the biodiversity of African forests using classic floristic inventories and long-term monitoring plots grouped into two main networks, the African Tropical Rainforest Observation Network (AfriTRON, http://www.afritron.org/) (accessed on 10 February 2022)and the Africa program of the Forest Global Earth Observatory Network (ForestGEO, https://forestgeo.si.edu/, accessed on 10 February 2022). In forest inventories, the species are identified merely on the basis of morphological characters, and this is challenging even for expert botanists. Often, up to 30% of the individuals in the plots remain unidentified for years [7] due to the absence during field surveys of flowers and fruits that are needed to achieve accurate identifications [8].

Biological identification through "DNA barcode" was proposed, first in the animal kingdom [9,10] and later on for land plants [11,12] as a molecular method that could supplement morphological identifications. DNA barcodes are short and standardized fragments of DNA that should be easy to amplify and to sequence, and that can rapidly and reliably distinguish species from each other. DNA barcoding slowly gained ground in Africa, with over 21,000 vascular plants and 3000 animal records in the Barcode of Life Data System in 2019 [13,14], and has been used to elucidate the systematics and ecology of several plant groups, e.g., [15–17]. Existing African DNA barcodes for plants have been concentrated in forest ecosystems in Southern and West Africa [14,18] and more recently in savanna ecosystems [13]. Furthermore, inferences on the effectiveness of DNA barcode to identify African forest trees have been mostly based on lowlands. Whereas montane forests significantly differ floristically and structurally from lowland forests, the effectiveness of DNA barcoding in identifying tree species in these forests is still lacking.

We constructed a local DNA barcode database to aid the identification of tree species and reconstruct their community phylogeny in a 20.28 ha plot located in montane forest in Northeastern Nigeria. Here, we test the ability of this DNA barcode to identify the plot species and genera.

2. Materials and Methods

2.1. Study Site and Sampling

The tissue samples for DNA extraction were collected from the 20.28 hectares (260 × 780 m) Ngel Nyaki Forest Dynamics plot, where all trees with diameter at breast height (dbh) > 1 cm had previously been measured, mapped tagged and morphologically identified [19]. The plot (07°04005″ N, 11°03024″ E) is located within the Ngel Nyaki Forest Reserve on the Mambilla Plateau, Taraba State, Nigeria, with elevation ranging from 1588 m to 1690 m, and is part of the Forest Global Earth Observatory (ForestGEO) network [20]. The mean annual rainfall is 1800 mm while the mean annual temperature is 19 °C. The vegetation of the area is a mosaic of grassland and montane forest [21].

The morphological identifications of the trees were performed in the field by non-professional taxonomists, but were partially checked by the first author. The resulting checklist comprised 105 morphospecies including 74 (71%) identified to species level, 22 (21%) to genus, and 9 (9%) unidentified, even to family level. Of the 105 species (with dbh > 1 cm) recorded in the Ngel Nyaki plot, we sampled 99 belonging to 90 genera and 47 families. Two additional woody species growing in the vicinity of the plot, *Dracaena* cf. *deisteliana* Engl. (Asparagaceae) and *Pittosporum viridiflorum* Sims (Pittosporaceae), were added to our sample, making a total of 101 species in 92 genera and 48 families. We collected leaf tissue from 1 (for species that were represented by a single individual in the plot) to 4 individuals per species. The samples were collected in the field and were immediately dried in silica-gel. They consisted of 5–50 cm^2 of leaf tissue. Voucher specimens accompanying the leaf tissue were also collected and are deposited at the National Museum of Natural History in Washington.

2.2. DNA Extraction and Sequencing

All laboratory work was carried out at the Canadian Centre for DNA Barcoding (CCDB) (https://ccdb.ca/, accessed on 10 February 2022) and following their protocols. Total genomic DNA was extracted from silica dried leaf material using the CCDB protocol (https://ccdb.ca/site/wp-content/uploads/2016/09/CCDB_DNA_Extraction-Plants.pdf, accessed on 10 February 2022). DNA barcode regions *rbcLa*, *matK* and the *trnH-psbA* intergenic spacer were amplified using CCDB standard PCR primers and protocols (https://ccdb.ca/site/wp-content/uploads/2016/09/CCDB_Amplification-Plants.pdf, accessed on 10 February 2022) with the primers available at https://ccdb.ca/site/wp-content/uploads/2016/09/CCDB_PrimerSets-Plants.pdf (accessed on 10 February 2022)

Voucher details and GenBank accession numbers for all sequences are listed in BOLD (http://www.boldsystems.org/) (accessed on 10 February 2022).

2.3. Testing the DNA Barcode Accuracy

Prior to evaluating the identification success of the two barcode regions, we used the Basic Local Alignment Search Tool (BLAST) [22] to compare our sequences to those available in GenBank (https://www.ncbi.nlm.nih.gov/genbank/, accessed on 10 February 2022), with the aim of confirming our identifications and identifying our unknowns. After matching our sequences in GenBank, the morphospecies names were updated only after comparison of their voucher specimens to either the type specimens available online or to other herbarium specimens and photographs in Tropicos (https://www.tropicos.org/, accessed on 10 February 2022).

To test the barcode efficiency, we followed [18]. Our DNA barcoding refence database (assumed to be exhaustive in terms of species) had 274 individuals and was used to assign individual trees to species or genera. The test was performed on species represented by at least two individuals in the database, so that we could have a query and a matching sample. The coding genes *matK* and *rbcLa* were aligned and manually adjusted using ClustalW in the Molecular Evolutionary Genetic Analysis software version 7.0.26 [23]. After the global alignment, we computed pairwise genetic distances among all sequences in the dataset using the Kimura's 2-parameter model [24]. The analysis was also performed in Mega7. In the resulting matrix, each sample (query) was assigned to a species or a genus of the sample (matching) from which it is separated with the least genetic distance (excluding itself). The identification was (1) correct when the query sample matched only the samples of its species of genus; (2) multiple if the query sample matched several species or genera including its correct one; and (3) wrong when the query sample matched species or genera different from its own [18]. We were not able to align *trnH-psbA* because it was too variable among the 48 plant families in the study. Hence this locus was not used in the test of DNA barcode accuracy analyses.

3. Results

3.1. Sequencing Success

DNA sequencing success was tested on 274 individual trees, representing 101 species. Sequencing success was lowest for *matK* and highest for *rbcLa*. Reliable contigs were obtained for only 78.9% of all individuals sequenced for *matK*, 95.3% *trnH-psbA*, but 97.5% for *rbcLa*, which corresponded to all the species represented in the database for *rbcLa* and *trnH-psbA*, but only to 93.1% of the species for *matK* (Table 1). The percentages of species represented by at least two individuals for *matK*, *rbcLa* and *trnH-psbA* in the database were 70.3%, 87.2% and 84.2% respectively.

Table 1. Sequencing success of montane forest trees from Ngel Nyaki for *rbcLa*, *matK* and *trnH-psbA* barcode regions.

	matK	*rbcLa*	*trnH-psbA*
Number of individuals tested	274	274	274
Sequencing success: N ind. (% ind.)	216 (78.9)	267 (97.5)	261 (95.3)
Sequencing success: N sp. (% sp.)	94 (93.1)	101 (100)	101 (100)
N sp. with sequences $\geq$ 2 samples	71 (70.3)	88 (87.2)	85 (84.2)

3.2. Taxonomic Update Using BLAST

The identification of 13 morphospecies was updated using the heuristic search in GenBank. Of the nine morphospecies for which the family was unknown, seven were identified to species level and two placed in different plant families. Furthermore, the identification of four other morphospecies was updated. The first morphospecies placed in the *Argocoffeopsis* Lebrun (Rubiaceae) Lebrun was updated to *Psilanthus mannii* Hook. within the same family. A morphospecies thought to belong to the genus *Beilschmiedia* Nees

(Lauraceae) was transferred to the family Sapotaceae. The identification of a morphospecies thought to be *Lannea barteri* (Oliv.) Engl. (Anacardiaceae) was updated to *Brucea antidysenterica* J.F. Mill. (Simaroubaceae). Finally, the morphospecies *Hannoa klaineana* Pierre ex Engl. (Simaroubaceae) was updated to *Ekebergia capensis* Sparrm.

3.3. Barcode Accuracy

The accuracy of two of the three barcode markers (*matK* and *rbcLa*) in identifying montane forest trees is presented in Table 2. The analyses were performed on all available samples for each marker. When used individually, highest success for the identification of species was obtained with *matK* (98.3%). The two makers performed slightly better when combined. At genus level, the same trend was maintained, but with even better performances. Here, *matK* and the *rbcLa* + *matK* combination successfully identified all the samples to the genus, while *rbcLa* alone was successful to identify 98.4% of samples to genus (Table 2).

Table 2. Barcoding accuracy in identifying Ngel Nyaki Afromontane forest trees at species and genus level.

	Barcoding Accuracy			Query Samples		
	Correct ID	Multiple ID	Wrong ID	N. ind.	N. sp.	N. Gen.
Species identification						
rbcLa	93.8	6.15	0	244	92	92
matK	98.3	1.1	0.55	181	67	59
matK + *rbcLa*	98.9	0.5	0.54	186	72	63
Genus identification						
rbcLa	98.4	1.6	0	244	92	92
matK	100	0	0	181	67	59
matK + *rbcLa*	100	0	0	186	72	63

4. Discussion

4.1. Recoverability of DNA Barcode Used

The two DNA barcodes *rbcLa* and *matK* used in this study have long been recognized having sufficient variation to discriminate among land plant species [11,25,26]. Among the three barcodes, *matK* had the lowest rate of recovery (79%), consistent with prior studies [18,27,28]. In contrast, *rbcLa* and *trnH-psbA* had higher rates of recovery (above 95%). However, it is worth pointing out that the rates of recovery were in general higher than in prior studies, probably due to the efficiency of the Canadian Centre for DNA Barcoding that has optimized protocols for higher rates of recovery. For example, recovery rates around 70% have been reported for *matK* in several studies [8,27,29], while sequencing and amplification success for *rbcLa* and *trnH-psbA* is often below 94% e.g., [8,27,30].

4.2. Tree Species Identification Using DNA Barcode in Ngel Nyaki Montane Forest

The morphological identification of the trees in the Ngel Nyaki plot was almost entirely performed by non-professional taxonomists who however accurately identified to species 69% of all tree species occurring in the plot. Only four species were wrongly identified. The DNA barcode was instrumental in updating the identification of 12% of the species in the plot for which prior sequences were available in Genbank. Due to the lack of adequate library in Genbank, 21% of the species in the plot for which good quality barcode sequences were generated could still not be identified to species level. Hence, molecular techniques such as DNA barcode may not replace traditional taxonomic techniques as suggested by some studies [31], but can only supplement it.

This study showed the efficiency of the two barcode loci *rbcLa* and *matK* in accurately assigning Afromontane forest tree species to a correct species or genera. When used alone, best results for species identification were obtained with *matK* (98%) compared to *rbcLa* (94%). These values are slightly higher than those reported in most lowland forests [8,18,27,30]. The combination of the two markers *matK* + *rbcLa* improved the barcoding success to 99%, a result consistent with those in most lowland forests. Barcoding success was even better at genus level, *rbcLa* alone identifying 98% of all genera, while *matK* and the combination *matK* + *rbcLa* accurately identified all the samples to genus.

The genetic distance method that we used did not allow us to test the accuracy of the intergenic spacer *trnH-psbA*. This locus, easy to amplify and short, is known to be very variable among angiosperms and thus is widely used in plant species identification [32]. In general, *trnH-psbA* locus is more variable than *matK* and *rbcL* and we assume its performance in the identification of montane forest species would even be greater. *matK* and *rbcLa* were variable enough that their combination to *trnH-psbA* was no more relevant.

4.3. The Efficiency of DNA Barcoding in the Context of the Afromontane Flora

DNA barcode is a powerful tool for identifying tree species to genus level. However the identification to species level is not always reliable, especially in plant communities with speciose genera [18]. For example, the identification of tree species (with dbh $\geq$ 1 cm) in a 50-ha plot in the highly diverse Korup National Park, Cameroon using three DNA barcode markers showed a significant decrease in their performance with increasing number of species per clade (genus) [18]. In fact, the five most speciose genera in the Korup plot *Cola* Schott & Endl., *Diospyros* L., *Psychotrya* L., *Rinorea* Aubl. and *Garcinia* L. have 23, 14, 13, 13 and 10 species respectively [33]. Such closely related species are more likely to hybridize, have incomplete lineage sorting and share haplotypes, all of which can lessen the ability of barcode loci to discriminate among them. At the other end of the spectrum, 165 (33%) species in Korup are represented by a single species.

The Ngel Nyaki plot had a relatively low diversity, with only 105 species in 92 genera. The most speciose genera here are *Ficus* L. and *Psychotria* L., each having three species. Five other genera have two species each, while the remaining 85 species (81%) are represented each by a single species. This species-to-genus (S/G) ratio is not specific to the Ngel Nyaki montane forest. In fact, most Afromontane forests are characterized by a low diversity of trees and low S/G ratio. For example, in Woodbush–De Hoek montane forest in South Africa, 50 species of trees with dbh > 5 cm and dbh > 10 cm in 46 genera (S/G = 1.09) were recorded within 1.5 ha circular plots [34]. Similarly, [35] in a study on trees with dbh $\geq$ 5 cm in dry Afromontane forests of Awi Zone, northwestern Ethiopia, recorded 18 species in 18 genera, 21 species in 21 genera, 20 species in 20 genera, 16 species in 16 genera and 23 species in 23 genera in 0.6 ha of Bari, Apini, Dabkuli, Tsahare Kan, and Kahtasa forests respectively.

We further explored the relationship between the S/G ratio and elevation, by comparing the Ngel Nyaki data other African forest sites for trees with dbh $\geq$ 10 cm (Table A1). The S/G ratio decreases with increasing elevation, with a correlation coefficient of -0.722 (Figure 1A). The Lambi 2 and Ngovayang mid-elevation plots in Cameroon had the highest S/G ratio (1.55 and 1.51 respectively) while higher elevation plots Bwindi 1 and Bwindi 4 had the lowest. The Lambi the Ngovayang plots seem to be outliers in our dataset. In fact, a stronger relationship with $r = -0.80$ is shown when these plots are removed. Higher S/G ratio of 2.6 and 3 have been reported elsewhere in the Manu forest (Peru) and Yasuni forest (Ecuador) respectively for trees with the same diameter cutoff [36]. The S/G ratio increases when smaller diameter size classes are considered and the correlation with elevation is stronger ($r = -0.84$, p-value = 0.007). A highest S/G ratio of 1.64 is observed for the lowland Rabi plot and 1.15 for the Ngel Nyaki plot for all trees with dbh $\geq$ 1 cm were measured (Figure 1B). In fact, the understory of most African forests are stocked with speciose genera of small-statured trees that never attain large size diameter classes [37,38]. Several studies have shown the decrease in tree species diversity with elevation, e.g., [39,40]. Our data also shows a decrease of generic diversity with increasing elevation ($r = -0.84$). This means that the low diversity in higher elevations is also due to the decrease in the number of genera, but coupled with the decrease in the number of species per genera. This result is consistent with Jaccard's observations in the Alps [41], who noted that "with increasing altitude, the number of genera decreases less rapidly than the number of species".

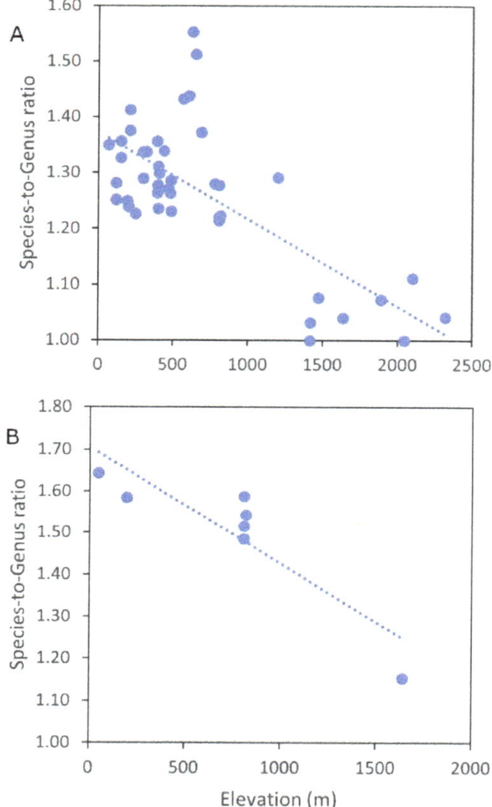

Figure 1. Correlation between the species-to-genus (S/G) ratio and elevation, (**A**) for trees with dbh > 10 cm in forty three 1-ha African forest plots, The correlation coefficient $r = -0.722$, p-value = 0.00000004635; (**B**) for trees with dbh > 1 cm in seven large (10–50-a) census plots, correlation coefficient $r = -0.88$, p-value = 0.007.

5. Conclusions

Our study highlighted how DNA barcoding can be efficient in identifying tree species in an Afromontane Forest. As in lowland forests, identification success is higher at genus than at species level. Identification success was higher than in lowland forest, due to the non-prevalence of highly diverse genera in this habitat. The comparison of species-to-genus among other sites with comparable data showed that Afromontane forests tend to have a low S/G ratio for tree species, which is an advantage for the use of DNA barcode in these forests.

Author Contributions: Conceptualization, D.K.; methodology, D.K., I.A. and H.C., investigation, D.K., I.A. and H.C.; data curation, D.K. and I.A.; writing—original draft preparation, D.K.; writing review and editing, D.K., I.A. and H.C.; funding acquisition, D.K. and H.C. All authors have read and agreed to the published version of the manuscript.

Funding: This study was made possible by the generous donation made by Retired General T.Y. Danjuma to the Nigerian Montane Forest Project (NMFP) through H. Chapman. Additional support towards the plot census and DNA barcoding was provided by the Forest Global Earth Observatory (ForestGEO) of the Smithsonian Tropical Research Institute. The Chester Zoo, England, and A.G. Leventis Foundation also provided additional financial assistance to the NMFP.

Data Availability Statement: Full census data for the Ngel Nyaki plot is available upon reasonable request from the ForestGEO data portal http://ctfs.si.edu/datarequest/ and the full plant DNA barcode library is available on BOLD (http://www.boldsystems.org/) (accessed on 10 February 2022).

Acknowledgments: We wish to thank the staff of the Ngel Nyaki Montane Forest Program, especially the ForestGEO local team who carried out the tree census and collected the DNA tissue and vouchers. We are also grateful to Douglas Sheil and Robert Bitariho for their authorization to use the Bwindi data, to Terry Sunderland for sharing the Takamanda data with us and finally to Daniel Zuleta for writing the R script to estimate the barcode success from the genetic distance matrix.

Conflicts of Interest: The authors declare no conflict of interest.

Appendix A

Table A1. Species-to-Genus ratio (S/G) among 43 African forest 1-ha plots for trees with dbh $\geq$ 10 cm. * denote large plots (10–50 ha) of the Forest Global Earth Observatory (ForestGEO) network. The data for each large plot was obtained by averaging the values in 1-ha subplots within the plot. S = number of species, G = number of genera. TEAM = Tropical Ecology Assessment and Monitoring.

Site	Country	Elevation (m)	S	G	S/G	Source
Bwindi-1	Burundi	1474	42	39	1.08	TEAM Network
Bwindi-2	Burundi	1419	28	28	1.00	TEAM Network
Bwindi-3	Burundi	1893	44	41	1.07	TEAM Network
Bwindi-4	Burundi	2049	27	27	1.00	TEAM Network
Bwindi-5	Burundi	2101	30	27	1.11	TEAM Network
Bwindi-6	Burundi	2321	25	24	1.04	TEAM Network
Bidjouka-1	Cameroon	392	99	73	1.36	[42]
Bidjouka-2	Cameroon	605	105	73	1.44	[42]
Korup 50-ha *	Cameroon	195	87.2	48.82	1.79	[33]
Lambi-1	Cameroon	396	106	83	1.28	[42]
Lambi-2	Cameroon	627	118	76	1.55	[42]
Ngovayang-1	Cameroon	650	121	80	1.51	[42]
Rumpi-hills-11	Cameroon	1450	32	31	1.03	[43]
Takamanda-10	Cameroon	210	108	78.5	1.38	[44]
Takamanda-11	Cameroon	210	113	80	1.41	[44]
Takamanda-12	Cameroon	150	105.5	79.5	1.33	[44]
Takamanda-13	Cameroon	150	118	87	1.36	[44]
Takamanda-14	Cameroon	120	87	69.5	1.25	[44]
Takamanda-15	Cameroon	120	91	71	1.28	[44]
Takamanda-6	Cameroon	320	103	77	1.34	[44]
Takamanda-7	Cameroon	400	97	74	1.31	[44]
Takamanda-8	Cameroon	780	64	50	1.28	[44]
Takamanda-9	Cameroon	1200	71	55	1.29	[44]
Dzanga-Sanga-1	Central African Republic	471	108	85	1.27	[45]
Dzanga-Sanga-2	Central African Republic	482	120	95	1.26	[45]
Dzanga-Sanga-3	Central African Republic	393	67	53	1.26	[45]
Dzanga-Sanga-4	Central African Republic	489	96	78	1.23	[45]
Dzanga-Sanga-5	Central African Republic	485	108	84	1.29	[45]
Edoro-1 (10-ha) *	DR Congo	808	65.4	53.6	1.22	[46]
Edoro-2 (10-ha) *	DR Congo	809	67.4	55.5	1.21	[46]
Lenda-1 (10-ha) *	DR Congo	808	60.4	47.3	1.28	[46]
Lenda-2 (10-ha) *	DR Congo	819	49.9	40.8	1.22	[46]

Table A1. *Cont.*

Site	Country	Elevation (m)	S	G	S/G	Source
Monts de Cristal-1	Gabon	400	89	72	1.24	[47]
Monts de Cristal-2	Gabon	300	89	69	1.29	[47]
Monts de Cristal-3	Gabon	300	99	74	1.34	[47]
Monts de Cristal-4	Gabon	200	88	71	1.24	[47]
Monts de Cristal-5	Gabon	250	108	88	1.23	[47]
Rabi 25-ha *	Gabon	47	84.6	62.68	1.35	[38]
Waka-10	Gabon	569	106	74	1.43	[48]
Waka-6	Gabon	438	83	62	1.34	[48]
Waka-7	Gabon	407	100	77	1.30	[48]
Waka-8	Gabon	687	107	78	1.37	[48]
Ngel Nyaki (20.28 ha) *	Nigeria	1639	41.1	39.5	1.04	[19]

References

1. Mittermeier, R.A.; Mittermeier, C.G.; Brooks, T.M.; Pilgrim, J.D.; Konstant, W.R.; da Fonseca, G.A.B.; Kormos, C. Wilderness and Biodiversity Conservation. *Proc. Natl. Acad. Sci. USA* **2003**, *100*, 10309–10313. [CrossRef] [PubMed]
2. Shapiro, A.C.; Grantham, H.S.; Aguilar-Amuchastegui, N.; Murray, N.J.; Gond, V.; Bonfils, D.; Rickenbach, O. Forest Condition in the Congo Basin for the Assessment of Ecosystem Conservation Status. *Ecol. Indic.* **2021**, *122*, 107268. [CrossRef]
3. Rudel, T.K. The National Determinants of Deforestation in Sub-Saharan Africa. *Philos. Trans. R. Soc. Lond. B Biol. Sci.* **2013**, *368*, 20120405. [CrossRef] [PubMed]
4. Lewis, S.L.; Sonké, B.; Sunderland, T.; Begne, S.K.; Lopez-Gonzalez, G.; van der Heijden, G.M.F.; Phillips, O.L.; Affum-Baffoe, K.; Baker, T.R.; Banin, L.; et al. Above-Ground Biomass and Structure of 260 African Tropical Forests. *Phil. Trans. R. Soc. B* **2013**, *368*, 20120295. [CrossRef] [PubMed]
5. Malhi, Y.; Adu-Bredu, S.; Asare, R.A.; Lewis, S.L.; Mayaux, P. African Rainforests: Past, Present and Future. *Philos. Trans. R. Soc. Lond. B Biol. Sci.* **2013**, *368*, 20120312. [CrossRef]
6. Brooks, T.M.; Mittermeier, R.A.; da Fonseca, G.A.B.; Gerlach, J.; Hoffmann, M.; Lamoreux, J.F.; Mittermeier, C.G.; Pilgrim, J.D.; Rodrigues, A.S.L. Global Biodiversity Conservation Priorities. *Science* **2006**, *313*, 58–61. [CrossRef]
7. Condit, R. *Tropical Forest Census Plots: Methods and Results from Barro Colorado Island, Panama and a Comparison with Other Plots*; Springer: Berlin/Heidelberg, Germany, 1998; ISBN 978-3-540-64144-5.
8. Gonzalez, M.A.; Baraloto, C.; Engel, J.; Mori, S.A.; Pétronelli, P.; Riéra, B.; Roger, A.; Thébaud, C.; Chave, J. Identification of Amazonian Trees with DNA Barcodes. *PLoS ONE* **2009**, *4*, e7483. [CrossRef]
9. Hebert, P.D.N.; Cywinska, A.; Ball, S.L.; deWaard, J.R. Biological Identifications through DNA Barcodes. *Proc. R. Soc. Lond. B Biol. Sci.* **2003**, *270*, 313–321. [CrossRef]
10. Taylor, H.R.; Harris, W.E. An Emergent Science on the Brink of Irrelevance: A Review of the Past 8 Years of DNA Barcoding. *Mol. Ecol. Resour.* **2012**, *12*, 377–388. [CrossRef]
11. CBOL Plant Working Group; Hollingsworth, P.M.; Forrest, L.L.; Spouge, J.L.; Hajibabaei, M.; Ratnasingham, S.; van der Bank, M.; Chase, M.W.; Cowan, R.S.; Erickson, D.L.; et al. A DNA Barcode for Land Plants. *Proc. Natl. Acad. Sci. USA* **2009**, *106*, 12794–12797. [CrossRef]
12. Kress, W.J.; Erickson, D.L. A Two-Locus Global DNA Barcode for Land Plants: The Coding RbcL Gene Complements the Non-Coding TrnH-PsbA Spacer Region. *PLoS ONE* **2007**, *2*, e508. [CrossRef] [PubMed]
13. Gill, B.A.; Musili, P.M.; Kurukura, S.; Hassan, A.A.; Goheen, J.R.; Kress, W.J.; Kuzmina, M.; Pringle, R.M.; Kartzinel, T.R. Plant DNA-Barcode Library and Community Phylogeny for a Semi-Arid East African Savanna. *Mol. Ecol. Resour.* **2019**, *19*, 838–846. [CrossRef] [PubMed]
14. Bezeng, B.S.; Davies, T.J.; Daru, B.H.; Kabongo, R.M.; Maurin, O.; Yessoufou, K.; van der Bank, H.; van der Bank, M. Ten Years of Barcoding at the African Centre for DNA Barcoding. *Genome* **2017**, *60*, 629–638. [CrossRef] [PubMed]
15. Boatwright, J.S.; Maurin, O.; van der Bank, M. Phylogenetic Position of Madagascan Species of Acacia s.l. and New Combinations in Senegalia and Vachellia (Fabaceae, Mimosoideae, Acacieae). *Bot. J. Linn. Soc.* **2015**, *179*, 288–294. [CrossRef]
16. Gere, J.; Yessoufou, K.; Daru, B.H.; Mankga, L.T.; Maurin, O.; van der Bank, M. Incorporating TrnH-PsbA to the Core DNA Barcodes Improves Significantly Species Discrimination within Southern African Combretaceae. *Zookeys* **2013**, *365*, 129–147. [CrossRef]
17. Daru, B.H.; Manning, J.C.; Boatwright, J.S.; Maurin, O.; Maclean, N.; Schaefer, H.; Kuzmina, M.; van der Bank, M. Molecular and Morphological Analysis of Subfamily Alooideae (Asphodelaceae) and the Inclusion of Chortolirion in Aloe. *Taxon* **2013**, *62*, 62–76. [CrossRef]
18. Parmentier, I.; Duminil, J.; Kuzmina, M.; Philippe, M.; Thomas, D.W.; Kenfack, D.; Chuyong, G.B.; Cruaud, C.; Hardy, O.J. How Effective Are DNA Barcodes in the Identification of African Rainforest Trees? *PLoS ONE* **2013**, *8*, e54921. [CrossRef]
19. Abiem, I.; Arellano, G.; Kenfack, D.; Chapman, H. Afromontane Forest Diversity and the Role of Grassland-Forest Transition in Tree Species Distribution. *Diversity* **2020**, *12*, 30. [CrossRef]

20. Davies, S.J.; Abiem, I.; Abu Salim, K.; Aguilar, S.; Allen, D.; Alonso, A.; Anderson-Teixeira, K.; Andrade, A.; Arellano, G.; Ashton, P.S.; et al. ForestGEO: Understanding Forest Diversity and Dynamics through a Global Observatory Network. *Biol. Conserv.* **2021**, *253*, 108907. [CrossRef]

21. Chapman, J.D.; Chapman, H.M. *The Forests of Taraba and Adamawa States, Nigeria: An Ecological Account and Plant Species Checklist*; University of Canterbury: Christchurch, New Zealand, 2001.

22. Altschul, S.F.; Gish, W.; Miller, W.; Myers, E.W.; Lipman, D.J. Basic Local Alignment Search Tool. *J. Mol. Biol.* **1990**, *215*, 403–410. [CrossRef]

23. Kumar, S.; Stecher, G.; Tamura, K. MEGA7: Molecular Evolutionary Genetics Analysis Version 7.0 for Bigger Datasets. *Mol. Biol. Evol.* **2016**, *33*, 1870–1874. [CrossRef] [PubMed]

24. Kimura, M. A Simple Method for Estimating Evolutionary Rates of Base Substitutions through Comparative Studies of Nucleotide Sequences. *J. Mol. Evol.* **1980**, *16*, 111–120. [CrossRef] [PubMed]

25. Newmaster, S.G.; Fazekas, A.J.; Ragupathy, S. DNA Barcoding in Land Plants: Evaluation of RbcL in a Multigene Tiered Approach. *Can. J. Bot.* **2006**, *84*, 335–341. [CrossRef]

26. Yu, J.; Xue, J.-H.; Zhou, S.-L. New Universal MatK Primers for DNA Barcoding Angiosperms. *J. Syst. Evol.* **2011**, *49*, 176–181. [CrossRef]

27. Kress, W.J.; Erickson, D.L.; Jones, F.A.; Swenson, N.G.; Perez, R.; Sanjur, O.; Bermingham, E. Plant DNA Barcodes and a Community Phylogeny of a Tropical Forest Dynamics Plot in Panama. *Proc. Natl. Acad. Sci. USA* **2009**, *106*, 18621–18626. [CrossRef]

28. Amandita, F.Y.; Rembold, K.; Vornam, B.; Rahayu, S.; Siregar, I.Z.; Kreft, H.; Finkeldey, R. DNA Barcoding of Flowering Plants in Sumatra, Indonesia. *Ecol. Evol.* **2019**, *9*, 1858–1868. [CrossRef]

29. Hollingsworth, P.M.; Graham, S.W.; Little, D.P. Choosing and Using a Plant DNA Barcode. *PLoS ONE* **2011**, *6*, e19254. [CrossRef] [PubMed]

30. Huang, X.; Ci, X.; Conran, J.G.; Li, J. Application of DNA Barcodes in Asian Tropical Trees—A Case Study from Xishuangbanna Nature Reserve, Southwest China. *PLoS ONE* **2015**, *10*, e0129295. [CrossRef]

31. Newmaster, S.G.; Ragupathy, S.; Janovec, J. A Botanical Renaissance: State-of-the-Art DNA Bar Coding Facilitates an Automated Identification Technology System for Plants. *Int. J. Comput. Appl. Technol.* **2009**, *35*, 50–60. [CrossRef]

32. Kress, W.J.; Wurdack, K.J.; Zimmer, E.A.; Weigt, L.A.; Janzen, D.H. Use of DNA Barcodes to Identify Flowering Plants. *Proc. Natl. Acad. Sci. USA* **2005**, *102*, 8369–8374. [CrossRef]

33. Kenfack, D.; Thomas, D.W.; Chuyong, G.; Condit, R. Rarity and Abundance in a Diverse African Forest. *Biodivers. Conserv.* **2007**, *16*, 2045–2074. [CrossRef]

34. Mensah, S.; Egeru, A.; Assogbadjo, A.E.; Glèlè Kakaï, R. Vegetation Structure, Dominance Patterns and Height Growth in an Afromontane Forest, Southern Africa. *J. For. Res.* **2020**, *31*, 453–462. [CrossRef]

35. Gebeyehu, G.; Soromessa, T.; Bekele, T.; Teketay, D. Species Composition, Stand Structure, and Regeneration Status of Tree Species in Dry Afromontane Forests of Awi Zone, Northwestern Ethiopia. *Ecosyst. Health Sustain.* **2019**, *5*, 199–215. [CrossRef]

36. Pitman, N.C.A.; Terborgh, J.W.; Silman, M.R.; Núñez, V.P.; Neill, D.A.; Cerón, C.E.; Palacios, W.A.; Aulestia, M. A Comparison of Tree Species Diversity in Two Upper Amazonian Forests. *Ecology* **2002**, *83*, 3210–3224. [CrossRef]

37. LaFrankie, J.V.; Ashton, P.S.; Chuyong, G.B.; Co, L.; Condit, R.; Davies, S.J.; Foster, R.; Hubbell, S.P.; Kenfack, D.; Lagunzad, D.; et al. Contrasting Structure and Composition of the Understory in Species-Rich Tropical Rain Forests. *Ecology* **2006**, *87*, 2298–2305. [CrossRef]

38. Memiaghe, H.R.; Lutz, J.A.; Korte, L.; Alonso, A.; Kenfack, D. Ecological Importance of Small-Diameter Trees to the Structure, Diversity and Biomass of a Tropical Evergreen Forest at Rabi, Gabon. *PLoS ONE* **2016**, *11*, e0154988. [CrossRef]

39. Bhat, J.A.; Kumar, M.; Negi, A.K.; Todaria, N.P.; Malik, Z.A.; Pala, N.A.; Kumar, A.; Shukla, G. Species Diversity of Woody Vegetation along Altitudinal Gradient of the Western Himalayas. *Glob. Ecol. Conserv.* **2020**, *24*, e01302. [CrossRef]

40. Song, X.; Cao, M.; Li, J.; Kitching, R.L.; Nakamura, A.; Laidlaw, M.J.; Tang, Y.; Sun, Z.; Zhang, W.; Yang, J. Different Environmental Factors Drive Tree Species Diversity along Elevation Gradients in Three Climatic Zones in Yunnan, Southern China. *Plant Divers.* **2021**, *43*, 433–443. [CrossRef]

41. Jaccard, P. The Distribution of the Flora in the Alpine Zone. *New Phytol.* **1912**, *11*, 37–50. [CrossRef]

42. Gonmadje, C.F.; Doumenge, C.; McKey, D.; Tchouto, G.P.M.; Sunderland, T.C.H.; Balinga, M.P.B.; Sonké, B. Tree Diversity and Conservation Value of Ngovayang's Lowland Forests, Cameroon. *Biodivers. Conserv.* **2011**, *20*, 2627–2648. [CrossRef]

43. Sainge, M.N.; Lyonga, N.M.; Mbatchou, G.P.T.; Kenfack, D.; Nchu, F.; Peterson, A.T. Vegetation, Floristic Composition and Structure of a Tropical Montane Forest in Cameroon. *Bothalia* **2019**, *49*, 12. [CrossRef]

44. Comiskey, J.A.; Sunderland, T.C.; Sunderland-Groves, J.L. *Takamanda: The Biodiversity of an African Rainforest*; Smithsonian Institution, SI/MAB Biodiversity Program: Washington, DC, USA, 2003; ISBN 1-893912-12-4.

45. Balinga, M.; Moses, S.; Fombod, E.; Sunderland, T.C.; Chantal, S.; Asaha, S. *A Preliminary Assessment of the Vegetation of the Dzanga Sangha Protected Area Complex, Central African Republic*; WWF: Gland, Switzerland; Smithsonian Institution: Washington, DC, USA; Forests, Resources and People: Limbe, Cameroon; CARPE: Kinshasa, Republic of the Congo, 2006; 124p.

46. Makana, J.R.; Terese, B.H.; Hibbs, D.E.; Condit, R.; Losos, E.; Leigh, E. Stand Structure and Species Diversity in the Ituri Forest Dynamics Plots: A Comparison of Monodominant and Mixed Forest Stands. *Trop. For. Divers. Dynamism* **2004**, 159–174.

47. Sunderland, T.; Walters, G.; Issembe, Y. A Preliminary Vegetation Assessment of the Mbé National Park, Monts de Cristal, Gabon. 2004. 47p. Available online: https://carpe.umd.edu/sites/default/files/documentsarchive/Monts_de_Cristal_final_report.pdf (accessed on 10 February 2022).
48. Balinga, M.P.B.; Sunderland, T.; Walters, G.; Issembe, Y.; Asaha, S.; Fombod, E. *A Vegetation Assessment of the Waka National Park, Gabon. CARPE Report*; Wildlife Conservation Society: Libreville, Gabon, 2006; 155p. Available online: https://carpe.umd.edu/sites/default/files/documentsarchive/SI_WakaNPGabon_Rpt_Nov2006.pdf (accessed on 10 February 2022).

Article

Gradients in the Diversity of Plants and Large Herbivores Revealed with DNA Barcoding in a Semi-Arid African Savanna

Patrick T. Freeman [1,2], Robert O. Ang'ila [3,4], Duncan Kimuyu [3,4], Paul M. Musili [5], David Kenfack [6], Peter Lokeny Etelej [4], Molly Magid [1,2], Brian A. Gill [1,2] and Tyler R. Kartzinel [1,2,4,*]

[1] Department of Ecology, Evolution, and Organismal Biology, Brown University, Providence, RI 02912, USA; patrick_freeman@alumni.brown.edu (P.T.F.); molly_magid@alumni.brown.edu (M.M.); gillbriana@gmail.com (B.A.G.)
[2] Institute at Brown for Environment and Society, Brown University, Providence, RI 02912, USA
[3] Department of Natural Resources, Karatina University, Karatina 10101, Kenya; robert.o.angila@gmail.com (R.O.A.); dkimuyu@karu.ac.ke (D.K.)
[4] Mpala Research Centre, Nanyuki 555-10400, Kenya; peterlokeny@gmail.com
[5] Botany Department, National Museums of Kenya, Nairobi 451660-0100, Kenya; pmutuku@museums.or.ke
[6] Smithsonian Tropical Research Institute, CTFS-ForestGEO, Washington, DC 20560, USA; kenfackd@si.edu
[*] Correspondence: tyler_kartzinel@brown.edu

Citation: Freeman, P.T.; Ang'ila, R.O.; Kimuyu, D.; Musili, P.M.; Kenfack, D.; Lokeny Etelej, P.; Magid, M.; Gill, B.A.; Kartzinel, T.R. Gradients in the Diversity of Plants and Large Herbivores Revealed with DNA Barcoding in a Semi-Arid African Savanna. *Diversity* **2022**, *14*, 219. https://doi.org/10.3390/d14030219

Academic Editors: Luc Legal and W. John Kress

Received: 31 January 2022
Accepted: 11 March 2022
Published: 17 March 2022

Publisher's Note: MDPI stays neutral with regard to jurisdictional claims in published maps and institutional affiliations.

Abstract: Do hotspots of plant biodiversity translate into hotspots in the abundance and diversity of large mammalian herbivores? A common expectation in community ecology is that the diversity of plants and animals should be positively correlated in space, as with the latitudinal diversity gradient and the geographic mosaic of biodiversity. Whether this pattern 'scales down' to landscape-level linkages between the diversity of plants or the activities of highly mobile megafauna has received less attention. We investigated spatial associations between plants and large herbivores by integrating data from a plant-DNA-barcode phylogeny, camera traps, and a comprehensive map of woody plants across the 1.2-km^2 Mpala Forest Global Earth Observatory (ForestGEO) plot, Kenya. Plant and large herbivore communities were strongly associated with an underlying soil gradient, but the richness of large herbivore species was negatively correlated with the richness of woody plants. Results suggest thickets and steep terrain create associational refuges for plants by deterring megaherbivores from browsing on otherwise palatable species. Recent work using dietary DNA metabarcoding has demonstrated that large herbivores often directly control populations of the plant species they prefer to eat, and our results reinforce the important role of megaherbivores in shaping vegetation across landscapes.

Keywords: behavioral ecology; DNA barcoding; elephant (*Loxodonta africana*); ForestGEO; geographic mosaic of species interactions; phylogenetic community ecology; landscape ecology; megaherbivores; phylogenetic signal; plant–herbivore interactions

1. Introduction

African savannas are home to the greatest extant concentration of wild large herbivores, many of which are facing extinctions that have far-reaching consequences for whole ecosystems, and plant communities in particular [1–3]. Although large herbivores can range widely, many exhibit habitat preferences that reflect different needs to access water, avoid predators, or forage on palatable plants [4–7]. Elucidating spatial relationships between plants and large herbivores is paramount to understanding and managing change in savanna ecosystems.

A common expectation in biodiversity science is that the diversity of plants and animals is positively correlated [8,9]. The latitudinal diversity gradient and the geographic mosaic of 'interaction biodiversity' are thought to be potential outcomes of this positive association [10]. Positive associations in plant and consumer diversity can emerge from similar responses to abiotic gradients [9], disturbance [11], and biotic interactions [12].

For example, resource gradients can generate overlapping gradients in plant and animal diversity across biomes (e.g., tundra vs. tropical rainforest) as well as within landscapes (e.g., mesic vs. xeric microhabitats) [13,14]. Whether positive biodiversity associations across trophic levels enable predictions about landscape-level linkages between the local diversity of plants and the activities of highly mobile herbivores that eat them is an open question, although recent dietary DNA-metabarcoding studies suggest the diversity of plants and large herbivores is more tightly linked than currently appreciated [15].

Analyses of local plant and animal communities can be used to test key hypotheses about how shared habitat associations may develop. The Resource Specialization Hypothesis posits that locally diverse plants support locally diverse consumers by providing a broader array of resources that can accommodate the dietary niches of more consumer species [16,17]. In contrast, the More Individuals Hypothesis posits that more diverse plant communities are often more productive, thereby increasing both the abundance of individual consumers and hence the number of consumer species [16,18]. If a positive correlation between the diversity of plants and large mammalian herbivores exists, it may arise from their generalist feeding strategies and requirements—consistent with the More Individuals Hypothesis. However, comparing these hypotheses to gain understanding about whether plant and animal diversities are locally linked requires consideration of what it means to be a 'specialist' and whether the evolutionary relationships of species reflect traits that determine where they occur [10,12,19]. Strong inferences can be made about how spatial distributions are structured by incorporating information on species' ecological and evolutionary similarities [20]. A relatively small number of species can represent a broad diversity of evolutionary lineages, and vice versa [21]. Within communities of large herbivores, for example, some species may consume relatively few plant taxa (taxonomic specialists) from many plant lineages (phylogenetic generalists) whereas others may consume many taxa from few lineages [19]. Thus, even 'generalized' large herbivores may exhibit feeding or habitat preferences that establish colocalized 'hotspots' of plant and animal diversity—consistent with the Resource Specialization Hypothesis and contrary to the More Individuals Hypothesis.

We combined a plant-DNA-barcode phylogeny, a comprehensive map of woody plants, and camera-trap data to evaluate whether the abundance and diversity of plants and large herbivores are correlated across the ForestGEO plot at Mpala Research Centre, Kenya. The Mpala ForestGEO plot spans soil and topographic gradients, making it possible to evaluate local covariation in the taxonomic and phylogenetic diversity of woody plants and large herbivores at scales of tens to hundreds of meters while controlling for the influence of large-scale biogeographic differences in species pools [22]. We used these data to test three hypotheses: (*i*) the diversity of woody plants and large herbivores is significantly different between habitats, such that habitats with high woody-plant density also have high taxonomic and phylogenetic diversity of both plants and animals; (*ii*) local herbivore diversity is strongly and positively correlated with local plant diversity, both within and between habitats that support different baseline resource availabilities (i.e., tree density) and diversities (i.e., taxonomic and phylogenetic); and (*iii*) the taxonomic and phylogenetic compositions of plant and herbivore communities are spatially congruent, consistent with a linkage between the diversity of plants and large mammals.

2. Materials and Methods

2.1. Study Location

Our analysis is based on the 1.2-km^2 Mpala ForestGEO plot (0°17′ N, 36°53′ E) [22]. The plot is in a semi-arid savanna (~600 mm annual rainfall) that supports at least 17 wild large herbivore species (>5 kg). The ForestGEO plot includes at least 62 woody plant species out of >460 plant species that occur in the region [22,23]. It spans three habitats characteristic of the Laikipia Highlands: (*i*) 'plateau' habitat on poorly drained and nutrient-rich clay vertisols (black-cotton soil; 1775–1792 m asl); (*ii*) 'low plain' habitat on well-drained, nutrient-poor, red sandy loams (red sands soil; 1669–1779 m asl); and (*iii*) a rocky 'slope' between the plateau and low plain (1679–1779 m asl).

We analyzed communities of woody plants and large herbivores at 33 sites within the ForestGEO plot (Figure 1). The sites were positioned at regular 100-m intervals across the plot, with 14 occurring in the plateau, 3 on the slope, and 16 in the low plain. Across the plot, we also measured fine-scale topographic habitat variables, including elevation, slope, and convexity, based on elevation data recorded in a 5 m × 5 m grid [22]. We associated our topographic values with a set of 20 m × 20-m vegetation quadrats using the *fgeo.analyze* package [24] in R [25]. After standardizing measures of elevation, slope, and convexity to a mean of 0 and standard deviation of 1, the three variables were used to classify the 3000 20 × 20-m quadrats into the three major habitats (plateau, slope, and low plain) using Ward hierarchical clustering (Figure 1). Finally, we calculated the topographic wetness index (TWI), which reflects the ability of a landscape to retain water and is a strong predictor of savanna wildlife distributions [26], by integrating the total water catchment area and slope of each grid cell using *build.layers* in the *dynatopmodel* package [27]. Cells with high TWI tend to be flat or concave.

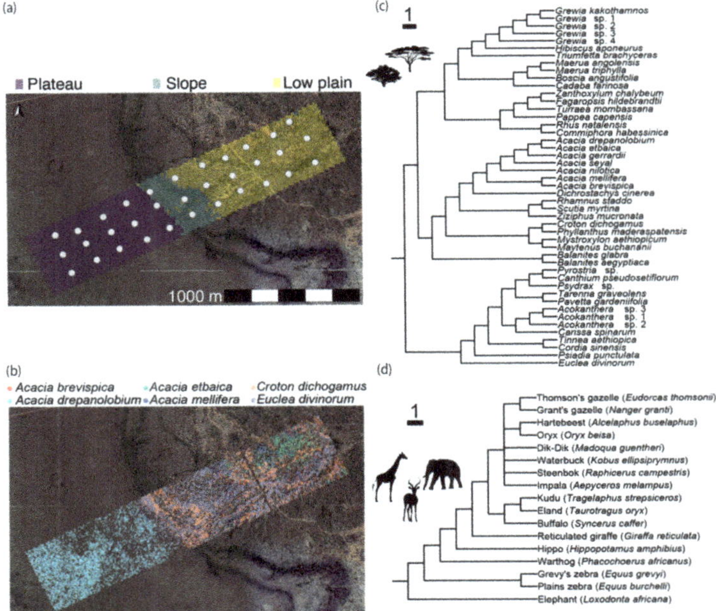

Figure 1. The Mpala ForestGEO ecosystem. (**a**) The map shows the extent of the study plot, with white circles representing 33 sampling sites. The three habitats are distinguished by color (purple = "plateau", teal = "slope", yellow = "low plain"). (**b**) The locations of six woody plant species comprising 80% of stems in the plot are shown with different color points. Phylogenies show relationships between (**c**) woody plants and (**d**) large mammalian herbivores; scale bars represent 1 MY. Note *Hibiscus aponeurus* in the phylogeny represents *Hibiscus* spp. in the ForestGEO data; see Table S1 in Supplementary Materials for description of grafted or substituted taxa.

2.2. Woody Plant Distribution and DNA Barcoding

The first comprehensive ForestGEO survey of woody plants began in 2010. It established a regular grid of 400-m² quadrats in which the main and auxiliary stems of woody trees and shrubs > 0.5 m tall were geolocated, tagged, and measured for diameter at knee height (dkh) [22]. Species were identified by researchers from the East African Herbarium at the National Museums of Kenya. We obtained the complete dataset from the ForestGEO portal (12 March 2019) [22]. It included 363,798 total stems and 139,078 main stems (henceforth 'individual trees') representing 67 species and 22 families. The branching architecture of shrubs such as *Croton* and *Euclea* can make it difficult to identify discrete individuals, but we assumed the data were internally consistent. The dataset was filtered to include only living stems > 2 cm dkh, species with >2 main stems in the plot, and sufficient identification for phylogenetic analysis. The filtered data retained 355,461 stems, 136,297 individual trees, and 55 morphospecies (Table S1). We extracted a dataset for analysis that focused on trees within 25 m^2 of our 33 grid sites.

Our plant phylogeny was based on an extensive plant DNA barcode library and phylogeny for Mpala, which was constructed using a supermatrix approach [15,23,28]. The full DNA barcode library includes high-quality data from 1760 specimens representing at least 438 species sequenced at up to 5 markers (*mat*K, *rbc*L, *psb*A-*trn*H, *trn*L-F, and ITS). A subset of species missing from the phylogeny were grafted in three complementary ways. First, we obtained new *trn*L and *rbc*L DNA barcodes from 5 species [23], and we used these data to determine how to graft them into the phylogeny (Table S1). Second, we represented taxa with substitutes that were already in the phylogeny (e.g., congeners, such as *Hibiscus aponeurus* used to represent *Hibiscus* spp.; Figure 1, Table S1). Third, we grafted remaining species based on the literature (see Table S1 for details).

2.3. Large Mammalian Herbivore Community Data

To assess herbivore distributions, we deployed camera traps from March 2018 to April 2019 (Bushnell, #11-9874C). We recorded date, time, and species using Wild-ID software [29]. Photos of large herbivores were extracted and filtered to independent detections (defined as >30 min apart) to reduce the impacts of temporal autocorrelation. For each species, we calculated a relative abundance index (RAI) as the total number of independent photographs divided by the total number of working camera days over the course of the survey. Simple RAI-based approaches yielded relative abundance estimates that correlated strongly with independent estimates of animal abundance for large mammals [30].

2.4. Community Diversity Indices

We compared species richness and two phylogenetic diversity metrics for the plants and herbivores across sites. Phylogenetic diversity metrics included the standard effect sizes of mean pairwise distance (*ses*MPD) and mean nearest-taxon distance (*ses*MNTD) in the package *picante* in R [31]. The *ses*MPD metric is sensitive to the phylogenetic diversity of deep-branching lineages, whereas *ses*MNTD is sensitive to diversity patterns close to the tips of the phylogeny [32]. These metrics use null models to determine whether observed communities contain taxa that are phylogenetically clustered or overdispersed [33]. For each site, we randomly generated 999 communities that shuffled the names of taxa across the phylogeny while holding the richness of the community constant.

To characterize turnover in plant and animal communities, we quantified dissimilarity across sites. We considered both the taxonomic and phylogenetic composition of communities using phylogenetic community dissimilarity (PCD) in *picante* [34]. The PCD metric accounts for taxonomic dissimilarity (i.e., PCDc; 'community' species differences) and evolutionary dissimilarity (PCDp; 'phylogenetic' differences between non-shared species), where PCD = PCDc × PCDp. If PCD = 1, two sites are no more or less similar than communities assembled at random from the species pool. If PCDp = 1, then PCD = PCDc, and communities reflect only taxonomic differences. However, if PCDc = 1, then differences are due entirely to the phylogenetic distances between nonshared species (i.e., PCDp). If all nonshared species are closely related, PCDp is low; if two sites have nonshared species from very divergent clades, then PCDp is high. Because the PCD metric is based on species presence/absence patterns, and because PCDp is undefined in comparisons of sites with identical species composition, we focused our analyses on PCD and PCDc and infer the relative importance of phylogeny by comparison.

2.5. Hypothesis Testing

To test hypothesis *i*, that the abundances and diversities of woody plants and large herbivores differs across habitats, we compared data from 33 sites. To estimate diversity, we used species richness, *ses*MPD, and *ses*MNTD at sites. To estimate abundance, we used the total number of plant stems and the summed RAI of all herbivore species. We focused on total stems as an indicator of thicket density and sightline obstruction, which inform large-herbivore habitat use [35]. However, different trees have different branching architectures—even within species, shorter trees can have more short stems—and heavy damage can promote stem proliferation where megaherbivores are abundant [36]. Thus, interpretations of total stem counts as a measure of thicket density and basal area as a measure of plant biomass may differ. Habitat comparisons were made using ANOVAs and Tukey's HSD. We also compared total plant-species richness across sites within each habitat using sample-based rarefaction based on the Bernoulli product model [37].

To test hypothesis *ii*, that local plant and herbivore diversities were positively correlated, we compared abundance and diversity both within and between habitats using ANCOVA. We constructed linear models in R using the herbivore-community characteristic as the dependent variable and the corresponding vegetation characteristic as the independent variable, with habitat types as the covariate (plateau vs. low plain). Because we had a small sample size of slope sites (N = 3), we only included the major plateau (N = 14) and low plain (N = 16) habitats in these linear models.

Finally, we evaluated hypothesis *iii*, that plant and animal communities are spatially linked. First, we tested for significant differences in PCD between habitats using permutational multivariate analyses of variance (perMANOVA) [38]. Second, we tested for significant correspondence between plant and animal PCD using Mantel tests. To account for the possibility that community similarities arise from spatial proximity, we used partial Mantel tests to evaluate correlations while accounting for distance between sites. Finally, to identify species-specific habitat associations, we performed indicator species analyses using the point biserial correlation coefficient based on Pearson's φ statistic with 999 bootstraps in *indicspecies* [39]. Pearson's φ ranges from −1 to 1, indicating strong avoidance or preference, respectively. The analysis was based on species' relative abundances using Hellinger transformation, corrected for unequal sampling across habitats, reported with *P*-values calculated independently across species.

3. Results

3.1. Ecological Characteristics of the Plot

The plateau was flat and topographically homogeneous compared to the slope and low plain (Figures 1 and S1, Table 1). Both slope and convexity were lowest in the plateau, intermediate in the low plain, and greatest on the slope (Figure S1). The plateau's flat, clay soils that provide little opportunity for rain infiltration or runoff produced the highest TWI (Figure S1). Plant communities at grid sites included 47 out of 55 woody plant species from the plot-wide data—species that together represented 98% of identified stems in the plot (Table S1). This included 7621 individual plants (median = 193 stems/site; range = 20–925; Table S2). Numerically dominant species were *Acacia drepanolobium*, *Croton dichogamus*, *A. brevispica*, *Euclea divinorum*, *A. mellifera*, and *A. etbaica* (Figure 1, Table S1). Herbivore communities were characterized by 8879 independent photographs of 17 species over 10,075 trap days (median = 329 days/site; Tables S3 and S4). The most common species were dik-dik, giraffe, plains zebra, impala, and elephant (Tables S3 and S4).

Table 1. Summary of ecological characteristics at 33 grid sites across three habitats.

Variable	Measure	Plateau	Slope	Low Plain
Abiotic characteristics	Number of sites	14	3	16
	Mean elevation (m)	1787	1754	1692
	Mean convexity	0.019	0.556	−0.022
	Mean slope (rad)	1.04	12.30	4.10
	Mean topographic water index (TWI)	10.6	7.9	9.0
Woody plants	Mean total stems	238	1602	855
	Mean richness	5.4	25.0	15.4
	Mean *ses*MPD	−1.7	0.9	−0.6
	Mean *ses*MNTD	−1.3	1.8	0.3
Large herbivores	Mean cumulative RAI	60.4	49.0	127.0
	Mean richness	11.4	6.3	9.8
	Mean *ses*MPD	1.1	0.2	−0.1
	Mean *ses*MPD	−0.6	−0.3	0.0

3.2. Hypothesis Testing

3.2.1. Hypothesis 1: Dense Habitats Should Have High Plant and Animal Diversity

Results were not consistent with our first hypothesis. Although high woody-plant abundances in the low plain and slope were generally associated with high plant taxonomic and phylogenetic diversities, the greatest herbivore abundances and diversities were not necessarily in these same habitats. Woody-plant density and richness was ~2–3-fold lower on the plateau compared to the slope and low plain (Table 1; Figure 2a,b). Similar patterns were found for phylogenetic diversity: the plateau had the lowest diversity and the low plain was intermediate (Figure 2c,d). In contrast, while herbivore abundance was also ~2-fold lower on the plateau than low plain, abundance on the slope was similar (Figure 2e) and richness on the slope was much lower (Figure 2f). Rarefaction revealed the greatest plant-species richness and lowest herbivore richness on the slope (Figure S2). Patterns of phylogenetic diversity also differed, with herbivore *ses*MPD being greatest on the plateau (Figure 2g) and *ses*MNTD not differing significantly across habitats (Figure 2h).

3.2.2. Hypothesis 2: Herbivore and Plant Diversity Should Be Positively Correlated

Contrary to our second hypothesis, there was not a significant positive correlation between the abundance and diversity of flora and fauna within and across habitats (Figure 3). In contrast, there was a significant negative correlation between the richness of flora and fauna within and between habitats (Figure 3b) and we found evidence only for overall habitat-level differences in the abundance and phylogenetic diversities of plants and animals (Figure 3a,c,d). There was an interesting contrast between herbivore phylogenetic diversity in the plateau and low plain: *ses*MPD, which is sensitive to deep patterns in the phylogeny, was greater in the plateau; *ses*MNTD, which is sensitive to variation near the tips, was greater in the low plain (Figure 3c,d). Thus, plateau sites hosted large herbivores that both represented species from disparate mammalian lineages and that were likely to include closely related pairs of species from those lineages.

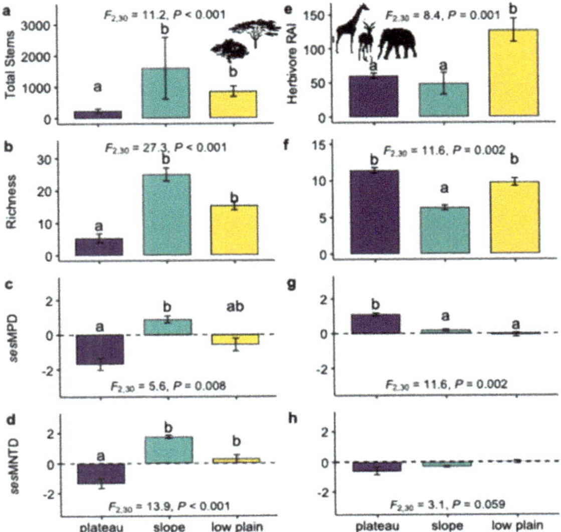

Figure 2. Abundance and diversity of (**a–d**) woody plants and (**e–h**) large herbivores across habitats (mean ± s.e.). Letters above each bar indicate significant differences based on Tukey's HSD following ANOVAs. Abundance is the (**a**) total plant stems/site and (**e**) total herbivore RIA/site. Richness is the mean count of (**b**) woody plant and (**f**) large herbivore species. Phylogenetic diversity was measured as (**c,g**) *ses*MPD and (**d,h**) *ses*MNTD, with positive values indicating overdispersion and negative values clustering.

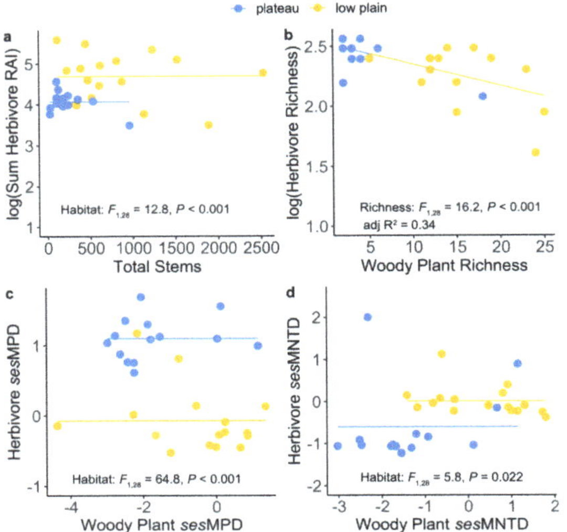

Figure 3. Relationships between herbivore (**a**) abundance, (**b**) taxonomic richness, (**c**) *ses*MPD, and (**d**) *ses*MNTD with corresponding measures for woody plants. For each pair of metrics, we constructed initial linear models with predictors including the woody-plant variable, habitat, and the plant variable × habitat interaction. We simplified models based on the subset of statistically significant variables. We report adjusted R^2 for the model that included a significant correlation (**b**). Horizontal lines show significant differences between habitats (**a,c,d**). Colors show habitat types.

3.2.3. Hypothesis 3: Spatial Links in the Composition of Plant and Animal Communities

Consistent with our third hypothesis, spatial turnover in woody-plant and large-herbivore communities was congruent. There was significant taxonomic dissimilarity between communities of plants and herbivores across habitats (Figure 4a,b). Taxonomic differentiation was strong across habitats, but incorporating information on phylogenetic variance eroded the signal of habitat associations (Figure 4b,d). There were significant positive correlations between local plant and herbivore community compositions, even after accounting for spatial proximity and phylogenetic variance, but these correlations were also strongest when accounting only for species composition (Figure 4e,f). About half of the tree species were significantly associated with or avoided a habitat (25/47, 53%; Figure S3a). Of these, most were negatively associated with the plateau (20/47 species, 43%) and only one had affinity for it (*Acacia drepanolobium*; Figure S3a). Eight tree species were significantly associated with the low plain, including several that were also negatively associated with the plateau (*Balanites glabra*, *Acacia gerrardii*, *A. etbaica*, *A. brevispica*, *Acokanthera* sp. 1, *Pyrostria* sp. and two *Grewia* spp.; Figure S3a). Many tree species had strong and positive affinities for the slope, especially those that also had positive affinities for the low plain, although slope sample sizes were small and results were not statistically significant (Figure S3a). In contrast to plants, nearly all large herbivore species exhibited significant habitat association or avoidance 94% (16/17; Figure S3b). Most species exhibited preferred either the plateau (9/17, 53%) or low plain (7/17, 41%), while none were associated with the slope and most avoided it (11/17, 65%; Figure S3b).

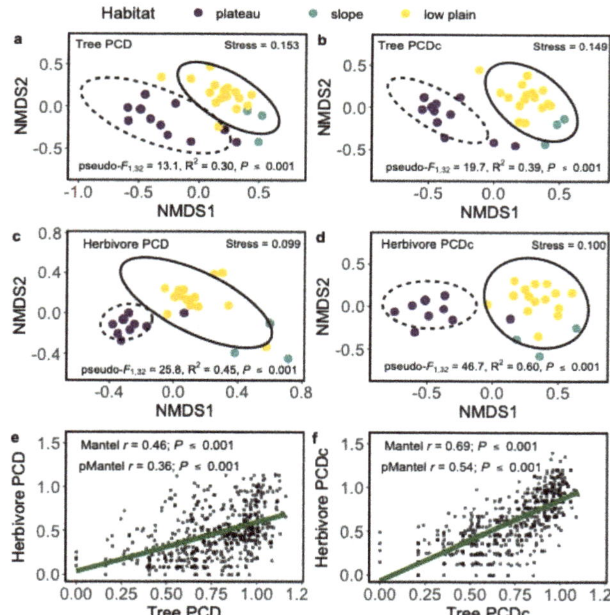

Figure 4. Taxonomic and phylogenetic differentiation in woody-plant and large-herbivore communities. Nonmetric multidimensional scaling (NMDS) revealed significant compositional dissimilarity in (**a**) tree PCD, (**b**) tree PDCc, (**c**) large-herbivore PCD, and (**d**) large-herbivore PCDc between habitats. Colors in (**a–d**) indicate habitat at each grid site. Results of perMANOVAs and stress values for corresponding NMDS plots are shown on each panel. We grouped the three slope sites with low plain sites for calculating perMANOVAs and 95% confidence ellipses. The (**e**) PCD and (**f**) PCDc of plant and animal communities were positively correlated based on Mantel and (p)artial Mantel tests; green trendlines were fit using generalized linear models.

4. Discussion

We used fine-scale data on woody-plant and large-herbivore communities to evaluate local biodiversity linkages across trophic levels. The flora with the most abundant woody stems also supported the greatest taxonomic and phylogenetic diversity of woody plants. However, this abundance and diversity of plants did not necessarily translate into a greater abundance or diversity of herbivores. Across habitats, the low plain had a high abundance of both plants and large herbivores, but the slope had a comparably high abundance of plants with a markedly lower abundance of herbivores (Figure 2). Both within and across habitats, there was also a negative relationship between local plant and herbivore-species richness (Figure 3). While plant and animal communities differed across habitats, this pattern was not reinforced by phylogeny (Figure 4). In contrast with prior studies reporting phylogenetic associations between plants and insect herbivores [40,41], our results reveal how compositional turnover in some plant and animal assemblages may be random with respect to phylogeny, even when habitat filtering is highly nonrandom with respect to plant and animal functional traits across disparate evolutionary lineages [20].

Results were not generally consistent with the More Individuals or Resource Specialization hypotheses. The More Individuals Hypothesis posits that greater resource availability enables more individuals to establish, and hence more species of consumers to co-occur [16]. In contrast to positive correlations between the abundance and diversity of some plant and herbivore assemblages that have been studied at local-to-global scales [14,42,43], our results revealed landscape-level contrasts between the abundance and diversity of plants and large herbivores, as well as surprising negative correlations in their species richness across 33 sites. Under the Resource Specialization Hypothesis, plant diversity should enable more specialized consumer species to establish in the community. For example, insect-species richness may increase with the richness and phylogenetic diversity of grassland and forest plants [42,43]. However, ungulate herbivores that require relatively large quantities of food are unlikely to be attracted to or excluded from a local community based on similar mechanisms involving narrow feeding specializations. The low plain had relatively high plant abundance and species richness compared to the plateau, but it only revealed higher average herbivore abundance and not higher herbivore richness. Further contradicting both hypotheses, the slope and low plain had comparable levels of plant abundance and species richness, yet the slope supported lower animal abundance and richness. We thus consider other non-mutually exclusive mechanisms to explain two striking patterns in our data: (*i*) the distinct flora and fauna across the gradient from heavy-clay vertisol ("black-cotton") soils of the plateau through the sloping transition to the sandy ("red") soils of the low plain and (*ii*) the negative correlation between plant and animal-species richness.

First, the flora of the plateau and low plain differed strikingly, but both were dominated by *Acacia* spp. (Figure 4). *Acacias* are a diverse suite of savanna trees that browsing ungulates eat extensively [19]. The clay soils of the plateau are dominated by *Acacia drepanolobium* whereas sandy soils of the low plain are dominated by *A. mellifera*, *A. etbaica*, and *A. brevispica* (Figure 1). Some herbivores were common in multiple habitats during the study period (e.g., giraffe, elephant, zebras, eland, and buffalo), whereas others were recorded almost exclusively in one habitat (hartebeest, oryx, and Grant's gazelle in plateau; dik-dik and waterbuck in low plain; Table S3, Figure S3) [44]. The subset of herbivores common across habitats spanned disparate lineages, whereas many of those with strong associations were close relatives from the Bovidae family (Figure 1). Thus, although the flora and fauna of each habitat is taxonomically different, taxa from diverse lineages share the functional abilities to occupy their shared habitats.

Second, there was a strong negative correlation between the richness of plants and animals together with contrasting patterns of plant and animal abundance and phylogenetic diversity across habitat types. Relatively high local plant abundance and diversity could be either a cause or consequence of relatively low herbivore abundance and diversity [45,46]. High plant abundance could cause of low animal abundance and diversity if thickets of woody plants are avoided by herbivores due to risk of predation. Many wild large herbivores perceive risk associated with the presence of predators and prefer good sight lines in habitats with low tree density [35,44]. Topographic features, particularly slope and convexity, further interrupt sightlines such that steep habitats with high tree and predator densities, coupled with challenging terrain, could deter herbivores from steep slopes, thereby alleviating top-down pressure on plants [4,36,47]. Rapid changes in topography can impose particularly strong mechanical and energetic constraints on megaherbivores such as elephants and giraffes, which avoided the steep slopes and have been shown to inflict markedly less damage on trees in this habitat [36].

Long-term herbivore-exclusion experiments at Mpala highlight the ability of megaherbivores to exert strong top-down effects on plant abundance and diversity. For example, elephants can reduce the availability of forbs that comprise a substantial portion of browsing mesoherbivore diets [19,48,49] while also increasing visibility and the availability of grasses for grazers [19,50]. The defensive strategies employed by abundant tree species on the slope suggest plant-herbivore sensitivity to top-down regulation that could contribute to their differential abundance across habitats. For example, some tree species that were relatively abundant on the steep slope invest heavily in antiherbivore defenses, including *Croton dichogamus* (Euphorbiaceae; produces noxious latex) and *Euclea divinorum* (Ebenaceae; produces tough, fibrous, and tannin-rich leaves) [19,51]. These defenses could deter consumption by the smaller-bodied ruminants that are abundant on the slope (e.g., dik-dik), but could be less effective against megaherbivores such as elephants and giraffes that otherwise frequently consume them elsewhere [52]. Whereas many recent studies focus on behaviorally mediated fear responses and trophic cascades [4,47], our results reinforce the important and spatially heterogeneous indirect effects that megaherbivores can have on vegetation [12,44,48–50,53].

The same physical habitat features that obstruct herbivore sightlines and could lead to avoidance of thickets can have methodological implications for camera-trap studies. We assumed approximately equivalent 25-m^2 radii detection ranges for camera traps at each grid site. However, reduced sightlines could lead to underestimates of the abundance and diversity of large herbivores in dense vegetation or on slopes [54]. This possibility suggests that the contrast between animal abundance and diversity observed in the plateau and low plain habitats could be conservative, since the low plain had both higher stem density and higher herbivore RIA (Table 1). However, the slope habitat had especially high stem density and complex topography, coupled with comparatively low herbivore RIA (Table 1), suggesting that further camera-trapping efforts may reveal additional animal use of plots in this habitat. It would be interesting to model seasonal variation in both detection rates and ungulate habitat associations across the plot: habitat associations might become stronger in the dry season when ungulates have the greatest need to monopolize the specific resources for which each is best able to compete, or alternatively these associations may be dampened if food depletion forces them to access the same reserves of riskier or less-preferred resources that accumulate on slopes and in thickets [4,5,55].

Our analyses combined a comprehensive woody-plant survey with DNA barcoding, which was methodologically useful both for refining plant identifications and modeling their phylogenetic relationships [23]. Data that support detailed investigations into ecological linkages between plants and herbivores—whether these linkages amount to spatial cooccurrence, trophic interactions, or both—are needed because these integrations have strong influences on the structure and function of both terrestrial and aquatic ecosystems worldwide [56,57]. Yet while the ecological and evolutionary processes that generate patterns of phylogenetic signal in relatively specialized species interactions have received substantial recent attention (e.g., host–microbiome [58], host–parasite [59]), less attention has been paid to phylogenetic structure in broader types of species interactions (e.g., plant–mycorrhiza [60], seed disperser networks [61], vertebrate trophic networks [19]) or habitat associations [62]. Because all species interactions and co-occurrence networks are subject to environmental and biological filters that act on species' functional traits, patterns of habitat use may generate phylogenetic signal in cooccurrence networks across trophic levels more frequently than currently realized. Our analysis focused on a uniquely fine-scale example of habitat associations across trophic levels, including phylogenetically diverse communities of woody plants and large herbivores. Results suggest no strong phylogenetic structure to the filtering of plant and animal traits that determine their habitat associations, since closely related species had divergent habitat associations and members of divergent clades often co-occurred.

Supplementary Materials: The following supporting information can be downloaded at: https://www.mdpi.com/article/10.3390/d14030219/s1, Figure S1: Maps of the study plot showing fine-scale major topographic characteristics, including (**a**) elevation, (**b**) convexity, (**c**) slope, and (**d**) topographic wetness index (TWI); Figure S2: Sample-based rarefaction of (**a**) woody-plant and (**b**) large-herbivore communities to compare total species richness within habitats based on our 33 grid sites (the actual numbers of sampling units in each habitat are shown as large points); Figure S3: Indicator species analysis for (**a**) woody plant and (**b**) large herbivore species across habitats. Points are colored by habitat and closed circles correspond to habitat associations (positive) or avoidance (negative) that differ significantly from 0; Table S1: Summary of plant species and DNA barcodes. The table matches names of species in our phylogeny with names used in the ForestGEO dataset, includes information on new DNA barcodes presented in this paper, descriptions of how species were added to the existing phylogeny [23], and the abundance of species across the plot as well as within our 33 grid sites; Table S2: Summary of plant abundances across 33 grid sites in this analysis. The table includes grid information (corresponding camera ID number, habitat, location in UTM) with site-specific plant abundance (total stems, individual trees) and diversity values (richness, *ses*MPD, *ses*MNTD); the *p*-values for site-specific phylogenetic diversity metrics are provided (significant positive values indicate phylogenetic overdispersion and negative values represent clustering). The site × species matrix based on the count of individual trees is then provided; Table S3: Summary of large-herbivore RAI values across 33 camera sites. The table includes grid information (camera ID number, habitat, location in UTM) with site-specific animal abundance (summed RAI) and diversity values (richness, *ses*MPD, *ses*MNTD), and the *P*-values for site-specific phylogenetic diversity are provided (significant positive values indicate phylogenetic overdispersion and negative values represent clustering). The site × species matrix based on RAI values is then provided; Table S4: Raw large-herbivore camera-trap records for the study period. For each photo, the table includes grid location, herbivore species, camera deployment date, photo-capture date and time, season (binned "wet" and "dry"), and the total camera-trap deployment days (in both wet and dry seasons). File S1: Nexus-formatted phylogeny for the 47 woody plant species present at the grid sites and used in analyses. References: [23,63–65] are cited in Supplementary Materials.

Author Contributions: Conceptualization, P.T.F. and T.R.K.; methodology, P.T.F., P.M.M. and D.K. (David Kenfack); P.T.F. and T.R.K.; investigation, P.T.F., R.O.A., D.K. (Duncan Kimuyu), P.L.E., M.M. and T.R.K.; data curation, P.T.F., B.A.G. and T.R.K.; writing—original draft preparation, P.T.F. and T.R.K.; writing—review and editing, all authors; funding acquisition, D.K. and T.R.K. All authors have read and agreed to the published version of the manuscript.

Funding: This research was funded by NSF DEB-1930820.

Institutional Review Board Statement: Not applicable.

Data Availability Statement: New plant DNA barcodes are available on Genbank (Accessions: *trn*L-F, OM238064-OM238067; *rbc*La, OM238062-OM238063), the full plant DNA barcode library is available on BOLD ('DS-UHURUR3', https://doi.org/10.5883/DS-UHURUR3), and the plant phylogeny is available in File S1.

Acknowledgments: We thank the Government of Kenya for granting permission to conduct this research, Mpala Research Centre for its staff and logistical support, and the Institute at Brown for Environment and Society for supporting student participation in the work. The Mpala ForestGEO plot is a collaborative project of the National Museums of Kenya and the Mpala Research Centre in partnership with the Forest Global Earth Observatory (ForestGEO) of the Smithsonian Tropical Research Institute (STRI). Funding for the first census was provided by ForestGEO. Smithsonian Tropical Research Institute, ForestGEO, and International Foundation for Science (D/5455-2) provided additional funding support for this work.

Conflicts of Interest: The authors declare no conflict of interest.

References

1. Galetti, M.; Moleón, M.; Jordano, P.; Pires, M.M.; Guimarães, P.R.; Pape, T.; Nichols, E.; Hansen, D.; Olesen, J.M.; Munk, M.; et al. Ecological and evolutionary legacy of megafauna extinctions. *Biol. Rev. Camb. Philos. Soc.* **2018**, *93*, 845–862. [CrossRef] [PubMed]
2. Ripple, W.J.; Newsome, T.M.; Wolf, C.; Dirzo, R.; Everatt, K.T.; Galetti, M.; Hayward, M.W.; Kerley, G.I.H.; Levi, T.; Lindsey, P.A.; et al. Collapse of the world's largest herbivores. *Sci. Adv.* **2015**, *1*, e1400103. [CrossRef]
3. Estes, J.A.; Terborgh, J.; Brashares, J.S.; Power, M.E.; Berger, J.; Bond, W.J.; Carpenter, S.R.; Essington, T.E.; Holt, R.D.; Jackson, J.B.C.; et al. Trophic downgrading of planet earth. *Science* **2011**, *333*, 301–306. [CrossRef] [PubMed]
4. Ford, A.T.; Goheen, J.R.; Otieno, T.O.; Bidner, L.; Isbell, L.A.; Palmer, T.M.; Ward, D.; Woodroffe, R.; Pringle, R.M. Large carnivores make savanna tree communities less thorny. *Science* **2014**, *346*, 346–349. [CrossRef] [PubMed]
5. Owen-Smith, N.; Roux, E.L.; Macandza, V. Are relatively rare antelope narrowly selective feeders? A sable antelope and zebra comparison. *J. Zool.* **2013**, *291*, 163–170. [CrossRef]
6. Riginos, C. Climate and the landscape of fear in an African savanna. *J. Anim. Ecol.* **2015**, *84*, 124–133. [CrossRef] [PubMed]
7. Thouless, C.R. Long distance movements of elephants in northern Kenya. *Afr. J. Ecol.* **1995**, *33*, 321–334. [CrossRef]
8. Loreau, M.; Naeem, S.; Inchausti, P.; Bengtsson, J.; Grime, J.P.; Hector, A.; Hooper, D.U.; Huston, M.A.; Raffaelli, D.; Schmid, B.; et al. Biodiversity and ecosystem functioning: Current knowledge and future challenges. *Science* **2001**, *294*, 804–808. [CrossRef]
9. Mittelbach, G.G.; Schemske, D.W.; Cornell, H.V.; Allen, A.P.; Brown, J.M.; Bush, M.B.; Harrison, S.P.; Hurlbert, A.H.; Knowlton, N.; Lessios, H.A.; et al. Evolution and the latitudinal diversity gradient: Speciation, extinction and biogeography. *Ecol. Lett.* **2007**, *10*, 315–331. [CrossRef]
10. Thompson, J.N. Evolutionary ecology and the conservation of biodiversity. *Trends Ecol. Evol.* **1996**, *11*, 300–303. [CrossRef]
11. Wootton, J.T. Effects of disturbance on species diversity: A multitrophic perspective. *Am. Nat.* **1998**, *152*, 803–825. [CrossRef] [PubMed]
12. Zhang, J.; Qian, H.; Girardello, M.; Pellissier, V.; Nielsen, S.E.; Svenning, J.-C. Trophic interactions among vertebrate guilds and plants shape global patterns in species diversity. *Proc. R. Soc. B* **2018**, *285*, 20180949. [CrossRef] [PubMed]
13. Cromsigt, J.P.G.M.; Prins, H.H.T.; Olff, H. Habitat heterogeneity as a driver of ungulate diversity and distribution patterns: Interaction of body mass and digestive strategy. *Divers. Distrib.* **2009**, *15*, 513–522. [CrossRef]
14. Jetz, W.; Kreft, H.; Ceballos, G.; Mutke, J. Global associations between terrestrial producer and vertebrate consumer diversity. *Proc. R. Soc. B* **2009**, *276*, 269–278. [CrossRef] [PubMed]
15. Kartzinel, T.R.; Chen, P.A.; Coverdale, T.C.; Erickson, D.L.; Kress, W.J.; Kuzmina, M.L.; Rubenstein, D.I.; Wang, W.; Pringle, R.M. DNA metabarcoding illuminates dietary niche partitioning by African large herbivores. *Proc. Nat. Acad. Sci. USA* **2015**, *112*, 8019–8024. [CrossRef]
16. Haddad, N.M.; Crutsinger, G.M.; Gross, K.; Haarstad, J.; Knops, J.M.H.; Tilman, D. Plant species loss decreases arthropod diversity and shifts trophic structure. *Ecol. Lett.* **2009**, *12*, 1029–1039. [CrossRef]
17. Hutchinson, G.E. Homage to Santa Rosalia or why are there so many kinds of animals? *Am. Nat.* **1959**, *93*, 145–159. [CrossRef]
18. Srivastava, D.; Lawton, J. Why more productive sites have more species: An experimental test of theory using tree-hold communities. *Am. Nat.* **1998**, *152*, 510–529. [CrossRef]
19. Kartzinel, T.R.; Pringle, R.M. Multiple dimensions of dietary diversity in large mammalian herbivores. *J. Anim. Ecol.* **2020**, *89*, 1482–1496. [CrossRef]
20. Swenson, N.G.; Stegen, J.C.; Davies, S.J.; Erickson, D.L.; Forero-Montaña, J.; Hurlbert, A.H.; Kress, W.J.; Thompson, J.; Uriarte, M.; Wright, S.J.; et al. Temporal turnover in the composition of tropical tree communities: Functional determinism and phylogenetic stochasticity. *Ecology* **2012**, *93*, 490–499. [CrossRef]
21. Cavender-Bares, J.; Kozak, K.H.; Fine, P.V.; Kembel, S.W. The merging of community ecology and phylogenetic biology. *Ecology Lett.* **2009**, *12*, 693–715. [CrossRef] [PubMed]
22. Mutuku, P.M.; Kenfack, D. Effect of local topographic heterogeneity on tree species assembly in an *Acacia*-dominated African savanna. *J. Trop. Ecol.* **2019**, *35*, 46–56. [CrossRef]
23. Gill, B.A.; Musili, P.M.; Kurukura, S.; Hassan, A.A.; Goheen, J.R.; Kress, W.J.; Kuzmina, M.; Pringle, R.M.; Kartzinel, T.R. Plant DNA-barcode library and community phylogeny for a semi-arid East African savanna. *Mol. Ecol. Resour.* **2019**, *19*, 838–846. [CrossRef] [PubMed]
24. Lepore, M.; Arellano, G.; Condit, R.; Detto, M.; Harms, K.; Lao, S.; Russo, S.; Zuleta, D. fgeo.analyze: Analyze ForestGEO Data, R Package Version 1.1.14. 2020. Available online: https://CRAN.R-project.org/package=fgeo.analyze (accessed on 12 December 2021).
25. R Core Team. *A Language and Environment for Statistical Computing*; R Foundation for Statistical Computing: Vienna, Austria, 2021.
26. Hopcraft, J.G.C.; Anderson, T.M.; Pérez-Vila, S.; Mayeba, E.; Olff, H. Body size and the division of niche space: Food and predation differentially shape the distribution of Serengeti grazers. *J. Anim. Ecol.* **2012**, *81*, 201–213. [CrossRef] [PubMed]
27. Metcalfe, P.; Beven, K.; Freer, J. Dynamic TOPMODEL: A new implementation in R and its sensitivity to time and space steps. *Environ. Model. and Softw.* **2015**, *72*, 155–172. [CrossRef]
28. Kress, W.J.; Erickson, D.L.; Swenson, N.G.; Thompson, J.; Uriarte, M.; Zimmerman, J.K. Advances in the use of DNA barcodes to build a community phylogeny for tropical trees in a Puerto Rican forest dynamics plot. *PLoS ONE* **2010**, *5*, e15409. [CrossRef]

29. Bolger, D.T.; Morrison, T.A.; Vance, B.; Lee, D.; Farid, H. Development and application of a computer-assisted system for photographic mark-recapture analysis. *Methods Ecol. Evolut.* **2012**, *3*, 813–822. [CrossRef]

30. Palmer, M.S.; Swanson, A.; Kosmala, M.; Arnold, T.; Packer, C. Evaluating relative abundance indices for terrestrial herbivores from large-scale camera trap surveys. *Afr. J. Ecol.* **2018**, *56*, 791–803. [CrossRef]

31. Kembel, S.W.; Cowan, P.D.; Helmus, M.R.; Cornwell, W.K.; Morlon, H.; Ackerly, D.D.; Blomberg, S.P.; Webb, C.O. Picante: R tools for integrating phylogenies and ecology. *Bioinformatics* **2010**, *26*, 1463–1464. [CrossRef]

32. Mazel, F.; Davies, T.J.; Gallien, L.; Renaud, J.; Groussin, M.; Münkemüller, T.; Thuiller, W. Influence of tree shape and evolutionary time-scale on phylogenetic diversity metrics. *Ecography* **2016**, *39*, 913–920. [CrossRef]

33. Webb, C.O.; Ackerly, D.D.; McPeek, M.A.; Donoghue, M.J. Phylogenies and community ecology. *Annu. Rev. Ecol. Syst.* **2002**, *31*, 475–505. [CrossRef]

34. Ives, A.R.; Helmus, M.R. Phylogenetic metrics of community similarity. *Am. Nat.* **2010**, *176*, E128–E142. [CrossRef] [PubMed]

35. Riginos, C.; Grace, J.B. Savanna tree density, herbivores, and the herbaceous community: Bottom-up vs. top-down effects. *Ecology* **2008**, *89*, 2228–2238. [CrossRef] [PubMed]

36. Kimuyu, D.M.; Kenfack, D.; Musili, P.M.; Ang'ila, R.O. Fine-scale habitat heterogeneity influences browsing damage by elephant and giraffe. *Biotropica* **2021**, *53*, 86–96. [CrossRef]

37. Colwell, R.K.; Chao, A.; Gotelli, N.J.; Lin, S.-Y.; Mao, C.X.; Chazdon, R.L.; Longino, J.T. Models and estimators linking individual-based and sample-based rarefaction, extrapolation and comparison of assemblages. *J. Plant Ecol.* **2012**, *5*, 3–21. [CrossRef]

38. Oksanen, J.; Blanchet, F.G.; Friendly, M.; Kindt, R.; Legendre, P.; McGlinn, D.; Minchin, P.R.; O'Hara, R.B.; Simpson, G.L.; Solymos, P.; et al. *Vegan: Community Ecology Package*; R Package Version 2.5-7:2020; R Core Team: Vienna, Austria, 2020.

39. De Caceres, M.; Legendre, P. Associations between species and groups of sites: Indices and statistical inference. *Ecology* **2009**, *90*, 3566–3574. [CrossRef]

40. Pellissier, L.; Ndiribe, C.; Dubuis, A.; Pradervand, J.-N.; Salamin, N.; Guisan, A.; Rasmann, S. Turnover of plant lineages shapes herbivore phylogenetic beta diversity along ecological gradients. *Ecol. Lett.* **2013**, *16*, 600–608. [CrossRef]

41. Kemp, J.E.; Linder, H.P.; Ellis, A.G. Beta diversity of herbivorous insects is coupled to high species and phylogenetic turnover of plant communities across short spatial scales in the Cape Floristic Region. *J. Biogeogr.* **2017**, *44*, 1813–1823. [CrossRef]

42. Dinnage, R.; Cadotte, M.W.; Haddad, N.M.; Crutsinger, G.M.; Tilman, D. Diversity of plant evolutionary lineages promotes arthropod diversity. *Ecol. Lett.* **2012**, *15*, 1308–1317. [CrossRef]

43. Schuldt, A.; Ebeling, A.; Kunz, M.; Staab, M.; Guimarães-Steinicke, C.; Bachmann, D.; Buchmann, N.; Durka, W.; Fichtner, A.; Fornoff, F.; et al. Multiple plant diversity components drive consumer communities across ecosystems. *Nat. Commun.* **2019**, *10*, 1460. [CrossRef]

44. Wells, H.B.M.; Crego, R.D.; Odepal, Ø.H.; Khasoha, L.M.; Alston, J.M.; Reed, C.G.; Weiner, S.; Kurukura, S.; Hassan, A.A.; Namoni, M.; et al. Experimental evidence that effects of megaherbivores on mesoherbivore space use are influenced by species' traits. *J. Anim. Ecol.* **2021**, *90*, 2510–2522. [CrossRef] [PubMed]

45. Hairston, N.G.; Smith, F.E.; Slobodkin, L. Community structure, population control, and competition. *Am. Nat.* **1960**, *879*, 421–425. [CrossRef]

46. Polis, G.A. Why are parts of the world green? Multiple factors control productivity and the distribution of biomes. *Oikos* **1999**, *86*, 3–15. [CrossRef]

47. Beschta, R.L.; Ripple, W.J. Large predators and trophic cascades in ecosystems of the western United States. *Biol. Conserv.* **2009**, *142*, 2401–2414. [CrossRef]

48. Coverdale, T.C.; Kartzinel, T.R.; Grabowski, K.L.; Shriver, R.K.; Hassan, A.A.; Goheen, J.R.; Palmer, T.M.; Pringle, R.M. Elephants in the understory: Opposing direct and indirect effects of consumption and ecosystem engineering by megaherbivores. *Ecology* **2016**, *97*, 3219–3230. [CrossRef] [PubMed]

49. Coverdale, T.C.; O'Connell, R.D.; Hutchinson, M.C.; Savagian, A.; Kartzinel, T.R.; Palmer, T.M.; Goheen, J.R.; Augustine, D.J.; Sankaran, M.; Tarnita, C.E.; et al. Large herbivores suppress liana infestation in an African savanna. *Proc. Nat. Acad. Sci. USA* **2021**, *118*, e2101676118. [CrossRef]

50. Goheen, J.R.; Augustine, D.J.; Veblen, K.E.; Kimuyu, D.M.; Palmer, T.M.; Porensky, L.M.; Pringle, R.M.; Ratnam, J.; Riginos, C.; Sankaran, M.; et al. Conservation lessons from large-mammal manipulations in East African savannas: The KLEE, UHURU, and GLADE experiments. *Ann. N. Y. Acad. Sci.* **2018**, *1429*, 31–49. [CrossRef]

51. Hattas, D.; Hjältén, J.; Julkunen-Tiitto, R.; Scogings, P.F.; Rooke, T. Differential phenolic profiles in six African savanna woody species in relation to antiherbivore defense. *Phytochemistry* **2011**, *72*, 1796–1803. [CrossRef]

52. Kartzinel, T.R.; Hsing, J.C.; Musili, P.M.; Brown, B.R.P.; Pringle, R.M. Covariation of diet and gut microbiome in African megafauna. *Proc. Nat. Acad. Sci. USA* **2019**, *116*, 23588–23593. [CrossRef]

53. Kimuyu, D.M.; Veblen, K.E.; Riginos, C.; Chira, R.M.; Githaiga, J.M.; Young, T.P. Influence of cattle on browsing and grazing wildlife varies with rainfall and presence of megaherbivores. *Ecol. Appl.* **2017**, *27*, 786–798. [CrossRef]

54. MacKenzie, D.I.; Nichols, J.D.; Hines, J.E.; Knutson, M.G.; Franklin, A.B. Estimating site occupancy, colonization, and local extinction when a species is detected imperfectly. *Ecology* **2003**, *84*, 2200–2207. [CrossRef]

55. Becker, J.A.; Hutchinson, M.C.; Potter, A.B.; Park, S.; Guyton, J.A.; Abernathy, K.; Americo, V.F.; da Conceição, A.; Kartzinel, T.R.; Kuziel, L.; et al. Ecological and behavioral mechanisms of density-dependent habitat expansion in a recovering African ungulate population. *Ecol. Monogr.* **2021**, *91*, e01476. [CrossRef]

56. Milchunas, D.G.; Lauenroth, W.K. Quantitative effects of grazing on vegetation and soils over a global range of environments. *Ecol. Monogr.* **1993**, *63*, 327–366. [CrossRef]

57. Wood, K.A.; O'Hare, M.T.; McDonald, C.; Searle, K.R.; Daunt, F.; Stillman, R.A. Herbivore regulation of plant abundance in aquatic ecosystems. *Biol. Rev.* **2017**, *92*, 1128–1141. [CrossRef] [PubMed]

58. Brooks, A.W.; Kohl, K.D.; Brucker, R.M.; van Opstal, E.J.; Bordenstein, S.R. Phylosymbiosis: Relationships and functional effects of microbial communities across host evolutionary history. *PLoS Biol.* **2017**, *15*, e1002587. [CrossRef]

59. Park, A.W.; Farrell, M.J.; Schmidt, J.P.; Huang, S.; Dallas, T.A.; Pappalardo, P.; Drake, J.M.; Stephens, P.R.; Poulin, R.; Nunn, C.L.; et al. Characterizing the phylogenetic specialism–generalism spectrum of mammal parasites. *Proc. R. Soc. B* **2018**, *285*, 20172613. [CrossRef]

60. Shefferson, R.P.; Bunch, W.; Cowden, C.C.; Lee, Y.I.; Kartzinel, T.R.; Yukawa, T.; Downing, J.; Jiang, H. Does evolutionary history determine specificity in broad ecological interactions? *J. Ecol.* **2019**, *107*, 1582–1593. [CrossRef]

61. Corro, E.J.; Villalobos, F.; Lira-Noriega, A.; Guevara, R.; Guimarães, P.R.; Dáttilo, W. Annual precipitation predicts the phylogenetic signal in bat-fruit interaction networks across the Neotropics. *Biol. Lett.* **2021**, *17*, 20210478. [CrossRef]

62. Staab, M.; Liu, X.; Assmann, T.; Bruelheide, H.; Buscot, F.; Durka, W.; Erfmeier, A.; Klein, A.-M.; Ma, K.; Michalski, S.; et al. Tree phylogenetic diversity structures multitrophic communities. *Funct. Ecol.* **2020**, *35*, 521–534. [CrossRef]

63. Simth, G.F.; Figueiredo, E. Conserving Acacia Mill. with a conserved type: What happenedin Melbourne? *Taxonomy* **2018**, *60*, 1504–1506. [CrossRef]

64. Antonellia, A.; Nylanderb, J.A.A.; Perssona, C.; Sanmartin, I. Tracing the impact of the Andean uplift on Neotropical plant evolution. *Proc. Nat. Acad. Sci. USA* **2009**, *106*, 9749–9754. [CrossRef] [PubMed]

65. Janssens, S.; Couvreur, T.L.; Mertens, A.; Dauby, G.; Dagallier, L.-P.; Abeele, S.V.; Vandelook, F.; Mascarello, M.; Beeckman, H.; Sosef, M.; et al. A large-scale species level dated angiosperm phylogeny for evolutionary and ecological analyses. *Biodivers. Data J.* **2020**, *8*, e39677. [CrossRef] [PubMed]

Article

Assessment of ITS1, ITS2, 5′-ETS, and *trnL-F* DNA Barcodes for Metabarcoding of Poaceae Pollen

Denis O. Omelchenko [1,*,†], Anastasia A. Krinitsina [2,*,†], Artem S. Kasianov [1], Anna S. Speranskaya [2,3], Olga V. Chesnokova [2], Svetlana V. Polevova [2] and Elena E. Severova [2,4]

[1] Laboratory of Plant Genomics, Institute for Information Transmission Problems, 127051 Moscow, Russia; artem.kasianov@gmail.com

[2] Department of Higher Plants, Faculty of Biology, Lomonosov Moscow State University, 119991 Moscow, Russia; hanna.s.939@gmail.com (A.S.S.); ChesnokovaOV@my.msu.ru (O.V.C.); polevova@mail.bio.msu.ru (S.V.P.); eseverova@hse.ru (E.E.S.)

[3] Group of Genomics and Postgenomic Technologies, Central Research Institute of Epidemiology, 111123 Moscow, Russia

[4] Faculty of Biology and Biotechnologies, National Research University Higher School of Economics, 101000 Moscow, Russia

* Correspondence: omelchenkodo@iitp.ru (D.O.O.); krinitsina@msu-botany.ru (A.A.K.)

† These authors contributed equally to this work.

Citation: Omelchenko, D.O.; Krinitsina, A.A.; Kasianov, A.S.; Speranskaya, A.S.; Chesnokova, O.V.; Polevova, S.V.; Severova, E.E. Assessment of ITS1, ITS2, 5′-ETS, and *trnL-F* DNA Barcodes for Metabarcoding of Poaceae Pollen. *Diversity* **2022**, *14*, 191. https://doi.org/10.3390/d14030191

Academic Editors: W. John Kress, Morgan Gostel and Michael Wink

Received: 30 January 2022
Accepted: 2 March 2022
Published: 5 March 2022

Abstract: Grass pollen is one of the major causes of allergy. Aerobiological monitoring is a necessary element of the complex of anti-allergic measures, but the similar pollen morphology of Poaceae species makes it challenging to discriminate species in airborne pollen mixes, which impairs the quality of aerobiological monitoring. One of the solutions to this problem is the metabarcoding approach employing DNA barcodes for taxonomical identification of species in a mix by high-throughput sequencing of the pollen DNA. A diverse set of 14 grass species of different genera were selected to create a local reference database of nuclear ITS1, ITS2, 5′-ETS, and plastome *trnL-F* DNA barcodes. Sequences for the database were Sanger sequenced from live field and herbarium specimens and collected from GenBank. New Poaceae-specific primers for 5′-ETS were designed and tested to obtain a 5′-ETS region less than 600 bp long, suitable for high-throughput sequencing. The DNA extraction method for single-species pollen samples and mixes was optimized to increase the yield for amplification and sequencing of pollen DNA. Barcode sequences were analyzed and compared by the barcoding gap and intra- and interspecific distances. Their capability to correctly identify grass pollen was tested on artificial pollen mixes of various complexity. Metabarcoding analysis of the artificial pollen mixes showed that nuclear DNA barcodes ITS1, ITS2, and 5′-ETS proved to be more efficient than the plastome barcode in both amplification from pollen DNA and identification of grass species. Although the metabarcoding results were qualitatively congruent with the actual composition of the pollen mixes in most cases, the quantitative results based on read-counts did not match the actual ratio of pollen grains in the mixes.

Keywords: grass pollen; metabarcoding; plant barcodes

1. Introduction

Grass pollen is one of the major causes of allergy, affecting 10–30% of the population around the globe [1,2]. There are over 400 species of grass in Europe, and their pollen is recognized as the leading cause of pollinosis [3]. About 100 species of grass could be found in the European part of Russia [4], flowering periods of which often overlap, and their pollen allergenicity is estimated to be from moderate to very high [5]. Aerobiological monitoring is a necessary element of the complex of anti-allergic measures allowing for tracking and predicting the dynamics of the concentration of major allergens in the air and adjusting the therapy and lifestyle of patients with pollinosis. The standard method of pollen identification in the air samples is light microscopy. However, one of the major disadvantages of pollen light microscopy analysis is that similar pollen morphology of Poaceae species makes it challenging to discriminate species in airborne pollen mixes, which impairs the quality of aerobiological monitoring [6,7]. DNA metabarcoding is an alternative approach that has been actively developing recently, allowing qualitative (to the level of species or genus for some taxa) and quantitative (to some extent) composition analysis of complex biological mixes. It employs high-throughput sequencing (HTS) and comparative analysis of specific DNA sequences called "DNA barcodes" to discriminate species present in the mix. DNA barcoding has been widely used in various areas of botanical research; for example, the phylogeny of wild cherry [8], archaeobotany of grapevine [9], authentication and identification of medicinal [10] and poisonous [11] plants, and plant species composition of honey [12,13].

Choosing the correct DNA barcode for the target taxa is one of the main problems of plant barcoding [14,15]. The resolution capacity of each of the primary chloroplast markers (first a combination of *matK* and *rbcL* recommended by the CBOL group [16] and later the nuclear ribosomal internal transcribed spacer (ITS) regions and several intergenic spacers) vary significantly between different taxa (for a review, see [17]). Many studies focused on the DNA barcoding of plants; note, that the identification at the high-rank taxa (order, family) is successful in more than 90% of cases, while insufficient data on reference DNA barcode sequences prevents determination to the level of genus or species [13,18]. Therefore, the right choice of DNA barcode and the primers to amplify them is the key to successful species identification.

The regions of the chloroplast genome *rbcL*, *matK*, *trnL*, *trnH-psbA*, and nuclear ITS2 are most often used as plant DNA barcodes. Some of these barcodes have been used with varying success for metabarcoding pollen (airborne or from food products such as honey). However, only *rbcL*, *matK*, ITS2, and *trnL* barcodes have been studied compared to the palynological analysis for assessing qualitative and quantitative consistency [19]. In particular, a comprehensive study of ITS2 and *rbcL* has shown their usability in metabarcoding of pollen for the construction of pollinator networks and qualitative analysis of pollen mixes. Though, the quantitative relativity of the metabarcoding results and real pollen abundance of mixture components has been low [20,21]. Another study has assessed *trnL* and ITS1 for quantitative pollen analysis using metabarcoding and concluded that *trnL* demonstrates the best sequence-to-pollen prediction [22]. Furthermore, comparative studies have shown a good capability of *trnL* intron and *trnL-trnF* (*trnL-F*) intergenic spacers, ITS region, and their combinations to resolve grass species [23–26]. Indel and SNP patterns of the *trnL-F* intergenic spacer and ITS region have been employed for infrageneric classification and phylogeny study of *Chascolytrum* and *Festuca* genera [27,28].

External transcribed spacer (ETS) is another nuclear DNA barcode closely related to the ITS region in rDNA, but it is less frequently used than ITS. However, ETS is regarded as a promising DNA barcode as the taxon-specific informativity of the ETS sequence has proved to be the highest among nuclear and plastid barcodes in several studies [29–31].

Many published studies report the species identification of different grasses using only some of these barcodes and focusing on a particular plant taxon (e.g., [32,33]). In this study, we have compared the plastome *trnL-F* and nuclear ITS1 and ITS2 barcodes with the 5′-ETS barcode and assessed their capability to identify the pollen of a diverse set of 14 grass species of different genera from the Poaceae family. New Poaceae-specific primers were designed to amplify the 5′-ETS fragment suitable for the HTS sequencing as its length is less than 600 bp for all species in the study (maximum length for Illumina paired-end sequencing at present). Additionally, we have optimized the protocol for DNA extraction from pollen grains to obtain high-quality DNA for amplification and sequencing. To identify the pollen composition, we have created a local barcode sequence database for the reference Poaceae species using Sanger sequenced *trnL-F*, *ITS1*, *ITS2*, and *5′-ETS* sequences of the live field samples, herbarium specimens, and available GenBank records.

All four barcode sequences were tested by their capacity to resolve the composition of the grass pollen mixes using artificial pollen mixes of various complexity.

2. Materials and Methods

2.1. Plant Material

To assess the nuclear (ITS1, ITS2, and 5′-ETS) and plastome (*trnL-F*) barcodes' capability to identify grass pollen and create a local reference database of barcode sequences, a broad spectrum of Poaceae species widespread in Central Russia were selected: *Alopecurus pratensis* L., *Arrhenatherum elatius* (L.) P.Beauv. ex J.Presl & C.Presl., *Briza media* L., *Bromus inermis* (Leyss.) Holub (syn. *Bromopsis inermis*), *Calamagrostis epigeios* (L.) Roth, *Dactylis glomerata* L., *Elymus repens* (L.) Gould (syn. *Elytrigia repens*), *Festuca pratensis* Huds., *Lolium perenne* L., *Phleum pratense* L., *Poa annua* L., *Poa pratensis* L., *Poa supina* Schrad., and *Poa trivialis* L. Additionally, *Festuca arundinacea* Schreb. and *Poa palustris* L. were collected for the ETS primers' design. Fresh leaf material of the morphologically identified grass plants was sampled in the field (Moscow region) and from the Moscow State University Herbarium collection specimens.

Pure single-species pollen for a subset of the selected reference species was manually collected to make artificial pollen mixes: *Calamagrostis epigeios, Phleum pratense, Bromus inermis, Festuca pratensis, Elymus repens, Alopecurus pratensis,* and *Lolium perenne.* These species pollinate in abundance and are easy to collect in enough quantities to create pollen mixes of various complexity. Therefore, pollen was collected during summer in the active pollination time of these species. The collected pollen was weighed, and a sample of 10 mg of each species was suspended in 100 µL TE-buffer. From each sample, 2 µL of the suspension was analyzed using light microscopy, and pollen grains were counted to estimate the number of pollen grains for each species. Each sample was diluted in TE buffer to achieve an equal pollen count per mL based on the observed number of pollen grains. Then single-species pollen samples were mixed by volume to create a 100 µL pollen mix that contained pollen from different species in equal abundance (approx. 10,000 pollen grains per mix in total). The species composition of each artificial mix is presented in Table 1.

Table 1. Pollen artificial mixes' composition.

Artificial Pollen Mix Name	Species
Pollen mixes with two species, 50% pollen of each type	
am1	*Calamagrostis epigeios, Phleum pretense*
am2	*Bromus inermis, Festuca pratensis*
am3	*Alopecurus pratensis, Lolium perenne*
am4	*Calamagrostis epigeios, Lolium perenne*
am5	*Phleum pratense, Alopecurus pratensis*
am6	*Phleum pratense, Elymus repens*
Pollen mixes with three species, 33.3% pollen of each type	
am7	*Calamagrostis epigeios, Phleum pratense, Bromus inermis*
am8	*Phleum pratense, Bromus inermis, Festuca pratensis*
am9	*Bromus inermis, Festuca pratensis, Elymus repens*
am10	*Phleum pratense, Lolium perenne, Elymus repens*
Pollen mixes with four species, 25% pollen of each type	
am11	*Calamagrostis epigeios, Phleum pratense, Bromus inermis, Festuca pratensis*
am12	*Phleum pratense, Calamagrostis epigeios, Elymus repens, Lolium perenne*
am13	*Phleum pratense, Festuca pratensis, Lolium perenne, Elymus repens*

Table 1. *Cont.*

Artificial Pollen Mix Name	Species
Pollen mixes with five species, 20% pollen of each type	
am14	*Calamagrostis epigeios, Phleum pratense, Bromus inermis, Festuca pratensis, Elymus repens*
am15	*Phleum pratense, Calamagrostis epigeios, Festuca pratensis, Lolium perenne, Elymus repens*
am16	*Phleum pratense, Festuca pratensis, Bromus inermis, Lolium perenne, Elymus repens*
Pollen mixes with six species, 16.7% pollen of each type	
am17	*Phleum pratense, Calamagrostis epigeios, Festuca pratensis, Bromus inermis, Lolium perenne, Elymus repens*
am18	*Calamagrostis epigeios, Phleum pratense, Bromus inermis, Festuca pratensis, Elymus repens, Alopecurus pratensis*

2.2. DNA Extraction

DNA from herbarium samples and fresh leaf material was extracted using the sorbent-based DiamondDNA Plant kit (ABT, Barnaul, Russia), according to the manufacturer's protocol with subsequent additional purification by magnetic silica beads, as described elsewhere [34].

Pollen DNA extraction protocol was optimized using pollen grains of *Phleum pratense* and *Bromus inermis*. The pollen sample (10 mg) was suspended in 100 µL TE-buffer and homogenized using a Precellys Bacteria lysing kit CK01 (Bertin Technologies, Montigny-le-Bretonneux, France) and MiniLys homogenizer (Bertin Technologies, Montigny-le-Bretonneux, France) at the maximum speed in two runs of 240 s each. Lysis efficiency was tested using three variants of the lysis buffer: (1) only CTAB-lysis buffer (2% CTAB, 0.1 M Tris-HCl pH 8.0, 1.4 M NaCl, 20 mM EDTA pH 8.0, 1% PVP, 0.2% 2-mercaptoethanol, 0.1 mM DTT); (2) CTAB-lysis buffer with 0.04% SDS; and (3) CTAB-lysis buffer with 0.4% SDS. Additionally, two variants of proteinase K concentration (0.2 mg and 0.4 mg per sample) and lysis incubation time (1 and 2 h) at 65 °C were tested for all variants of lysis buffer. DNA from the homogenized and lysed samples were extracted using 1 v/v chloroform: isoamyl alcohol 24: 1. Then, DNA was precipitated at -20 °C for 1 h with 0.1 v/v of 3 M sodium acetate and 1 v/v of isopropanol.

According to our observations with the light microscope, 10 mg of pollen contains ~150,000 pollen grains. Therefore, to check the minimum amount of pollen grains required to extract enough DNA for further analysis and HTS, the pollen DNA extraction efficiency from different amounts of pollen was also tested: 150,000, 37,500, 9375, 2344, 586, and 150 pollen grains. Test samples were created by $4\times$ serial dilution of the initial 10 mg pollen sample.

DNA extraction from these samples was performed using the best extraction protocol determined at the previous step: CTAB-buffer with 0.04% SDS, 0.2 mg per sample proteinase K, lysis incubation for 2 h at 65 °C. DNA from artificial pollen mixes was also extracted according to the optimized protocol.

The purity of the DNA samples was assessed by the A260/280 and A260/230 ratios on a NanoPhotometer N60-Touch (Implen, Munich, Germany), and the concentration was measured by fluorescence intensity using a Qubit dsDNA HS Assay Kit (Invitrogen, Waltham, MA, USA) and Qubit 3.0 fluorometer (Invitrogen, Waltham, MA, USA).

2.3. PCR and Primer Design

Primers for nuclear DNA barcodes ITS1 and ITS2 were designed to anneal to conservative regions of plant rDNA selectively and not fungi (ITS1-F 5′-GGAAGGAGAAGTCGTAACAAGG-3′, ITS1-R 5′-AGATATCCGTTGCCGAGAGT-3′ [35]; ITS2-F 5′-ATCGAGTYTTTGAACGCAAGTTG-3′, ITS2-R 5′-TCCTCCATGCTCTATTG-3′ not published). Primers for the chloroplast intergenic spacer *trnL-F* barcode were obtained from [36] (trnL_F 5′-GGTTCAAGTCCCTCTATCCC-3′; trnF_R 5′-ATTTGAACTGGTGACACGAG-3′).

Based on the alignment of all available 3′ ends of the 26S and 5′ ends of the 18S rDNA sequences of Poaceae plants from the GenBank database, two pairs of primers were designed for amplification of the complete rDNA intergenic region (IGS) for subsequent Sanger sequencing of 5′-ETS (26S_end_F 5′-GATCCACTGAGATCCAGCCC-3′; 18S_start_R 5′-CTGGCAGGATCAACCAGGTA-3′). Amplification was carried out on DNA from the leaves of the field plants collected during the vegetation season. In addition, the ETS region sequences of the Poaceae species were also collected from GenBank to create a MAFFT alignment of all ETS fragments available. New Poaceae-specific 5′-ETS primers were designed based on this alignment.

A schematic representation of the primer binding sites is presented in Figure 1.

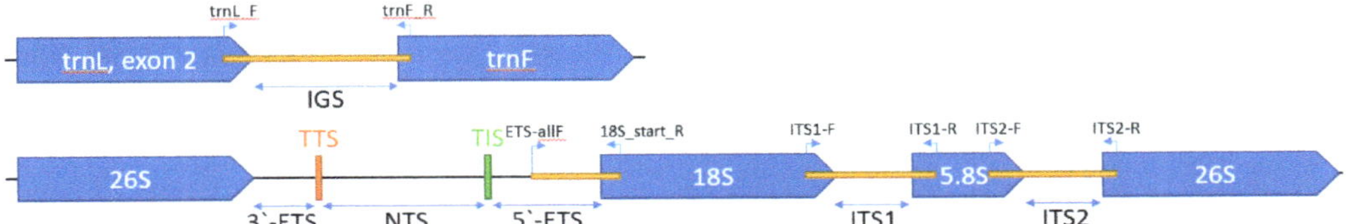

Figure 1. Primers' binding sites scheme. IGS—intergenic spacer; NTS—non-transcribed part of rDNA; TIS—transcription initiation site; TTS—transcription termination site.

The PCR for the Sanger sequencing of DNA from the herbarium specimen and field samples and DNA library indexation PCR for HTS were performed using NEBNext Ultra II Q5 Master Mix (NEB, Ipswich, MA, USA) containing high-fidelity Q5 polymerase. The PCR of barcodes from DNA of artificial pollen mixes was performed using the Encyclo Plus PCR Kit (Evrogen, Moscow, Russia) containing a mix of high-fidelity and high-processivity polymerases.

2.4. Library Preparation and Sequencing

A simplified two-step PCR using primers for DNA barcodes fused with Illumina adaptor sequences was performed for DNA library preparation as described elsewhere [35,37]. Products of the first PCR for each barcode were mixed equimolar (or by volume if product concentration was below detection level) for each sample, indexed in the second PCR, and sequenced on the MiSeq platform with the MiSeq Reagent Kit v3 for 600 cycles (2 × 300 nt paired-end) (Illumina, San Diego, CA, USA).

2.5. Local Barcode Reference Database Construction

A local reference database was created using ITS1, ITS2, *trnL-F*, and 5′-ETS sequences of the reference Poaceae species from herbarium and field samples Sanger sequenced at the Evrogen company (Moscow, Russia). Sanger-sequenced barcode sequences were trimmed from both ends by the quality and aligned using MAFFT v7.490 (FFT-NS-I algorithm). In addition, sequences of the studied DNA barcodes of the reference species have been retrieved from the GenBank database (if available) and added to the alignment if sequences overlapped and showed similarity by more than 90% with our sequences. Detailed information on the corresponding MSU Herbarium voucher numbers, field samples, and the GenBank accessions can be found in Supplementary Table S1. All barcode sequences were used to construct a local reference BLAST database as described elsewhere [37].

2.6. Data Analysis and Taxonomical Identification

Sanger sequencing results were manually reviewed and processed using CLC Genomics Workbench 8.5 software (Qiagen, Hilden, Germany), and all obtained sequences were submitted to the GenBank database.

Intra- and interspecific distances between the barcode sequences were calculated in MEGA v11.0.9 [38]. The best DNA/Protein models (ML) search function determined the best substitution model for each barcode alignment. The selected best substitution model for the alignment was used to calculate the distances. Analyses were conducted using the Tamura 3-parameter model [39] with a gamma distribution (shape parameter = 0.51), and all positions containing gaps and missing data were eliminated (complete deletion parameter) for the trnL-F barcode; and the Kimura 2-parameter model [40] with a gamma distribution and complete deletion for ITS1, ITS2, and 5′-ETS barcode (gamma distribution shape parameter: 0.77, 0.76, and 1.11, respectively).

Raw sequencing reads were trimmed using the trimmomatic software v.0.38 [41] with the parameters "LEADING: 3 TRAILING: 3 SLIDINGWINDOW: 4: 10 MINLEN: 40". Taxonomic classification of the reads was carried out using the BLAST-based bioinformatic pipeline described elsewhere [37]. Taxons that demonstrated abundance less than 1% for all barcodes in each sample were discarded from the analysis. Spearman's rank-order correlation was used to calculate the correlation between the mapped reads' abundance per species and actual pollen abundance in the artificial pollen mixes.

Analysis results were aggregated and plotted using Python with the Pandas [42], Matplotlib [43], and Seaborn [44] packages.

3. Results

3.1. ETS Primers Design

We aimed to design primers to amplify the 5′-ETS fragment up to 600 bp in length so it could be fully sequenced using second-generation high-throughput sequencing (2 × 300 bp maximum length for the Illumina paired-end sequencing). Agarose gel-electrophoresis of the full ETS amplification products showed 1000–5000 bp length bands for most of the reference species, and a 5′-ETS fragment adjacent to 18S rRNA were Sanger sequenced.

We have aligned the 5′-ETS region of the Sanger-sequenced samples and sequences from the GenBank database to find a region suitable for the Poaceae universal ETS primers. Unfortunately, we have not found a consecutive conservative region with a length sufficient to design one universal primer for all Poaceae species. Therefore, we have chosen the least discontinuous conserved region with a degenerate sequence ETS-allF 5′-GCYDTTGGTYYHGGATG-3′ for the 5′-ETS forward primer, with a reasonable Tm range and desired amplicon size less than 600 bp. According to the alignment of the reference Poaceae species, there are seven unique sequences for the forward primer (Table 2). Therefore, only these seven variants were synthesized and then mixed equimolar as a forward primer (ETS-Fmix) for subsequent PCR of the 5′-ETS barcode, to reduce possible nonspecific amplification.

Table 2. 5′-ETS forward primers.

Name	5′-3′	Tm (Q5), °C	Binomial Species Name
ETS-allF	GCYDTTGGTYYHGGATG	53–70	
ETS-1F	GCTATTGGTCTCGGATG	59	*Poa palustris*
ETS-2F	GCTGTTGGTCTCGGATG	63	*Poa trivialis, Poa pratensis, Alopecurus pratensis, Lolium perenne, Festuca pratensis, Festuca arundinacea, Poa annua, Poa supina, Elymus repens*
ETS-3F	GCCGTTGGTCTCGGATG	66	*Phleum pratense*
ETS-4F	GCTTTTGGTCTAGGATG	56	*Bromus inermis*
ETS-5F	GCTGTTGGTTTCGGATG	61	*Briza media*
ETS-6F	GCTGTTGGTTTTGGATG	58	*Calamagrostis epigeios, Arrhenatherum elatius*
ETS-7F	GCCGTTGGTCCTGGATG	66	*Dactylis glomerata*

The PCR test with ETS-Fmix and 18S_start_R primer on DNA from herbarium specimens was successful for all species. Gel electrophoresis analysis showed one product band per species (Supplementary Figure S1) with a 200–300 bp length product for most reference species, except *Poa supina*, *Poa annua*, and *Bromus inermis*. Their PCR product length is ~500 bp for *Poa supina* and *Poa annua* and ~450 bp for *Bromus inermis*. Thus, Poaceae-specific primers have been designed to amplify the 5′-ETS fragment that fits into the desired limit of 600 bp, suitable for high-throughput sequencing on the Illumina platform.

3.2. Pollen DNA Extraction Optimization

The largest quantity of DNA extracted from 10 mg of pollen has been achieved using CTAB lysis buffer containing 0.04% SDS and 0.2 mg per sample of proteinase K. The average concentration of the extracted DNA was 16.57 and 13.62 ng $* \mu L^{-1}$ for *Poa pratense* and *Bromus inermis* pollen, respectively. The purity of the extracted DNA was in the range of 1.883–2.006 OD 260/230 and 2.095–2.142 OD 260/280 regardless of the extraction protocol. An increase in proteinase K concentration in the lysis buffer led to a lower extracted DNA yield, and an increase in lysis time led to a slight increase in the yield in most cases (Table 3). Thus, we have chosen a protocol with a lysis incubation time of 2 h in the CTAB lysis buffer with 0.04% SDS and 0.2 mg per sample proteinase K for all further extractions.

Table 3. Pollen DNA extraction lysis-buffer optimization results.

Lysis Buffer	Proteinase K mg per Sample	1 H Lysis			2 H Lysis		
		Yield, ng $* \mu L^{-1}$	260/280	260/230	Yield, ng $* \mu L^{-1}$	260/280	260/230
				P. pratense DNA			
CTAB	0.2	12.50 ± 0.66	2.06 ± 0.01	1.99 ± 0.05	13.79 ± 0.49	2.03 ± 0.03	2.00 ± 0.01
CTAB	0.4	7.34 ± 0.22	1.99 ± 0.11	1.99 ± 0.06	6.29 ± 0.29	2.01 ± 0.05	1.96 ± 0.06
CTAB + 0.04% SDS	0.2	16.57 ± 0.61	2.03 ± 0.004	1.95 ± 0.09	16.88 ± 0.19	2.05 ± 0.05	1.99 ± 0.09
CTAB + 0.04% SDS	0.4	9.51 ± 0.88	2.07 ± 0.01	1.99 ± 0.01	9.08 ± 0.96	2.04 ± 0.10	1.98 ± 0.08
CTAB + 0.4% SDS	0.2	4.41 ± 1.04	2.01 ± 0.03	1.92 ± 0.09	4.43 ± 0.93	1.97 ± 0.06	1.99 ± 0.03
CTAB + 0.4% SDS	0.4	2.91 ± 0.11	2.01 ± 0.01	1.89 ± 0.05	2.78 ± 0.6	1.99 ± 0.03	2.00 ± 0.03
				B. inermis DNA			
CTAB	0.2	9.84 ± 0.34	2.01 ± 0.03	1.89 ± 0.09	11.41 ± 0.42	2.05 ± 0.05	2.00 ± 0.01
CTAB	0.4	7.72 ± 0.44	2.01 ± 0.18	1.99 ± 0.003	8.31 ± 0.31	2.06 ± 0.09	1.89 ± 0.09
CTAB + 0.04% SDS	0.2	13.62 ± 0.64	2.04 ± 0.02	1.95 ± 0.08	15.55 ± 0.59	2.09 ± 0.10	1.99 ± 0.02
CTAB + 0.04% SDS	0.4	6.46 ± 0.53	1.99 ± 0.01	1.97 ± 0.09	9.69 ± 0.80	2.00 ± 0.02	1.99 ± 0.05
CTAB + 0.4% SDS	0.2	4.15 ± 0.53	2.00 ± 0.05	1.93 ± 0.08	5.36 ± 0.19	2.00 ± 0.05	1.99 ± 0.09
CTAB + 0.4% SDS	0.4	3.34 ± 0.91	2.01 ± 0.04	1.99 ± 0.02	3.54 ± 0.49	1.95 ± 0.05	1.98 ± 0.04

The quantity of DNA extracted from 4 × serial dilutions of pollen suspension steadily decreased along with the pollen count and became undetectable (measured by fluorometric method) starting from a sample with 2350 pollen grains (Table 4). Thus, we have chosen 10,000 pollen grains for artificial pollen mixes creation.

Table 4. Pollen DNA extraction test results.

Pollen Amount, Dilution Factor	Approximate Pollen Grain Count	Concentration, ng $* \mu L^{-1}$
10 mg	150,000	14.7 ± 0.7
1:4	37,500	3.88 ± 0.28
1:16	10,000	1.07 ± 0.12
1:64	2350	too low
1:256	600	too low
1:1024	150	too low

3.3. 5.′-ETS, ITS1, ITS2, and trnL-F Barcodes Comparison

All four barcodes were amplified from DNA of herbarium specimens of 14 reference Poaceae species, Sanger sequenced, and submitted to the GenBank database. The obtained sequences were aligned with the corresponding GenBank sequences of these barcodes and used to construct a local reference database. The length and GC content of the barcode sequences varies slightly within each marker, except for the length of 5′-ETS: 307–363 bp, GC content 29–33% for *trnL-F*; 175–509 bp, GC content 50–59% for 5′-ETS; 190–204 bp, GC content 55–67% for ITS1; 193–207 bp, GC content 59–68% for ITS2. Evaluation of intra- and interspecific variability showed that while all barcodes have low intraspecific distances, the 5′-ETS barcode has the highest interspecific distance closely followed by ITS2 (Table 5). Plastome barcode *trnL-F* showed the lowest intra- and interspecific distances compared to the nuclear barcodes.

Table 5. Intra- and interspecific distance statistics.

Barcode	Interspecific			Intraspecific		
	Minimum	Maximum	Median	Minimum	Maximum	Median
ITS1	0.0000	0.4112	0.1514	0.0000	0.0756	0.0000
ITS2	0.0000	0.4549	0.2345	0.0000	0.0726	0.0000
ETS	0.0000	0.6142	0.2897	0.0000	0.0431	0.0000
trnL-F	0.0000	0.1278	0.0698	0.0000	0.0254	0.0000

However, the difference between the barcode sequences is low for the species of the same genus (*Poa* in this study). For example, all barcodes of *Poa annua* and *Poa supina* have identical sequences, which means that these species will be impossible to distinguish. Other possible misidentification sources with barcoding gap less than 1% could be *Arrhenatherum elatius* vs. *Calamagrostis epigeios* and *Alopecurus pratensis*, *Lolium perenne* vs. *Festuca pratensis* (barcoding gap equals −0.008, 0.001, and −0.0001, respectively) for the ITS1 barcode; *Poa pratensis* vs. *Phleum pratense* (−0.0142), *Calamagrostis epigeios* vs. *Briza media*, *Poa pratensis*, and *Phleum pratense* (−0.002, 0.007, and 0.008, respectively) for ITS2; *Poa annua* vs. *Poa pratensis*, *Alopecurus pratensis*, and *Phleum pratense*, *Lolium perenne* vs. *Festuca pratensis* (−0.004, 0.009, 0.009 and 0.0000, respectively) for *trnL-F*. Barcoding gaps for all four barcodes are present in Figure 2. Additionally, barcode intra- and interspecific distances per species are present in Supplementary Figure S2.

3.4. Metabarcoding Analysis of the Artificial Pollen Mixes

Using the optimized protocol for pollen DNA extraction, we have obtained DNA of 1.2–1.5 ng $* \mu L^{-1}$ from artificial pollen mixes (am). The quality of obtained DNA was the same as we have obtained for the *Poa pratense* and *Bromus inermis* single-species pollen at the optimization step. Amplification was successful for all barcodes and all samples of artificial pollen mixes, though the amplification efficiency differs significantly between the barcodes and decreases as follows: ITS2 > ITS1 > 5′-ETS > *trnL-F* (confidence intervals for amplified barcode concentrations are 20.02 ± 4.44, 13.04 ± 3.47, 8.32 ± 1.69, and 0.43 ± 0.07 ng $* \mu L^{-1}$, respectively).

The species composition of the artificial pollen mixes determined by HTS analysis is congruent with the actual pollen species content in 10 out of 18 artificial mixes. The most frequent erroneous identification has occurred in mixes containing either *Lolium* or *Festuca* pollen. In these mixes, the erroneous presence of *Lolium*, where only *Festuca* is present, and vice versa, was detected. However, the abundance of the erroneously identified species is often low (less or close to 1%). This issue is common for all barcodes in the study, especially for the plastome *trnL-F* barcode (1.7–4.9% *Festuca/Lolium* errors). Nuclear barcodes show fewer errors of this type, minimal for ITS2, for which abundance of erroneously identified *Lolium* or *Festuca* is close to 0 in all cases.

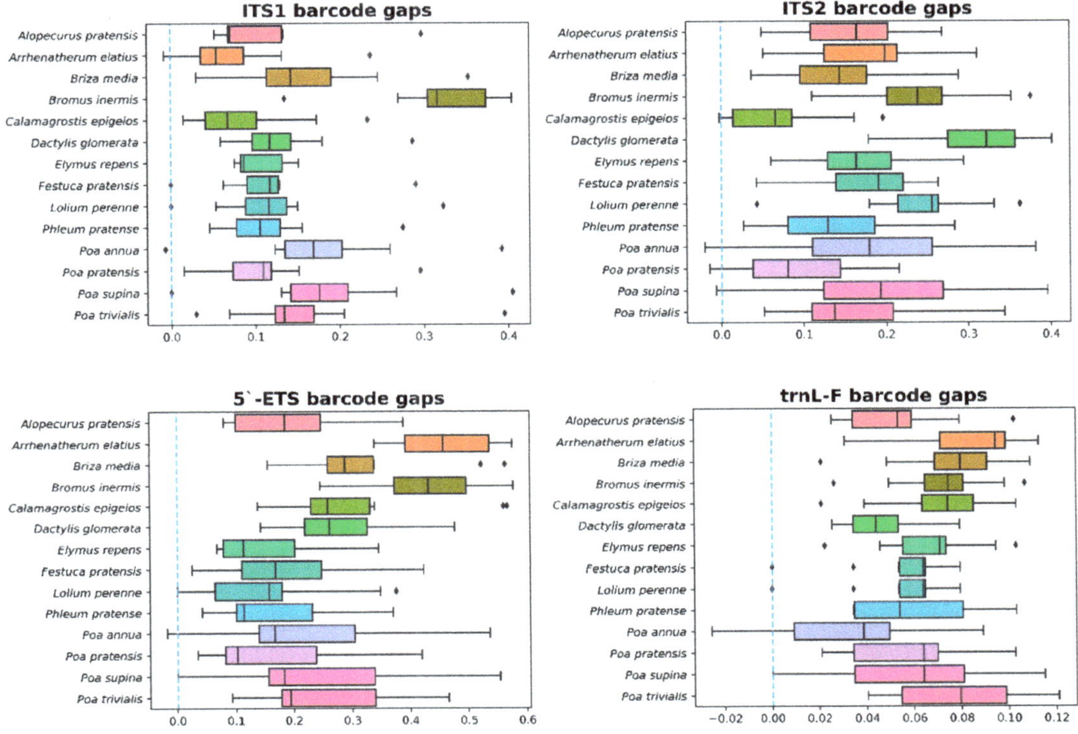

Figure 2. Barcoding gaps for all four DNA barcodes in the study per species.

Spearman's correlation coefficient between HTS determined the abundance and true abundance of each species in the artificial pollen mix decreases as follows: ITS1 > ITS2 > trnL-F > 5'-ETS (0.8, 0.78, 0.63, and 0.59, respectively). For the 5'-ETS barcode, the *Bromus inermis* abundance in all the mixes is significantly lower than for the other barcodes and actual mix composition (0.41–3.16%). As the complexity of the artificial pollen mix increases, the abundance of the detected 5'-ETS of *Bromus inermis* decreases. The low representation of the 5'-ETS barcode of *Bromus inermis* is most likely related to the length of the amplified 5'-ETS fragment (444 bp vs. 220 bp in average for other reference species in the database, except for 509 bp of *Poa supina* and *Poa annua*), which could lead to a lower amplification efficiency of the 5'-ETS of *Bromus inermis* when in the mix with other species.

Overall, the nuclear barcodes proved to be the most effective in the amplification and species classification. The plastome *trnL-F* barcode has demonstrated a lower amplification efficiency and a higher rate of erroneously identified species than the nuclear barcodes. Though the mix composition could be determined well qualitatively by metabarcoding analysis, quantitative results for each pollen species, determined by read counts, is rarely congruent with the actual abundance of pollen species in the mix. Most of the congruent quantitative results were achieved using ITS1 and ITS2 barcodes (Figure 3).

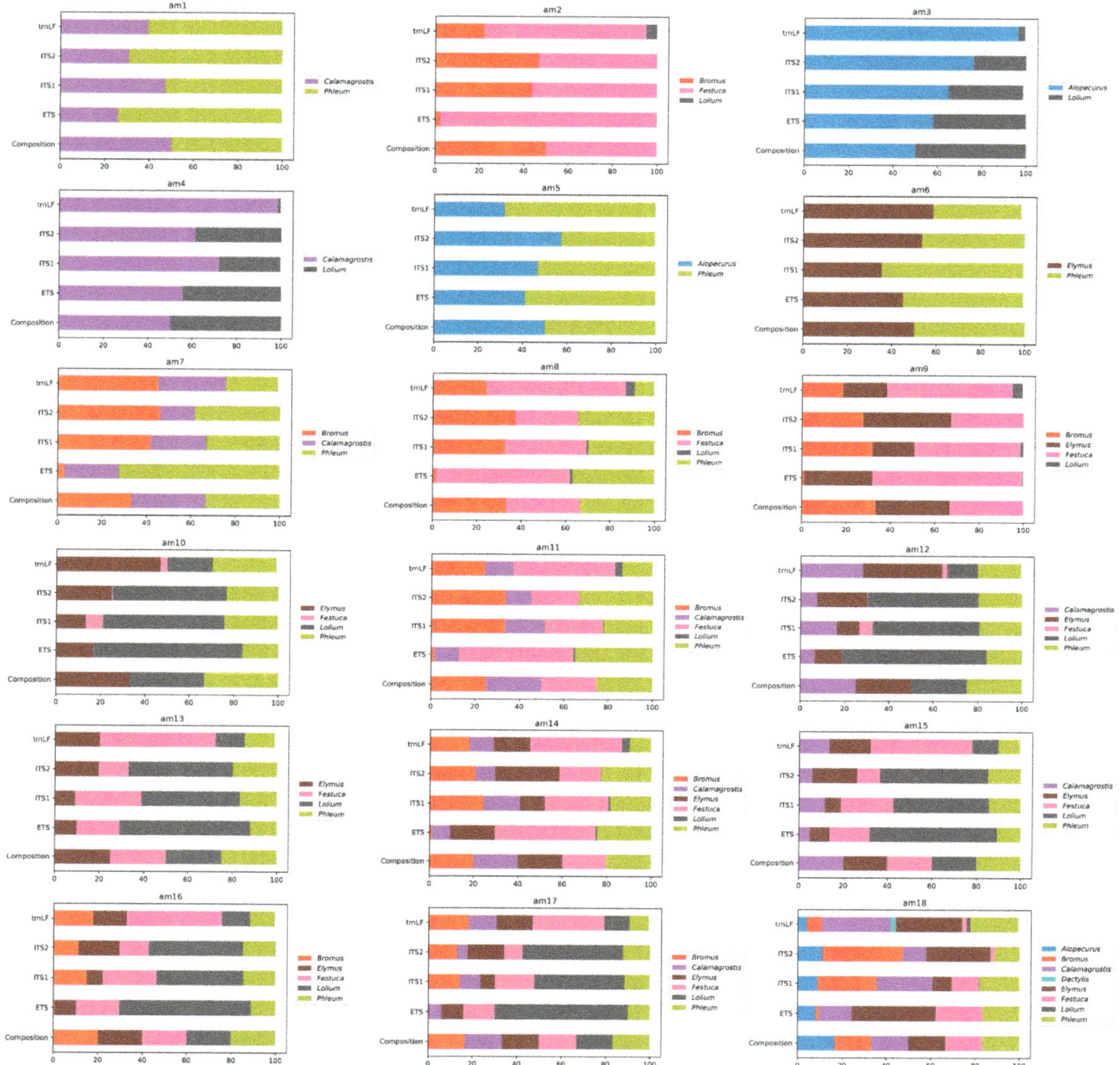

Figure 3. Metabarcoding results for the artificial pollen mixes.

4. Discussion

Previously designed primers for ETS amplification for some of the Poaceae genera and species could amplify a fragment of ~500–900 bp [30,45,46], but we aimed to obtain a shorter amplicon for a broad spectrum of Poaceae species of diverse genera that could be sequenced entirely using HTS and is suitable for metabarcoding analysis. Unfortunately, we could not find a region for a universal forward primer for all Poaceae species in the study due to the lack of long consecutive conservative regions. However, we have found a region that allowed us to design 7 primers, an equimolar mix of which proved to be efficient for specific amplification of all species in the study. New Poaceae-specific primers (degenerate 5′-ETS forward and universal 18S reverse) amplify the 5′-ETS fragment less than 600 bp, which is ~300 bp shorter than other published primers could amplify for the same species in this study.

The effectiveness of the protocols for sample preparation for HTS highly depends, among other things, on the quality and quantity of the DNA. Various methods for pollen DNA extraction involve using commercial solutions such as column-based and DNA binding with magnetic beads purification methods after preliminary homogenization of the pollen sample with metal beads [47,48]. We propose a protocol based on a classical CTAB-lysis extraction method [49] with modifications that achieve results similar to commercial kits for pollen DNA extraction. The addition of a small amount of SDS, which helped increase the DNA extraction efficiency from fossil pollen of *Abies* spp. from Pleistocene peat [42], showed increased extraction efficiency from Poaceae pollen as well. The DNA yield from the samples is associated with the lysis efficiency of the pollen grains. In different plant species, the structure of the shells of the pollen grains can vary greatly. The use of methods of mechanical destruction, such as grinding with metal balls [47] or the use of a bullet blender [24], increases the DNA yield. In this case, the yield becomes comparable with the one we have achieved using tubes with fine sand for grinding pollen.

Festuca and *Lolium* genera form a phylogenetic complex where *Lolium* is a subgroup of the *Festuca* genus according to several phylogenetic studies that employed restriction fragment length polymorphism (RFLP), random amplified polymorphic DNA (RAPD), as well as rDNA (ITS region) and cpDNA sequences data for analysis [50,51]. It was also pointed out that *Festuca pratensis* is the most closely related to *Lolium* species in *Festuca*/*Lolium* complex, and suggested that the closeness of *Festuca pratensis* ITS to *Lolium* ITS sequences could represent a reticulate evolutionary event [50,51]. The closeness of these species is also supported by the fact that *Festuca* species readily cross with *Lolium* species in nature or synthetically form *Festulolium* hybrids (e.g., *F. pratensis* × *L. perenne*, or *L. multiflorum* × *F. pratensis*) [52]. Furthermore, species of these genera display a high level of sequence similarity for orthologous genes (>91% identity) and conservation of gene family content, as showed by the transcriptome analysis [53]. Several plastome barcodes have been used to untangle the relationships within complex and construct phylogenetic trees [54,55], though nuclear barcode ITS2 showed better results than plastome barcodes [56]. We have found that the 5′-ETS barcode has 6 SNPs, ITS1 has 3 SNPs, and ITS2 has 8 SNPs between *Festuca pratensis* and *Lolium perenne* sequences. In contrast, *trnL-F* 5′-ETS has only 2 short insertions (4 and 5 bp long) and no SNP in *Lolium perenne* sequences compared to *Festuca pratensis*. Thus, nuclear barcodes resolve these species better than the *trnL-F* plastome barcode, and ITS2 shows the least errors in distinguishing these species due to more SNPs than other nuclear barcodes have.

Plant pollen taxon identification using the *trnL* barcode showed promising results and a pollen-to-read quantitative correlation [22]. However, it was also shown that this barcode could give incorrect taxon predictions, e.g., *Lolium/Festuca* and *Arrenatherum/Poa* [24]. In this study, we have assessed the taxon identification capabilities of the adjacent plastome region *trnL-F* intergenic spacer, but it has also shown a high error rate in resolving *Lolium* and *Festuca* species. Moreover, we have observed a low amplification efficiency of this barcode for pollen-extracted DNA. The low efficiency of amplification of the plastome area may be caused by the fact that during the development of the pollen grain, chloroplasts, which can be found in both vegetative and generative cells, are destroyed, and thus cpDNA can be severely damaged [57].

5. Conclusions

ITS1 and ITS2 proved to be the most effective qualitatively and quantitatively, and we recommend using them for Poaceae pollen analysis. Another nuclear barcode, 5′-ETS, showed good qualitative results, but due to variability in fragment length, failed to show good quantitative results. We suggest that 5′-ETS could be successfully used in phylogenetic studies or direct PCR detection of certain species due to the highest genetic distance between species among all barcodes in the study, if not for the metabarcoding of pollen. Plastome *trnL-F* showed the lowest amplification efficiency, intra- and interspecific distances, and the highest error rate for pollen identification, especially in resolving *Lolium* and *Festuca* sequences. In general, we can say that the barcodes used in this study allow efficient amplification and metabarcoding analysis of Poaceae pollen of various genera, and we suggest that nuclear barcodes are better for this task than plastome ones.

Supplementary Materials: The following supporting information can be downloaded at: https://www.mdpi.com/article/10.3390/d14030191/s1, Figure S1: PCR test results with primers ETS-Fmix + 18S_start_R; Figure S2: Barcodes' intra- and interspecific distances per species; Table S1: Accession numbers and basic characteristics of DNA barcodes used to create local reference database for grass pollen metabarcoding.

Author Contributions: Conceptualization, D.O.O. and A.A.K.; methodology, A.A.K. and A.S.S.; software, A.S.K. and D.O.O.; validation, D.O.O. and A.A.K.; formal analysis, A.S.K. and D.O.O.; investigation, A.A.K., D.O.O., O.V.C., S.V.P. and E.E.S.; resources, A.A.K. and E.E.S.; data curation, A.S.K., A.A.K. and D.O.O.; writing—original draft preparation, D.O.O. and A.A.K.; writing—review and editing, D.O.O., A.A.K. and E.E.S.; visualization, D.O.O.; supervision, E.E.S.; project administration, E.E.S.; funding acquisition, E.E.S. All authors have read and agreed to the published version of the manuscript.

Funding: This research was funded by Russian Foundation for Basic Research, project 19-05-50035.

Institutional Review Board Statement: Not applicable.

Informed Consent Statement: Not applicable.

Data Availability Statement: All the sequenced data are deposited in the public GenBank database. Accession numbers are present in Supplementary Table S1.

Acknowledgments: The authors would like to thank Maria D. Logacheva (Skolkovo Institute of Science and Technology, Moscow, Russia) for valuable advice on the high-throughput sequencing procedures and access to the Illumina MiSeq platform to perform the sequencing for this study. The authors would also like to thank Margarita V. Remizowa and Dmitry D. Sokoloff (Lomonosov Moscow State University, Moscow, Russia) for help with the morphological identification of the collected plants.

Conflicts of Interest: The authors declare no conflict of interest.

References

1. García-Mozo, H. Poaceae Pollen as the Leading Aeroallergen Worldwide: A Review. *Allergy* **2017**, *72*, 1849–1858. [CrossRef] [PubMed]
2. Damialis, A.; Traidl-Hoffmann, C.; Treudler, R. Climate Change and Pollen Allergies. In *Biodiversity and Health in the Face of Climate Change*; Marselle, M.R., Stadler, J., Korn, H., Irvine, K.N., Bonn, A., Eds.; Springer International Publishing: Cham, Switzerland, 2019; pp. 47–66. ISBN 978-3-030-02318-8.
3. D'Amato, G.; Cecchi, L.; Bonini, S.; Nunes, C.; Annesi-Maesano, I.; Behrendt, H.; Liccardi, G.; Popov, T.; Cauwenberge, P.V. Allergenic Pollen and Pollen Allergy in Europe. *Allergy* **2007**, *62*, 976–990. [CrossRef]
4. Mayevsky, P.F. *Flora of the Middle Zone of the European Part of Russia*; KMK Scientific Press Ltd.: Moscow, Russia, 2014.
5. Tree and Plant Allergy Info for Research—Allergen and Botanic Reference Library. Available online: http://www.pollenlibrary.com/ (accessed on 16 March 2020).
6. Erdtman, G. *Pollen Morphology and Plant Taxonomy: Angiosperms*; E.J. Brill: Leiden, The Netherlands, 1986; ISBN 978-90-04-08122-2.
7. Joly, C.; Barillé, L.; Barreau, M.; Mancheron, A.; Visset, L. Grain and Annulus Diameter as Criteria for Distinguishing Pollen Grains of Cereals from Wild Grasses. *Rev. Palaeobot. Palynol.* **2007**, *146*, 221–233. [CrossRef]
8. Ünsal, S.G.; Çiftçi, Y.Ö.; Eken, B.U.; Velioğlu, E.; Di Marco, G.; Gismondi, A.; Canini, A. Intraspecific Discrimination Study of Wild Cherry Populations from North-Western Turkey by DNA Barcoding Approach. *Tree Genet. Genomes* **2019**, *15*, 16. [CrossRef]
9. Gismondi, A.; Di Marco, G.; Martini, F.; Sarti, L.; Crespan, M.; Martínez-Labarga, C.; Rickards, O.; Canini, A. Grapevine Carpological Remains Revealed the Existence of a Neolithic Domesticated Vitis Vinifera L. Specimen Containing Ancient DNA Partially Preserved in Modern Ecotypes. *J. Archaeol. Sci.* **2016**, *69*, 75–84. [CrossRef]
10. Techen, N.; Parveen, I.; Pan, Z.; Khan, I.A. DNA Barcoding of Medicinal Plant Material for Identification. *Curr. Opin. Biotechnol.* **2014**, *25*, 103–110. [CrossRef] [PubMed]
11. Bruni, I.; De Mattia, F.; Galimberti, A.; Galasso, G.; Banfi, E.; Casiraghi, M.; Labra, M. Identification of Poisonous Plants by DNA Barcoding Approach. *Int. J. Leg. Med.* **2010**, *124*, 595–603. [CrossRef]

12. Bruni, I.; Galimberti, A.; Caridi, L.; Scaccabarozzi, D.; De Mattia, F.; Casiraghi, M.; Labra, M. A DNA Barcoding Approach to Identify Plant Species in Multiflower Honey. *Food Chem.* **2015**, *170*, 308–315. [CrossRef]

13. Prosser, S.W.J.; Hebert, P.D.N. Rapid Identification of the Botanical and Entomological Sources of Honey Using DNA Metabarcoding. *Food Chem.* **2017**, *214*, 183–191. [CrossRef]

14. Taylor, H.R.; Harris, W.E. An Emergent Science on the Brink of Irrelevance: A Review of the Past 8 Years of DNA Barcoding. *Mol. Ecol. Resour.* **2012**, *12*, 377–388. [CrossRef]

15. Coissac, E.; Riaz, T.; Puillandre, N. Bioinformatic Challenges for DNA Metabarcoding of Plants and Animals. *Mol. Ecol.* **2012**, *21*, 1834–1847. [CrossRef]

16. CBOL Plant Working Group A DNA Barcode for Land Plants. *Proc. Natl. Acad. Sci. USA* **2009**, *106*, 12794–12797. [CrossRef] [PubMed]

17. Shneyer, V.S.; Rodionov, A.V. Plant DNA Barcodes. *Biol. Bull Rev.* **2019**, *9*, 295–300. [CrossRef]

18. Beng, K.C.; Tomlinson, K.W.; Shen, X.H.; Surget-Groba, Y.; Hughes, A.C.; Corlett, R.T.; Slik, J.W.F. The Utility of DNA Metabarcoding for Studying the Response of Arthropod Diversity and Composition to Land-Use Change in the Tropics. *Sci. Rep.* **2016**, *6*, 24965. [CrossRef] [PubMed]

19. Bell, K.L.; de Vere, N.; Keller, A.; Richardson, R.T.; Gous, A.; Burgess, K.S.; Brosi, B.J. Pollen DNA Barcoding: Current Applications and Future Prospects. *Genome* **2016**, *59*, 629–640. [CrossRef]

20. Bell, K.L.; Fowler, J.; Burgess, K.S.; Dobbs, E.K.; Gruenewald, D.; Lawley, B.; Morozumi, C.; Brosi, B.J. Applying Pollen DNA Metabarcoding to the Study of Plant–Pollinator Interactions. *Appl. Plant Sci.* **2017**, *5*, 1600124. [CrossRef]

21. Bell, K.L.; Burgess, K.S.; Botsch, J.C.; Dobbs, E.K.; Read, T.D.; Brosi, B.J. Quantitative and Qualitative Assessment of Pollen DNA Metabarcoding Using Constructed Species Mixtures. *Mol. Ecol.* **2019**, *28*, 431–455. [CrossRef]

22. Baksay, S.; Pornon, A.; Burrus, M.; Mariette, J.; Andalo, C.; Escaravage, N. Experimental Quantification of Pollen with DNA Metabarcoding Using ITS1 and TrnL. *Sci. Rep.* **2020**, *10*, 1–9. [CrossRef]

23. Peterson, P.M.; Romaschenko, K.; Soreng, R.J. A Laboratory Guide for Generating DNA Barcodes in Grasses: A Case Study of Leptochloa s.l. (Poaceae: Chloridoideae). *Webbia* **2014**, *69*, 1–12. [CrossRef]

24. Kraaijeveld, K.; de Weger, L.A.; García, M.V.; Buermans, H.; Frank, J.; Hiemstra, P.S.; Dunnen, J.T. den Efficient and Sensitive Identification and Quantification of Airborne Pollen Using Next-Generation DNA Sequencing. *Mol. Ecol. Resour.* **2015**, *15*, 8–16. [CrossRef]

25. Naciri, Y.; Caetano, S.; Salamin, N. Plant DNA Barcodes and the Influence of Gene Flow. *Mol. Ecol. Resour.* **2012**, *12*, 575–580. [CrossRef] [PubMed]

26. Columbus, J.; Cerros-Tlatilpa, R.; Kinney, M.; Siqueiros-Delgado, M.E.; Bell, H.; Griffith, M.; Refulio-Rodriguez, N. Phylogenetics of Chloridoideae (Gramineae): A Preliminary Study Based on Nuclear Ribosomal Internal Transcribed Spacer and Chloroplast TrnL–F Sequences. *Aliso J. Syst. Evol. Bot.* **2007**, *23*, 565–579. [CrossRef]

27. Lloyd, K.; Hunter, A.; Orlovich, D.; Draffin, S.; Stewart, A.; Lee, W. Phylogeny and Biogeography of Endemic Festuca (Poaceae) from New Zealand Based on Nuclear (ITS) and Chloroplast (TrnL–TrnF) Nucleotide Sequences. *Aliso J. Syst. Evol. Bot.* **2007**, *23*, 406–419. [CrossRef]

28. Da Silva, L.N.; Essi, L.; Iganci, J.R.V.; Souza-Chies, T.T.D. Advances in the Phylogeny of the South American Cool-Season Grass Genus Chascolytrum (Poaceae, Pooideae): A New Infrageneric Classification. *Bot. J. Linn. Soc.* **2019**, *192*, 97–120. [CrossRef]

29. Wang, A.; Gopurenko, D.; Wu, H.; Lepschi, B. Evaluation of Six Candidate DNA Barcode Loci for Identification of Five Important Invasive Grasses in Eastern Australia. *PLoS ONE* **2017**, *12*, e0175338. [CrossRef]

30. Alonso, A.; Bull, R.D.; Acedo, C.; Gillespie, L.J. Design of Plant-Specific PCR Primers for the ETS Region with Enhanced Specificity for Tribe Bromeae and Their Application to Other Grasses (Poaceae). *Botany* **2014**, *92*, 693–699. [CrossRef]

31. Logacheva, M.D.; Valiejo-Roman, C.M.; Degtjareva, G.V.; Stratton, J.M.; Downie, S.R.; Samigullin, T.H.; Pimenov, M.G. A Comparison of NrDNA ITS and ETS Loci for Phylogenetic Inference in the Umbelliferae: An Example from Tribe Tordylieae. *Mol. Phylogenetics Evol.* **2010**, *57*, 471–476. [CrossRef]

32. Cai, Z.-M.; Zhang, Y.-X.; Zhang, L.-N.; Gao, L.-M.; Li, D.-Z. Testing Four Candidate Barcoding Markers in Temperate Woody Bamboos (Poaceae: Bambusoideae). *J. Syst. Evol.* **2012**, *50*, 527–539. [CrossRef]

33. Su, X.; Liu, Y.P.; Chen, Z.; Chen, K.L. Evaluation of Candidate Barcoding Markers in Orinus (Poaceae). *Genet. Mol. Res. GMR* **2016**, *15*. [CrossRef]

34. Krinitsina, A.A.; Sizova, T.V.; Zaika, M.A.; Speranskaya, A.S.; Sukhorukov, A.P. A Rapid and Cost-Effective Method for DNA Extraction from Archival Herbarium Specimens. *Biochemistry* **2015**, *80*, 1478–1484. [CrossRef]

35. Omelchenko, D.O.; Speranskaya, A.S.; Ayginin, A.A.; Khafizov, K.; Krinitsina, A.A.; Fedotova, A.V.; Pozdyshev, D.V.; Shtratnikova, V.Y.; Kupriyanova, E.V.; Shipulin, G.A.; et al. Improved Protocols of ITS1-Based Metabarcoding and Their Application in the Analysis of Plant-Containing Products. *Genes* **2019**, *10*, 122. [CrossRef] [PubMed]

36. Taberlet, P.; Gielly, L.; Pautou, G.; Bouvet, J. Universal Primers for Amplification of Three Non-Coding Regions of Chloroplast DNA. *Plant Mol. Biol.* **1991**, *17*, 1105–1109. [CrossRef] [PubMed]

37. Speranskaya, A.S.; Khafizov, K.; Ayginin, A.A.; Krinitsina, A.A.; Omelchenko, D.O.; Nilova, M.V.; Severova, E.E.; Samokhina, E.N.; Shipulin, G.A.; Logacheva, M.D. Comparative Analysis of Illumina and Ion Torrent High-Throughput Sequencing Platforms for Identification of Plant Components in Herbal Teas. *Food Control* **2018**, *93*, 315–324. [CrossRef]

38. Tamura, K.; Stecher, G.; Kumar, S. MEGA11: Molecular Evolutionary Genetics Analysis Version 11. *Mol. Biol. Evol.* **2021**, *38*, 3022–3027. [CrossRef] [PubMed]

39. Tamura, K. Estimation of the Number of Nucleotide Substitutions When There Are Strong Transition-Transversion and G+C-Content Biases. *Mol. Biol. Evol.* **1992**, *9*, 678–687. [CrossRef] [PubMed]

40. Kimura, M. A Simple Method for Estimating Evolutionary Rates of Base Substitutions through Comparative Studies of Nucleotide Sequences. *J. Mol. Evol.* **1980**, *16*, 111–120. [CrossRef]

41. Bolger, A.M.; Lohse, M.; Usadel, B. Trimmomatic: A Flexible Trimmer for Illumina Sequence Data. *Bioinformatics* **2014**, *30*, 2114–2120. [CrossRef]

42. McKinney, W. Data Structures for Statistical Computing in Python. In Proceedings of the 9th Python in Science Conference, Austin, TX, USA, 28 June–3 July 2010; Volume 445, pp. 56–61.

43. Hunter, J.D. Matplotlib: A 2D Graphics Environment. *Comput. Sci. Eng.* **2007**, *9*, 90–95. [CrossRef]

44. Waskom, M.L. Seaborn: Statistical Data Visualization. *J. Open Source Softw.* **2021**, *6*, 3021. [CrossRef]

45. Gillespie, L.J.; Soreng, R.J.; Paradis, M.; Bull, R.D. *Phylogeny and Reticulation in Subtribe Poinae and Related Subtribes (Poaceae) Based on NrITS, ETS, and TrnTLF Data. Diversity, Phylogeny, and Evolution in the Monocotyledons*; Aarhus University Press: Aarhus, Denmark, 2010; p. 29.

46. Consaul, L.L.; Gillespie, L.J.; Waterway, M.J. Evolution and Polyploid Origins in North American Arctic Puccinellia (Poaceae) Based on Nuclear Ribosomal Spacer and Chloroplast DNA Sequences. *Am. J. Bot.* **2010**, *97*, 324–336. [CrossRef]

47. Leontidou, K.; Vernesi, C.; De Groeve, J.; Cristofolini, F.; Vokou, D.; Cristofori, A. DNA Metabarcoding of Airborne Pollen: New Protocols for Improved Taxonomic Identification of Environmental Samples. *Aerobiologia* **2018**, *34*, 63–74. [CrossRef]

48. Ghitarrini, S.; Pierboni, E.; Rondini, C.; Tedeschini, E.; Tovo, G.R.; Frenguelli, G.; Albertini, E. New Biomolecular Tools for Aerobiological Monitoring: Identification of Major Allergenic Poaceae Species through Fast Real-Time PCR. *Ecol. Evol.* **2018**, *8*, 3996–4010. [CrossRef] [PubMed]

49. Doyle, J.J.; Doyle, J.L. A Rapid DNA Isolation Procedure for Small Quantities of Fresh Leaf Tissue. *Phytochem. Bull.* **1987**, *19*, 11–15.

50. Charmet, G.; Ravel, C.; Balfourier, F. Phylogenetic Analysis in the Festuca-Lolium Complex Using Molecular Markers and ITS RDNA. *Theor. Appl. Genet.* **1997**, *94*, 1038–1046. [CrossRef]

51. Gaut, B.S.; Tredway, L.P.; Kubik, C.; Gaut, R.L.; Meyer, W. Phylogenetic Relationships and Genetic Diversity among Members of TheFestuca-Lolium Complex (Poaceae) Based on ITS Sequence Data. *Pl. Syst. Evol.* **2000**, *224*, 33–53. [CrossRef]

52. Ghesquière, M.; Humphreys, M.W.; Zwierzykowski, Z. Festulolium. In *Fodder Crops and Amenity Grasses*; Boller, B., Posselt, U.K., Veronesi, F., Eds.; Handbook of Plant Breeding; Springer: New York, NY, USA, 2010; pp. 288–311. ISBN 978-1-4419-0760-8.

53. Czaban, A.; Sharma, S.; Byrne, S.L.; Spannagl, M.; Mayer, K.F.; Asp, T. Comparative Transcriptome Analysis within the Lolium/Festuca Species Complex Reveals High Sequence Conservation. *BMC Genom.* **2015**, *16*, 249. [CrossRef]

54. Loera-Sánchez, M.; Studer, B.; Kölliker, R. DNA Barcode TrnH-PsbA Is a Promising Candidate for Efficient Identification of Forage Legumes and Grasses. *BMC Res. Notes* **2020**, *13*, 35. [CrossRef]

55. Cheng, Y.; Zhou, K.; Humphreys, M.W.; Harper, J.A.; Ma, X.; Zhang, X.; Yan, H.; Huang, L. Phylogenetic Relationships in the Festuca-Lolium Complex (Loliinae; Poaceae): New Insights from Chloroplast Sequences. *Front. Ecol. Evol.* **2016**, *4*, 89. [CrossRef]

56. Wu, S.; Yin, L.; Deng, Z.; Chen, Q.; Fu, Y.; Xue, H. Using DNA Barcoding to Identify the Genus Lolium. *Not. Bot. Horti Agrobot. Cluj-Napoca* **2015**, *43*, 536–541. [CrossRef]

57. Sodmergen; Suzuki, T.; Kawano, S.; Nakamura, S.; Tano, S.; Kuroiwa, T. Behavior of Organelle Nuclei (Nucleoids) in Generative and Vegetative Cells during Maturation of Pollen InLilium Longiflorum AndPelargonium Zonale. *Protoplasma* **1992**, *168*, 73–81. [CrossRef]

Article

Patterns of Herbivory in Neotropical Forest Katydids as Revealed by DNA Barcoding of Digestive Tract Contents

Christine M. Palmer [1,*], **Nicole L. Wershoven** [1], **Sharon J. Martinson** [2,3,4], **Hannah M. ter Hofstede** [2,3,5], **W. John Kress** [2,6] and **Laurel B. Symes** [2,3,4]

[1] Natural Sciences Department, Castleton University, 233 South Street, Castleton, VT 05735, USA; nicole.wershoven@med.uvm.edu

[2] Department of Biological Sciences, Dartmouth College, 78 College Street, Hanover, NH 03755, USA; wyomingirl@gmail.com (S.J.M.); hannah.ter.hofstede@dartmouth.edu (H.M.t.H.); kressj@si.edu (W.J.K.); symes@cornell.edu (L.B.S.)

[3] Smithsonian Tropical Research Institute, Balboa, Ancón, Apartado 0843-03092, Panama

[4] K. Lisa Yang Center for Conservation Bioacoustics, Cornell Lab of Ornithology, Cornell University, 159 Sapsucker Woods Road, Ithaca, NY 14850, USA

[5] Graduate Program in Ecology, Evolution, Environment and Society, Dartmouth College, 64 College Street, Suite 102, Hanover, NH 03755, USA

[6] Department of Botany, National Museum of Natural History, Smithsonian Institution, P.O. Box 37012, Washington, DC 20013, USA

* Correspondence: christine.palmer@castleton.edu

Abstract: Many well-studied animal species use conspicuous, repetitive signals that attract both mates and predators. Orthopterans (crickets, katydids, and grasshoppers) are renowned for their acoustic signals. In Neotropical forests, however, many katydid species produce extremely short signals, totaling only a few seconds of sound per night, likely in response to predation by acoustically orienting predators. The rare signals of these katydid species raises the question of how they find conspecific mates in a structurally complex rainforest. While acoustic mechanisms, such as duetting, likely facilitate mate finding, we test the hypothesis that mate finding is further facilitated by colocalization on particular host plant species. DNA barcoding allows us to identify recently consumed plants from katydid stomach contents. We use DNA barcoding to test the prediction that katydids of the same species will have closely related plant species in their stomach. We do not find evidence for dietary specialization. Instead, katydids consumed a wide mix of plants within and across the flowering plants (27 species in 22 genera, 16 families, and 12 orders) with particular representation in the orders Fabales and Laurales. Some evidence indicates that katydids may gather on plants during a narrow window of rapid leaf out, but additional investigations are required to determine whether katydid mate finding is facilitated by gathering at transient food resources.

Keywords: trophic interactions; diet specialization; DNA barcoding; bush cricket; Barro Colorado Island; Panama; katydid; tropical trees

Citation: Palmer, C.M.; Wershoven, N.L.; Martinson, S.J.; ter Hofstede, H.M.; Kress, W.J.; Symes, L.B. Patterns of Herbivory in Neotropical Forest Katydids as Revealed by DNA Barcoding of Digestive Tract Contents. *Diversity* **2022**, *14*, 152. https://doi.org/10.3390/d14020152

Academic Editor: Luc Legal

Received: 27 January 2022
Accepted: 17 February 2022
Published: 21 February 2022

1. Introduction

Animals experience intense pressure to find food and mates while avoiding predation [1–3]. For many species, mate finding relies on signals that allow one sex to locate the other, and can integrate a variety of sensory modalities including visual, acoustic, electrical, and olfactory channels [4]. The same signals that increase detectability by mates can increase detectability by predators as well, and predators can impose intense selection both on the individuals doing the signaling, as well as on the individuals who are searching for mates [5–8]. Consequently, there is often intense selection for strategies that facilitate mate finding while minimizing exposure to predation.

Many Orthopterans, including crickets, katydids, and grasshoppers, are known for their conspicuous signals [9,10]. These species often use energetically expensive acoustic signals to attract mates [11,12], with signals repeated again and again for a large portion of the day or night. Orthopteran signals often attract predators as well [13,14]. In the Neotropics, eavesdropping gleaning bats such as *Trachops cirrhosis* and *Lophostoma silvicolum* hunt katydids and other small animals by eavesdropping on the sounds that they produce [15–17]. Likely as a result of this acoustically targeted predation, many Neotropical forest katydids produce vanishingly little sound. In a survey of 16 phaneropterine katydid species (Tettigoniidae), none produced more than five seconds of sound per night, with most species emitting infrequent calls of 20–200 ms in duration [18]. Many of these katydid species have calls with carrier frequencies that are in the high audible or ultrasonic range [19], characteristics that would also cause a call to attenuate quickly [20], particularly in dense vegetation. Beyond Phaneropterinae, other katydid subfamilies also have instances of low calling rates in Neotropical forests, with some conocephaline and pseudophylline species producing less than 30 s of sound per night [21]. While many Neotropical forest katydids produce little sound, there are examples of species that call substantially more, such as *Ischnomela pulchripennis*, a species that is associated with spiny bromeliads that provide protection from predator attack [17]. However, with key exceptions [22], relatively little is known about possible associations between particular species of katydids and their host plants and how those associations may interact with signal structure or signaling strategy.

For species that produce so few acoustic signals, one of the inescapable questions is how they are able to encounter mates in a dense and structurally complex rainforest. There are multiple mechanisms that could facilitate mate finding in species that produce only seconds of long-distance advertisement signal each night. Phaneropterine katydids engage in mating duets, where the female replies to the male signal with a short tick, providing information about her presence and receptivity (reviewed in [23]). Duetting alone may be enough to allow katydids to find each other, if the female replies incentivize short-term risk taking and elevate male signaling rate. Males will also produce ticks that resemble the female reply, likely as a competitive mechanism that confounds interception by other males [24]. If males are producing sounds to jam other males, it suggests that it is common for multiple individuals to be present and interacting during mating, again reflecting effective strategies for co-localizing with conspecifics, rather than rare chance encounters between pairs in the forest.

An additional mechanism that could further facilitate mate encounter while minimizing conspicuousness to predators is host plant specificity. In some species, animals mate on or near their food resources, streamlining the encounter process [25,26]. Animals that find mates on a food resource can reduce travel time and associated predation risk, and species that gather at food resources may also be able to use less conspicuous signals that enable them to compete for nearby mates without attracting distant predators. If katydids gather on particular host plants and search for mates where they are gathered, this food-based aggregation strategy could dramatically lower the hurdle to mate finding, reducing female travel costs and predation risk and allowing effective pairing, even with rare, short duration, rapidly attenuating signals.

Host plant specialization provides opportunities and challenges. Mature tropical forests contain a diversity of vegetation, much of which is heavily protected by secondary compounds and chemically defended against most herbivores [27–29]. Herbivores respond to plant defenses with a diversity of strategies, including extreme host generalization, where they eat small quantities of many plants to minimize the impacts of each type of toxin [30,31], or host plant specialization, where they evolve to tolerate or even repurpose a particular type of chemical defense [32,33]. In most habitats, herbivorous Orthopterans consume a wide range of plants [34]. However, there are cases where Orthopterans specialize on a particular food source, with some displaying strong associations and genetic differentiation based on diet [35,36], while others demonstrate strong preferences for specific plants but accept other plant species when preferred options are not available [37].

We test the hypothesis that many Neotropical forest phaneropterines are host plant specialists, facilitating pairing and reducing the demands associated with eating a large diversity of highly defended rainforest plants. The diet specialization hypothesis predicts that all katydids of a particular species will consistently consume the same or closely related plant species across space and time. An alternative hypothesis is that katydids are more generalized in their diets, a hypothesis that would be supported by sampling multiple individuals of the same katydid species and finding that they had been eating taxonomically diverse plant species.

Katydids are renowned for their camouflage and finding them in the forest can be very challenging, particularly because many species occur in the forest canopy [38]. Because it is so difficult to observe these animals in the wild, it is also difficult to determine if they are dietary specialists. A detailed review of the literature on these katydid species [19] did not yield any published records of diet. Fortunately, DNA sequencing approaches make it possible to identify plants that are part of the diets by extracting plant material from the stomachs of captured katydids and amplifying plant DNA barcodes [39–42]. These plant DNA barcodes have been developed specifically for the research site on Barro Colorado Island in Panama and have been previously utilized to generate a community phylogeny of trees and assign plant species to fine roots collected in the soil [43,44]. Here we employ these barcodes to test diet specialization of katydids.

2. Materials and Methods

2.1. Katydid Capture and Sample Preparation

For gut content characterization, katydids (family Tettigoniidae, primarily Phaneropterinae) were captured at building lights on Barro Colorado Island, Panama (9.16491, −79.83734). Katydids were collected during nighttime hours between late December and early March of 2015–2018, with 82% of the samples collected between 29 December and 31 January. During this season, katydids are actively calling and mating. Lights were checked two times per night, at approximately 23:00 and 04:30. When analyses were reported by year, the data are grouped by collecting season (e.g., '2017' represents data collected from late December of 2016 to early March of 2017). For each katydid, we recorded the location of the light and time of capture, then immediately froze the insect to interrupt digestion. Katydids were identified to species in accordance with available resources [19,45], with published morphospecies names used to identify three species. Light trapping captured many different katydid species, with relatively even representation across species, rather than a few dominant species. In the current study, we focused on the most commonly captured species in order to represent multiple individuals of the same species. Plant DNA was obtained by dissecting the katydid and isolating the digestive tract, which was placed in a microcentrifuge tube for DNA extraction [40].

2.2. DNA Extraction and Amplification

DNA was purified and amplified as described in Symes et al. [40]. Briefly, DNA was purified from dissected digestive tracts following the manufacturer's instructions using the QIAGEN DNeasy Plant Mini Kit (Qiagen 69104; Qiagen, Hilden, Germany) with the following modification. After initial homogenization with micropestles, 50 μL of AP1 buffer was added and the sample was further crushed before adding the 350 μL of AP1 buffer. Primers were utilized to amplify three conserved regions of the plastid genome following the procedure described in [43] (Table S1). Of the three regions, the *rbcL* region has the highest sequence conservation across plant species and is easy to sequence, the *psbA/trnH* region is intermediate in variability, but sometimes difficult to interpret, and the *matK* region is highly variable, but often difficult to amplify. PCR reactions were prepared using 5 μL template, 20 μL GoTaq Green PCR mastermix (Promega M7122; Madison, WI, USA), 0.5 μM F and 0.5 μM R primer, and water to a final volume of 40 μL. Thermocycler (ABI Veriti 96-well, model 9902) conditions for *rbcLa* and *psbA-trnH* were: 95 °C for 3 min, 35 cycles (95 °C for 30 s, 55 °C for 30 s, 72 °C for 1 min), 72 °C for 10 min and for *matK* were: 95 °C for 3 min, 40 cycles (95 °C for 30 s, 50 °C for 30 s, 72 °C for 1 min), 72 °C for 10 min. Samples were analyzed with *rbcLa* and *psbA-trnH* primers first, and then PCR was conducted with *matK* primers if needed to achieve a minimum of two successful primer sets per sample.

For positive samples, the remaining 30 μL of PCR product was purified using a Wizard PCR Clean-Up System (Promega A9281; Madison, WI, USA) and resuspended in 50 μL nuclease-free water. 2 μL of purified PCR product was mixed with 25 ng of the appropriate forward primer for the amplicon (*rbc*, rbc_SI_For; *psb*, psbA3′f; *matK*, matKfor_KIM3F) in a 15 μL reaction volume and sequenced using Sanger sequencing through Genewiz (Genewiz, South Plainfield, NJ, USA). Sequences were trimmed using Geneious and a 0.1 error probability limit. Following editing, sequences were exported in FASTA format for analysis with BLAST. Sequence data is available in FASTA format with the Supplementary Materials.

2.3. Identification of Plant Species

Sequences were assigned to plant species using the BLAST tool in Geneious based on identity similarity and query coverage between sequences from the gut samples and a custom reference database containing BCI-specific plant barcodes (*rbcL, psbA/trnH, matK*) for trees ([43]), shrubs, and lianas (Kress, unpubl). (Table S2) For samples with successful amplification for two or three of the gene markers, plant species identification was assigned if there was a match for the highest sequence similarity across multiple gene markers. If there were multiple taxa with the same similarity, identification was made to genus or family level. If there was no overlap between two or three gene markers, the sample was labeled as "Conflict", indicating that different primer sets match different plant taxa. For samples with successful amplifications for only one gene marker, plant species identification was made to the species with the highest percent identity and query coverage. If more than one plant had similar identity and coverage, sequences were assigned to the level of concordance in the cluster (genus or family). Data visualizations were created with Microsoft Excel and Powerpoint.

2.4. Growth Height Assignment

Plants with species- or genus-level identification were assigned a growth height category based on the "plant growth form" data in the relevant overview page for each plant species from the Encyclopedia of Life [46]. Plants were binned into the following categories with the maximum height listed: Shrub (6 m), Understory tree (Tree 15 m, Tree 20 m, Tree 25 m), and Canopy Tree (Tree 30 m, Tree 40 m, Tree 70 m). In one case (genus *Ocotea*), genus-level identification was included because all species in that genus fell in the same growth category.

3. Results

Successful plant sequences were recovered from the stomach contents of 71 insects representing 17 Neotropical forest katydid species. These katydid species consumed a wide variety of plant families, with each katydid species consuming multiple families of plants (Figure 1), and multiple katydid species consuming the same plant species. Comparing 2016 to 2017, katydids sampled from the same locations were often eating different plant families in different years (Figure 2).

In total 27 species of host plants were determined from the gut contents of the sampled katydids by DNA barcodes (Table 1 and Table S3). These 27 species were distributed across 22 genera in 16 families and 12 orders and were phylogenetically spread across the 23 orders of flowering trees on Barro Colorado Island (Figure 3 and Figure S1). The largest numbers of species were found in the Fabales (six species), Sapindales (six species), and Laurales (three species). The remaining nine orders each had one or two species of host plants. Relatively few or no host species were detected in the speciose orders Gentianales, Rosales, Malpighiales, and Myrtales.

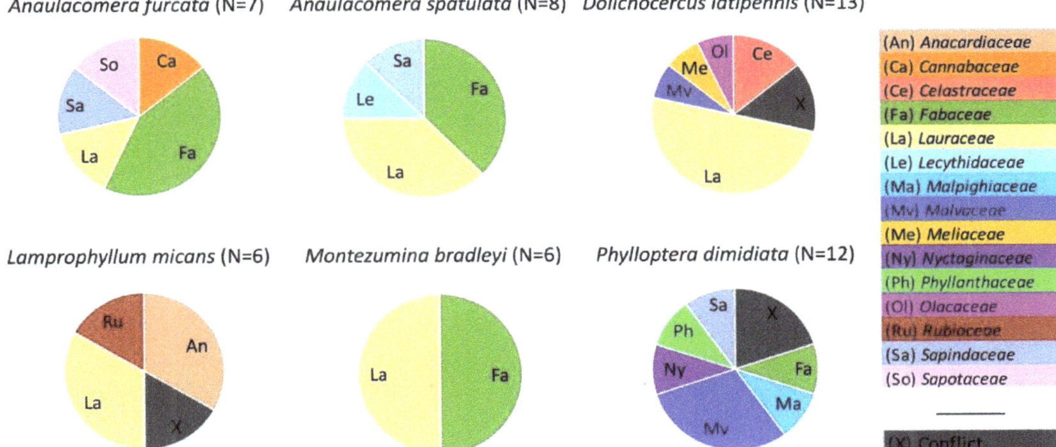

Figure 1. Plant families recovered from the stomach contents of six common katydid species on Barro Colorado Island, Panama. "Conflict" indicates that different primer sets show different plant species identifications for a given individual katydid.

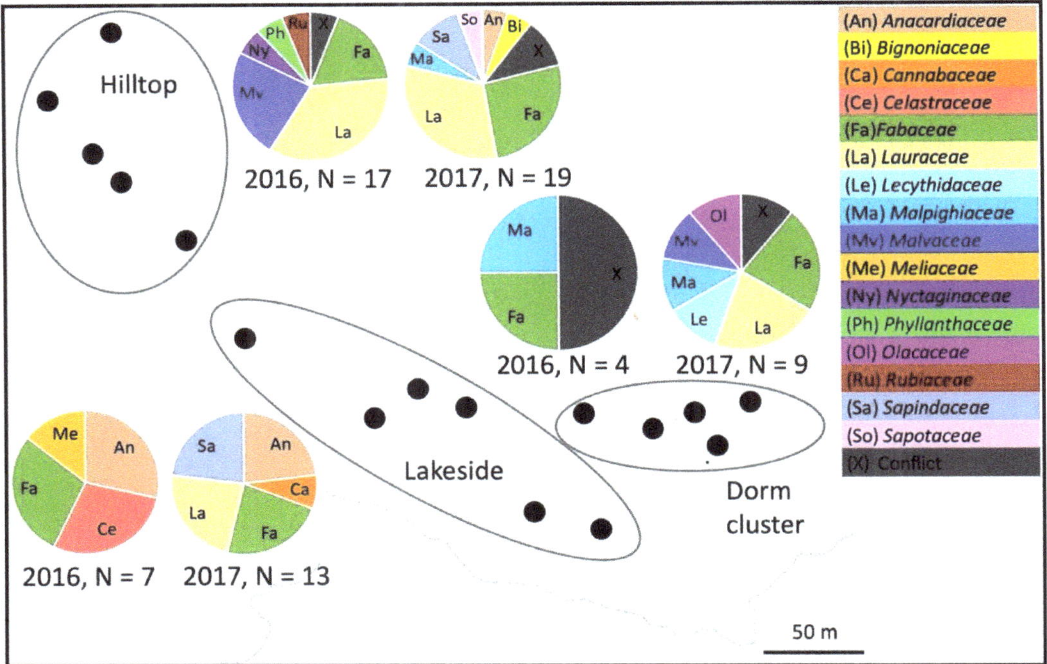

Figure 2. Map of katydid collection localities on Barro Colorado Island, with pie charts representing the plant families that were recovered from katydid stomachs by location and year. Each black dot represents a light capture location, with lights divided into three spatially and elevationally clustered zones. The inset pie charts represent the plant families that were sequenced in a given zone and year. "Conflict" indicates that different primer sets show different plant species identifications for a given individual katydid.

Table 1. Plant species identified by DNA barcoding from the digestive tracts of Neotropical katydids.

Katydid Species	Plant Order	Plant Family	Plant Genus	Plant Species	Growth Habit
Anaulacomera "wallace"	*Laurales (2 *)*	*Lauraceae (2 *)*	*Nectandra (2 *)*	*lineata (2 *)*	Understory tree (10–25 m)
	*Ericales **	*Sapotaceae **	*Pouteria **	*fossicola **	Understory tree (10–25 m)
			Inga	*sp*	
Anaulacomera furcata (v)	*Fabales (2 + 1 *)*	*Fabaceae (2 + 1 *)*	*Swartzia **	*simplex **	Understory tree (10–25 m)
			Tachigali	*versicolor*	Canopy tree (>25 m)
	*Laurales **	*Lauraceae **	*Nectandra **	*lineata **	Understory tree (10–25 m)
	*Rosales **	*Cannabaceae **	*Trema **	*micrantha **	Understory tree (10–25 m)
	Sapindales	*Sapindaceae*	*Cupania*	*cinerea*	Understory tree (10–25 m)
	Ericales	*Lecythidaceae*	*Gustavia*	*superba*	Understory tree (10–25 m)
			*Inga (2 *)*	*goldmanii*	Understory tree (10–25 m)
Anaulacomera spatulata	*Fabales (1 + 2 *)*	*Fabaceae (1+2 *)*		*punctata*	Understory tree (10–25 m)
			–	–	
	*Laurales (3 *)*	*Lauraceae (3 *)*	*Nectandra (3 *)*	*lineata (3 *)*	Understory tree (10–25 m)
	Sapindales	*Sapindaceae*	*Cupania*	*rufescens*	Understory tree (10–25 m)
	*Fabales **	*Fabaceae **	*Inga **	–	
Arota festae	*Malpighiales **	*Malpighiaceae **	*Malpighia **	*romeroana **	Shrub (<6 m)
	*Sapindales **	*Anacardiaceae **	*Anacardium **	*excelsum **	Canopy tree (>25 m)
Ceraia mytra	*Lamiales*	*Bignoniaceae*	*Jacaranda*	*copaia*	Canopy tree (>25 m)
Docidocercus gigliotosi	*Malpighiales*	*Malpighiaceae*	*Malpighia*	*romeroana*	Shrub (<6 m)
	*Celastrales (1 + 1 *)*	*Celastraceae (1 + 1 *)*	*Maytenus (1 + 1*)*	*schippii (1 + 1 *)*	Understory tree (10–25 m)
	*Laurales (6 + 1 *)*	*Lauraceae (6 + 1 *)*	*Nectandra (5 + 1*)*	*lineata (5 + 1 *)*	Understory tree (10–25 m)
Dolichocercus latipennis			*Ocotea*	–	Understory tree (10–25 m)
	Malvales	*Malvaceae*	–	–	
	Santalales	*Olacaceae*	*Heisteria*	*concinna*	Understory tree (10–25 m)
	Sapindales	*Meliaceae*	*Guarea*	–	Understory tree (10–25 m)
Euceraia insignis	*Sapindales (2 *)*	*Anacardiaceae **	*Anacardium**	*excelsum**	Canopy tree (>25 m)
		*Sapindaceae **	*Cupania**	–	Understory tree (10–25 m)
Hyperphrona irregularis	*Fabales **	*Fabaceae **	*Inga**	–	
Idiarthron major	*Fabales (2 *)*	*Fabaceae (2 *)*	*Dipteryx**	*oleifera*	Canopy tree (>25 m)
			*Inga**	–	
	Gentianales	*Rubiaceae*	*Chimarrhis*	*parviflora*	Understory tree (10–25 m)
Lamprophyllum micans	*Laurales (1 + 1 *)*	*Lauraceae (1 + 1 *)*	*Nectandra (1 + 1 *)*	*lineata (1 + 1 *)*	Understory tree (10–25 m)
	*Sapindales (1 + 1 *)*	*Anacardiaceae (1 + 1 *)*	*Anacardium (1 + 1 *)*	*excelsum (1 + 1 *)*	Canopy tree (>25 m)
Microcentrum "polka"	*Fabales **	*Fabaceae **	*Inga**	–	
	Laurales	*Lauraceae*	*Ocotea*	*puberula*	Understory tree (10–25 m)
Microcentrum championi	*Malvales*	*Malvaceae*	*Luehea*	*seemannii*	Canopy tree (>25 m)
	*Sapindales (1 + 1 *)*	*Anacardiaceae **	*Spondias **	*radlkoferi **	Canopy tree (>25 m)
		Sapindaceae	*Cupania*	*cinerea*	Understory tree (10–25 m)
Montezumina bradleyi	*Fabales (3)*	*Fabaceae (3)*	*Inga (3)*	–	
	Laurales (3)	*Lauraceae (3)*	*Nectandra (3)*	*lineata (3)*	Understory tree (10–25 m)

Table 1. *Cont.*

Katydid Species	Plant Order	Plant Family	Plant Genus	Plant Species	Growth Habit
Phylloptera dimidiata	*Caryophyllales*	*Nyctaginaceae*	*Guapira*	*standleyana*	Canopy tree (>25 m)
	Fabales *	*Fabaceae* *	*Swartzia* *	*simplex* *	Understory tree (10–25 m)
	*Malpighiales (1 + 1 *)*	*Malpighiaceae*	*Malpighia*	*romeroana*	Shrub (<6 m)
		Phyllanthaceae *	*Margaritaria* *	*nobilis* *	Understory tree (10–25 m)
	*Malvales (2 + 1 *)*	*Malvaceae (2 + 1 *)*	*Ceiba (2 + 1 *)*	*pentandra (2 + 1 *)*	Canopy tree (>25 m)
	Sapindales *	*Sapindaceae* *	*Cupania* *	*latifolia* *	Understory tree (10–25 m)
"Waxy" sp.	*Fabales*	*Fabaceae*	*Inga*	*sp*	
	Sapindales	*Anacardiaceae*	*Anacardium*	*excelsum*	Canopy tree (>25 m)

Plant species listed are supported by two or more primer sets. Plant species with * are supported by a single primer. The paranthetical number indicates the number of individual katydids associated with the identified plant.

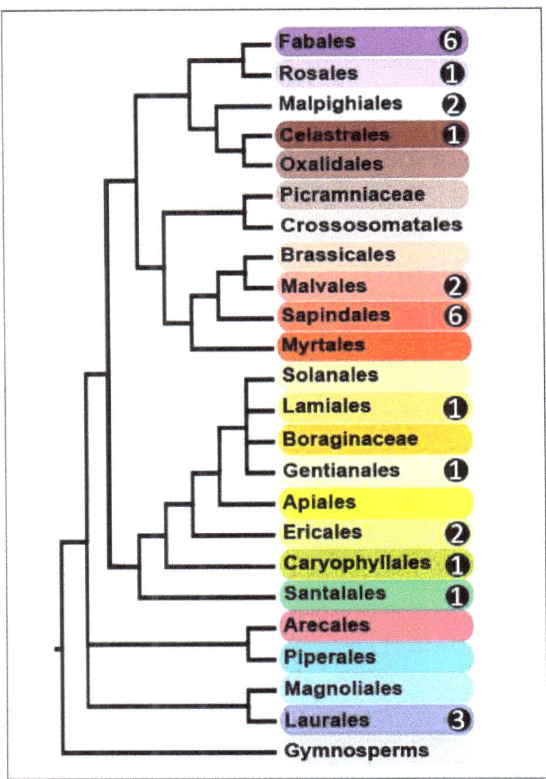

Figure 3. The phylogenetic distribution of plant species in the diets of katydids on BCI. The evolutionary relationships of the 23 orders of flowering plants found on BCI are represented in the branching diagram (modified from Figure 1 in [39]; see Figure S1 for a full representation of the species diversity of trees in the 50-ha forest dynamics plot). Circled numbers indicate the number of host plant species per order detected in the gut contents of katydids, as determined by DNA barcoding.

Most katydid species were consuming plant species that could grow into the canopy layer (Figure 4). However, some katydid species were foraging on plants that never grow higher than the understory.

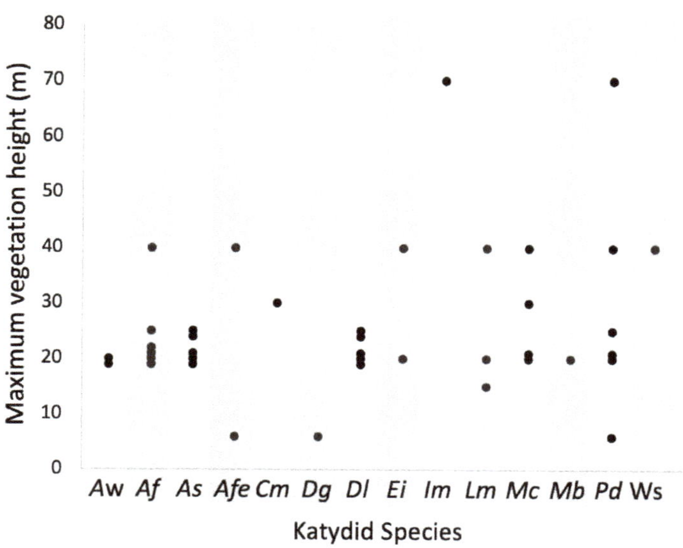

Figure 4. Maximum published growth height for the plants contained in the katydid diet, shown by katydid species. In the case of repeated plant height values, points are jittered slightly to show all data. Plants that could not be identified to species or genus level are not included due to variability of family-level maximum growth height. Katydid species abbreviations (*Aw: Anaulacomera* "wallace"; *Af: Anaulacomera furcata; As: Anaulacomera spatulata; Afe: Arota festae; Cm: Ceraia mytra; Dg: Docidocercus gigliotosi; Dl: Dolichocercus latipennis; Ei: Euceraia insignis; Im: Idiarthron major; Lm: Lamprophyllum micans; Mc: Microcentrum championi; Mb: Montezumina bradleyi; Pd: Phylloptera dimidiata;* Ws: Waxy sp.).

4. Discussion

The results of our study do not support the hypothesis that Neotropical katydid species specialize their diet by host plant. By extracting plant DNA from the digestive systems of katydids, we were able to use DNA barcoding to identify which plant species and/or families were recently consumed. Individual katydids of the same species often had multiple and different plant families in their digestive systems, indicating that katydid species were not specializing on single host plants. In addition, the use of multiple primers sometimes recovered different plant species from the same katydid, providing evidence that katydids would feed on a diversity of plants even at short timescales. Consequently, dietary specialization on a specific host plant is not providing these species with a means of facilitating mate localization in the face of intense predation on signaling males and mate-searching females.

In contrast, the katydid species studied here were consuming a wide variety of plant families (16) and orders (12) comprising over half the orders of plants found on BCI. While many plant families and orders appeared in the katydid diet, some were particularly well-represented. Within the katydid samples that yielded a single identified plant order, 27% of the samples were Laurales and 23% were Fabales. No other plant family or order comprised more than 8% of the samples (Table 1). The abundance of Laurales and Fabales in the diet appears to be consistent with the relative abundance of stems of these plant orders on Barro Colorado Island, although species diversity in several other lineages, such as Gentianales, Myrtales, Rosales, and Malpighiales, is equally high or higher, suggesting that katydids may avoid some lineages of plants in favor of others [47–49]. Furthermore, it should be noted that the katydid specimens were collected primarily in January when certain trees may be in the young leaf stage and hence easier to digest than other species.

If this is the case, then our results may in part be dependent on tree phenology. Further sampling at different times of the year in different seasons is certainly warranted.

The dietary composition of katydids provides one-directional information about the height at which katydids are foraging. Because even canopy emergent tree species start as saplings, the presence of canopy species in the katydid diet does not necessarily mean that the species was consumed in the canopy. In contrast, when katydids are eating shrubs and understory trees, it is strong evidence that the species is foraging low in the forest. While sample sizes are small for some katydid species, katydid species that are observed at ground level are in fact consuming understory vegetation (e.g., *Docidocercus gigliotosi* [22]) (Figure 4).

One notable finding was that many plant families were detected in only one year. In part, the year-to-year differences likely reflect the fact that there are many plant families and most were represented with relatively low frequency. However, the fact that some plant families are well-represented in one year and rare or absent in others suggests that there may be times when particular plant families or individual plants are especially palatable. For example, in a given sampling location on a given night, multiple katydids of multiple species had the same plant in their stomach (Table S3), perhaps reflecting a nearby feeding opportunity that attracted multiple species of katydid.

Transient feeding opportunities on particular plants would be consistent with what is known about the palatability and phenology of many tropical plants. Even though leaves may persist for several years, 25–70% of leaf damage occurs in the weeks when leaves are expanding [50,51]. In response to herbivore pressure, tropical plants have evolved a variety of strategies to minimize their window of vulnerability to herbivores [27]. Herbivore evasion strategies can include exceptionally rapid expansion of leaves, delayed greening, and synchronous flushing, strategies that minimize exposure to herbivory by compressing the window when leaves are maximally palatable [52,53].

The possibility of short-term feeding windows on specific plants is also supported by anecdotal field observations during this study, where katydids of multiple species were observed aggregated on a tree a few days before the tree produced substantial and obvious new growth (L. Symes, personal observation).

If katydids do aggregate to exploit transient feeding opportunities, co-localization on food sources might still facilitate mate finding, even for diet generalists. There are several avenues of investigation that could provide information on whether katydids aggregate to feed on plants during leafout. One strategy is to bait traps with plant volatiles such as benzyl nitrile, phenyl acetaldehyde, and/or 2-phenylethanol [54,55] to determine whether these chemical compounds are attractive to katydids, suggesting targeted feeding on vulnerable plants. A second strategy is to deploy long-term acoustic recorders and test for periods of time when a single location on the landscape has an elevated number of katydid calls, reflecting aggregation or one or more katydid species in an area for a short period of time. Understanding whether katydids aggregate on plants that are producing new leaves is important for understanding herbivore pressures on the forest vegetation, patterns of food availability for insect predators, and impacts of habitat patch size on insect populations and foraging effectiveness.

The results of DNA barcoding, when applied to Neotropical forest katydids, helps to define previously unseen connections between plants and herbivores, including connections that occur out of sight in the forest canopy. The generality of the katydid diet is more or less consistent with observations on other species in other locations, but underscores our lack of knowledge of how animals with rare short-range signals find each other in tropical forests. Our observations also suggest several avenues for future research.

Supplementary Materials: The following supporting information can be downloaded at: https://www.mdpi.com/article/10.3390/d14020152/s1, Table S1: Primers utilized for DNA barcode amplification; Table S2: List of accessions of DNA barcodes for BCI trees/shrubs/lianas used in BLAST searches; Table S3: List of katydid samples, plant BLAST result, and collection metadata. Figure S1: The phylogenetic distribution of plant species in the diets of katydids on BCI.

Author Contributions: Conceptualization, C.M.P. and L.B.S.; methodology, C.M.P. and L.B.S.; formal analysis, C.M.P. and L.B.S.; investigation, C.M.P., N.L.W., S.J.M. and L.B.S.; resources, C.M.P., H.M.t.H., W.J.K. and L.B.S.; data curation, C.M.P. and L.B.S.; writing—original draft preparation, C.M.P. and L.B.S.; writing—review and editing, C.M.P., H.M.t.H., W.J.K. and L.B.S.; visualization, C.M.P. and L.B.S.; supervision, C.M.P. and L.B.S.; project administration, C.M.P. and L.B.S.; funding acquisition, C.M.P., H.M.t.H. and L.B.S. All authors have read and agreed to the published version of the manuscript.

Funding: C.M.P. and N.L.W. were funded by Castleton University and the Vermont Biomedical Research Network (VBRN). Research reported in this publication was supported by an Institutional Development Award (IDeA) from the National Institute of General Medical Sciences of the National Institutes of Health, grant number P20GM103449. Its contents are solely the responsibility of the authors and do not necessarily represent the official views of NIGMS or NIH. L.B.S. was supported by the Neukom Institute of Dartmouth College. L.B.S., S.J.M. and H.M.t.H. were supported by funding from the Smithsonian Tropical Research Institute, Dartmouth College, and an Artificial Intelligence for Earth Innovation grant from Microsoft/National Geographic, grant number NG5-57246T-18. W.J.K. was supported by the Smithsonian Institution.

Institutional Review Board Statement: Not applicable.

Data Availability Statement: Sequence data for chloroplast DNA amplified from gut contents is available in the Supplementary Materials. Reference plant barcodes (accessions found in Table S2) can be found at https://www.ncbi.nlm.nih.gov/genbank/ (accessed on 10 August 2021).

Acknowledgments: The authors would like to thank Steier, J., Lehman, K., and Wright, J. for their help in generating the barcode sequence data for shrubs and lianas on Barro Colorado Island.

Conflicts of Interest: The authors declare no conflict of interest. The funders had no role in the design of the study; in the collection, analyses, or interpretation of data; in the writing of the manuscript, or in the decision to publish the results.

References

1. Ryan, M.J.; Tuttle, M.D.; Rand, A.S. Bat Predation and Sexual Advertisement in a Neotropical Anuran. *Am. Nat.* **1982**, *119*, 136–139. [CrossRef]
2. Langerhans, R.B.; Layman, C.A.; DeWitt, T.J. Male Genital Size Reflects a Tradeoff between Attracting Mates and Avoiding Predators in Two Live-Bearing Fish Species. *Proc. Natl. Acad. Sci. USA* **2005**, *102*, 7618–7623. [CrossRef] [PubMed]
3. Halfwerk, W.; Jones, P.L.; Taylor, R.C.; Ryan, M.J.; Page, R.A. Risky Ripples Allow Bats and Frogs to Eavesdrop on a Multisensory Sexual Display. *Science* **2014**, *343*, 413–416. [CrossRef] [PubMed]
4. Andersson, M. *Sexual Selection*; Princeton University Press: Princeton, NJ, USA, 1994; ISBN 9780691000572.
5. Zuk, M.; Kolluru, G.R. Exploitation of Sexual Signals by Predators and Parasitoids. *Q. Rev. Biol.* **1998**, *73*, 415–438. [CrossRef]
6. Wing, S.R. Cost of Mating for Female Insects: Risk of Predation in *Photinus collustrans* (Coleoptera: Lampyridae). *Am. Nat.* **1988**, *131*, 139–142. [CrossRef]
7. Ter Hofstede, H.; Voigt-Heucke, S.; Lang, A.; Römer, H.; Page, R.; Faure, P.; Dechmann, D. Revisiting Adaptations of Neotropical Katydids (Orthoptera: Tettigoniidae) to Gleaning Bat Predation. *Neotrop. Biodivers.* **2017**, *3*, 41–49. [CrossRef] [PubMed]
8. Heller, K.-G. Risk Shift between Males and Females in the Pair-Forming Behavior of Bushcrickets. *Naturwissenschaften* **1992**, *79*, 89–91. [CrossRef]
9. Carl Gerhardt, H.; Huber, F. *Acoustic Communication in Insects and Anurans: Common Problems and Diverse Solutions*; University of Chicago Press: Chicago, IL, USA, 2002; ISBN 9780226288338.
10. SINA Singing Insects of North America. Available online: https://sina.orthsoc.org/index.htm (accessed on 8 October 2021).
11. Symes, L.B.; Ayres, M.P.; Cowdery, C.P.; Costello, R.A. Signal Diversification in *Oecanthus* Tree Crickets Is Shaped by Energetic, Morphometric, and Acoustic Trade-Offs. *Evolution* **2015**, *69*, 1518–1527. [CrossRef]
12. Prestwich, K.N.; Walker, T.J. Energetics of Singing in Crickets: Effect of Temperature in Three Trilling Species (Orthoptera: Gryllidae). *J. Comp. Physiol. B* **1981**, *143*, 199–212. [CrossRef]
13. Walker, T.J. Experimental Demonstration of a Cat Locating Orthopteran Prey by the Prey's Calling Song. *Fla. Entomol.* **1964**, *47*, 163–165. [CrossRef]
14. Cade, W. Acoustically Orienting Parasitoids: Fly Phonotaxis to Cricket Song. *Science* **1975**, *190*, 1312–1313. [CrossRef]
15. Falk, J.J.; Ter Hofstede, H.M.; Jones, P.L.; Dixon, M.M.; Faure, P.A.; Kalko, E.K.V.; Page, R.A. Sensory-Based Niche Partitioning in a Multiple Predator–multiple Prey Community. *Proc. R. Soc. B* **2015**, *282*, 20150520. [CrossRef] [PubMed]
16. Jones, P.L.; Ryan, M.J.; Page, R.A. Population and Seasonal Variation in Response to Prey Calls by an Eavesdropping Bat. *Behav. Ecol. Sociobiol.* **2014**, *68*, 605–615. [CrossRef]

17. Belwood, J.J. The Influence of Bat Predation on Calling Behavior in Neotropical Forest Katydids (Insecta: Orthoptera: Tettigoniidae). Ph.D. Thesis, University of Florida, Gainesville, FL, USA, 1990.
18. Symes, L.B.; Robillard, T.; Martinson, S.J.; Dong, J.; Kernan, C.E.; Miller, C.R.; ter Hofstede, H.M. Daily Signaling Rate and the Duration of Sound per Signal Are Negatively Related in Neotropical Forest Katydids. *Integr. Comp. Biol.* **2021**, *61*, 877–899. [CrossRef]
19. Ter Hofstede, H.M.; Symes, L.B.; Martinson, S.J.; Robillard, T.; Faure, P.; Madhusudhana, S.; Page, R.A. Calling Songs of Neotropical Katydids (Orthoptera: Tettigoniidae) from Panama. *J. Orthoptera Res.* **2020**, *29*, 137. [CrossRef]
20. Romer, H.; Lewald, J. High-Frequency Sound Transmission in Natural Habitats: Implications for the Evolution of Insect Acoustic Communication. *Behav. Ecol. Sociobiol.* **1992**, *29*, 437–444. [CrossRef]
21. Symes, L.B.; Martinson, S.J.; Kernan, C.E.; ter Hofstede, H.M. Sheep in Wolves' Clothing: Prey Rely on Proactive Defences When Predator and Non-Predator Cues Are Similar. *Proc. Biol. Sci.* **2020**, *287*, 20201212. [CrossRef]
22. Lang, A.B.; Römer, H. Roost Site Selection and Site Fidelity in the Neotropical Katydid *Docidocercus gigliotosi* (Tettigoniidae). *Biotropica* **2008**, *40*, 183–189. [CrossRef]
23. Heller, K.-G.; Hemp, C.; Ingrisch, S.; Liu, C. Acoustic Communication in Phaneropterinae (Tettigonioidea)—A Global Review with Some New Data. *J. Orthop. Res.* **2015**, *24*, 7–18. [CrossRef]
24. Villarreal, S.M.; Gilbert, C. Male *Scudderia pistillata* Katydids Defend Their Acoustic Duet against Eavesdroppers. *Behav. Ecol. Sociobiol.* **2014**, *68*, 1669–1675. [CrossRef]
25. Xu, H.; Turlings, T.C.J. Plant Volatiles as Mate-Finding Cues for Insects. *Trends Plant. Sci.* **2018**, *23*, 100–111. [CrossRef]
26. Cocroft, R.B.; Rodríguez, R.L.; Hunt, R.E. Host Shifts and Signal Divergence: Mating Signals Covary with Host Use in a Complex of Specialized Plant-Feeding Insects. *Biol. J. Linn. Soc.* **2009**, *99*, 60–72. [CrossRef]
27. Kursar, T.A.; Coley, P.D. Convergence in Defense Syndromes of Young Leaves in Tropical Rainforests. *Biochem. Syst. Ecol.* **2003**, *31*, 929–949. [CrossRef]
28. Sedio, B.E.; Rojas Echeverri, J.C.; Boya, P.C.A.; Wright, S.J. Sources of Variation in Foliar Secondary Chemistry in a Tropical Forest Tree Community. *Ecology* **2017**, *98*, 616–623. [CrossRef] [PubMed]
29. Coley, P.D. Herbivory and Defensive Characteristics of Tree Species in a Lowland Tropical Forest. *Ecol. Monogr.* **1983**, *53*, 209–234. [CrossRef]
30. Singer, M.S.; Bernays, E.A.; Carrière, Y. The Interplay between Nutrient Balancing and Toxin Dilution in Foraging by a Generalist Insect Herbivore. *Anim. Behav.* **2002**, *64*, 629–643. [CrossRef]
31. Pulliam, H.R. Diet Optimization with Nutrient Constraints. *Am. Nat.* **1975**, *109*, 765–768. [CrossRef]
32. Birnbaum, S.S.L.; Abbot, P. Insect Adaptations toward Plant Toxins in Milkweed–herbivores Systems—A Review. *Entomol. Exp. Appl.* **2018**, *166*, 357–366. [CrossRef]
33. Dobler, S.; Petschenka, G.; Pankoke, H. Coping with Toxic Plant Compounds—The Insect's Perspective on Iridoid Glycosides and Cardenolides. *Phytochemistry* **2011**, *72*, 1593–1604. [CrossRef]
34. Unsicker, S.B.; Oswald, A.; Köhler, G.; Weisser, W.W. Complementarity Effects through Dietary Mixing Enhance the Performance of a Generalist Insect Herbivore. *Oecologia* **2008**, *156*, 313–324. [CrossRef]
35. Smith, T.R.; Capinera, J.L. Host Preferences and Habitat Associations of Some Florida Grasshoppers (Orthoptera: Acrididae). *Environ. Entomol.* **2005**, *34*, 210–224. [CrossRef]
36. Sword, G.A.; Joern, A.; Senior, L.B. Host Plant-Associated Genetic Differentiation in the Snakeweed Grasshopper, *Hesperotettix viridis* (Orthoptera: Acrididae). *Mol. Ecol.* **2005**, *14*, 2197–2205. [CrossRef] [PubMed]
37. Valtonen, A.; Malinga, G.M.; Junes, P.; Opoke, R.; Lehtovaara, V.J.; Nyeko, P.; Roininen, H. The Edible Katydid *Ruspolia differens* is a Selective Feeder on the Inflorescences and Leaves of Grass Species. *Entomol. Exp. Appl.* **2018**, *166*, 592–602. [CrossRef]
38. Montealegre, Z.F.; Sarria, F.A.; Pimienta, M.C.; Mason, A.C. Lack of Correlation between Vertical Distribution and Carrier Frequency, and Preference for Open Spaces in Arboreal Katydids That Use Extreme Ultrasound, in Gorgona, Colombia (Orthoptera: Tettigoniidae). *Rev. Biol. Trop.* **2014**, *62*, 289–296. [CrossRef]
39. Kress, W.J. Plant DNA Barcodes: Applications Today and in the Future. *J. Syst. Evol.* **2017**, *55*, 291–307. [CrossRef]
40. Symes, L.B.; Wershoven, N.L.; Hoeger, L.-O.; Ralston, J.S.; Martinson, S.J.; Ter Hofstede, H.M.; Palmer, C.M. Applying and Refining DNA Analysis to Determine the Identity of Plant Material Extracted from the Digestive Tracts of Katydids. *PeerJ* **2019**, *7*, e6808. [CrossRef] [PubMed]
41. García-Robledo, C.; Erickson, D.L.; Staines, C.L.; Erwin, T.L.; John Kress, W. Tropical Plant–Herbivore Networks: Reconstructing Species Interactions Using DNA Barcodes. *PLoS ONE* **2013**, *8*, e52967. [CrossRef]
42. Kress, W.J.; García-Robledo, C.; Uriarte, M.; Erickson, D.L. DNA Barcodes for Ecology, Evolution, and Conservation. *Trends Ecol. Evol.* **2015**, *30*, 25–35. [CrossRef]
43. Kress, W.J.; Erickson, D.L.; Jones, F.A.; Swenson, N.G.; Perez, R.; Sanjur, O.; Bermingham, E. Plant DNA Barcodes and a Community Phylogeny of a Tropical Forest Dynamics Plot in Panama. *Proc. Natl. Acad. Sci. USA* **2009**, *106*, 18621–18626. [CrossRef]
44. Jones, F.A.; Erickson, D.L.; Bernal, M.A.; Bermingham, E.; Kress, W.J.; Herre, E.A.; Muller-Landau, H.C.; Turner, B.L. The Roots of Diversity: Below Ground Species Richness and Rooting Distributions in a Tropical Forest Revealed by DNA Barcodes and Inverse Modeling. *PLoS ONE* **2011**, *6*, e24506. [CrossRef]

45. Nickle, D.A. Katydids of Panama (Orthoptera: Tettigoniidae). In *Insect Panama Mesoamerica*; Oxford University Press: Oxford, UK, 1992; pp. 142–184.

46. Encyclopedia of Life. Available online: http://eol.org (accessed on 8 January 2022).

47. Condit, R.; Hubbell, S.P.; Lafrankie, J.V.; Sukumar, R.; Manokaran, N.; Foster, R.B.; Ashton, P.S. Species-Area and Species-Individual Relationships for Tropical Trees: A Comparison of Three 50-Ha Plots. *J. Ecol.* **1996**, *84*, 549–562. [CrossRef]

48. Harms, K.E.; Condit, R.; Hubbell, S.P.; Foster, R.B. Habitat Associations of Trees and Shrubs in a 50-Ha Neotropical Forest Plot. *J. Ecol.* **2001**, *89*, 947–959. [CrossRef]

49. Croat, T.B. *Flora of Barro Colorado Island*; Stanford University Press: Redwood City, CA, USA, 1978; ISBN 9780804709507.

50. Coley, P.D.; Kursor, T.A. Anti-Herbivore Defenses of Young Tropical Leaves: Physiological Constraints and Ecological Trade-Offs. In *Tropical Forest Plant Ecophysiology*; Mulkey, S.S., Chazdon, R.L., Smith, A.P., Eds.; Springer: Boston, MA, USA, 1996; pp. 305–336. ISBN 9781461311638.

51. Aide, T.M. Others Comparison of Herbivory and Plant Defenses in Temperate and Tropical Broad-Leaved Forests. In *Plant-Animal Interactions: Evolution Ecology in Tropical and Temperate Regions*; John Wiley and Sons: Hoboken, NJ, USA, 1991; pp. 25–49.

52. Aide, T.M. Patterns of Leaf Development and Herbivory in a Tropical Understory Community. *Ecology* **1993**, *74*, 455–466. [CrossRef]

53. Lieberman, D.; Lieberman, M. The Causes and Consequences of Synchronous Flushing in a Dry Tropical Forest. *Biotropica* **1984**, *16*, 193–201. [CrossRef]

54. El-Sayed, A.M.; Knight, A.L.; Byers, J.A.; Judd, G.J.R.; Suckling, D.M. Caterpillar-Induced Plant Volatiles Attract Conspecific Adults in Nature. *Sci. Rep.* **2016**, *6*, 37555. [CrossRef]

55. Szendrei, Z.; Rodriguez-Saona, C. A Meta-Analysis of Insect Pest Behavioral Manipulation with Plant Volatiles. *Entomol. Exp. Appl.* **2010**, *134*, 201–210. [CrossRef]

Article

Are Introduced Alien Species More Predisposed to Invasion in Recipient Environments If They Provide a Wider Range of Services to Humans?

Kowiyou Yessoufou * and Annie Estelle Ambani

Department of Geography, Environmental Management and Energy Studies (GEMES), Kingsway Campus, University of Johannesburg, Corner Kingsway and University Road, P.O. Box 524, Johannesburg 2006, South Africa; anniee.ambani@gmail.com
* Correspondence: kowiyouy@uj.ac.za

Abstract: The drivers of invasion success of alien species remain, to some extent, a matter of debate. Here, we suggest that the services (the benefits humans obtain from a species) provided by alien plants could predict their invasion status, such that alien species providing more services would be more likely to be invasive than not. The rationale for this expectation is that alien species providing multiple services stand a better chance of being introduced in various numbers and multiple times outside their native range (propagule pressure theory). We investigated this hypothesis on alien woody species in South Africa. First, we defined 12 services provided by all the 210 known naturalized alien woody plants in South Africa. Then, we tested for a phylogenetic signal in these services using a DNA barcode-based phylogeny. Finally, we tested for potential links between the services and invasion status by fitting GLM models with appropriate error families. We found a phylogenetic signal in most services, suggesting that closely related species tend to provide similar services. Counter-intuitively, we consistently found that alien non-invasive species tend to provide more services, or even unique services, in comparison to alien invasive species. Although alternative scenarios are plausible to explain this unexpected finding, we speculate that harvesting alien plants for human benefits may limit their invasion ability. This warrants further investigation.

Keywords: alien woody plants; horizon scanning; DNA barcode; predicting invasion success; environmental policy; propagule pressure

Citation: Yessoufou, K.; Ambani, A.E. Are Introduced Alien Species More Predisposed to Invasion in Recipient Environments If They Provide a Wider Range of Services to Humans? . *Diversity* **2021**, *13*, 553. https://doi.org/10.3390/d13110553

Academic Editors: W. John Kress, Morgan Gostel and Michael Wink

Received: 20 August 2021
Accepted: 26 October 2021
Published: 30 October 2021

Publisher's Note: MDPI stays neutral with regard to jurisdictional claims in published maps and institutional affiliations.

1. Introduction

Over the past four centuries, some alien woody plants were introduced intentionally into South Africa to meet the growing human demands for various goods and services (charcoal, timber production, ornaments, dune stabilization, medicine, etc.; [1–5]). The selection and use of plants by humans have been shown to be non-random, but this non-randomness has been widely demonstrated for native plants (e.g., [6–10]). However, it is increasingly shown that alien species intentionally selected and introduced into new environments for human use (e.g., medicine) are also non-random selections from local floras (e.g., [11]; see [12] for further references). While some of the introduced alien plants fail to establish a viable population in their recipient environments, many others have naturalized, and some of the naturalized species have become invasive [13]. Alien invasive plants are naturalized plants that produce reproductive offspring, often in very large numbers, at considerable distances from their points of introduction, and thus have the potential to spread over a considerable area [14]. Although some of the invasive species pose a severe ecological threat to their recipient systems [15–17], the levels of threat are not equal; some species are strong invaders and pose high ecological and economic threats, while others are weak invaders [12,18,19].

In the face of these threats, a massive research effort has focused on understanding the predisposition of alien plants to invasion success in a foreign environment [5,20–23]. The findings reported in these studies are, to some extent, contradictory. This is because the drivers of invasion of some taxa in an environment do not explain the invasion of other taxa or the same taxa in different habitats [23–27], thus revealing the environment-dependent nature of invasion and the need for case-specific management solutions. Here, we propose that the services that alien species provide, and which motivate their introduction into a new environment, should better predict the invasion status of these species (naturalized vs. invasive) in their recipient ranges. There are two reasons underlying this expectation.

Firstly, since functions (ecological or physiological) generate services, and functional traits correlate with the invasion status of species in South Africa [21], services should also predict the invasion status of alien species. Several studies have tested, albeit indirectly, this potential link between services and invasion by focusing only on the link between functional traits and invasion [21,28]. Secondly (and this is the most critical basis for our expectation), an alien species that provides a diverse array of services is more likely to be introduced independently multiple times and in various numbers into new environments than an alien species that provides only one or a few services [9,29]. However, we also acknowledge that multiple independent introductions in large numbers may not necessarily be due to a diverse array of services, but rather could be driven, for some species, by a single service of high use-value for local communities. These alternative scenarios match the prediction of the propagule pressure theory [30], also termed "introduction effort" [31], which is the number of individuals introduced into a new environment and how often the introduction events occur [30,32].

In the present study, our aim is to link the services of naturalized woody plants to their invasiveness status in South Africa. Specifically, we ask the following questions: Are plant species selected and used by humans a random selection with regard to the services they provide? Does the total number of services (used as a proxy for propagule pressure) of alien species predict their invasion status? We explored these questions using the alien woody flora recorded as intentionally introduced to South Africa [21].

2. Materials and Methods

2.1. Native and Alien Woody Flora of South Africa

The present study focuses on South Africa but literature across Southern Africa, which includes Botswana, Lesotho, Mozambique, Namibia, South Africa, Swaziland, and Zimbabwe, was also consulted for the purpose of a comprehensive documentation of information. For example, native and alien plants to South Africa and their uses have been reported in various sources across Southern Africa. The Southern Africa's woody flora comprises approximately 2200 native species [21,33,34]. Of this native flora, 1190 species are included in the present study. This list includes 210 intentionally introduced alien species now documented as naturalized in South Africa. In total, 1400 species are included in our study, of which 1190 species are native, and 210 species are naturalized alien species.

2.2. Categorization of Alien Species

First, naturalized alien species were categorized into invasive and non-invasive following Bezeng et al. [20,21] as their study provides the most recent and comprehensive record and categorization of alien woody species in South Africa. In South Africa, the NEMBA list of alien plants is the official list of species considered as invasive and non-invasive (naturalized) in South Africa. The list is generated, through a lengthy process, by the government of South Africa through the Department of Environment, Forestry and Fisheries (DEFF). The process through which the list of alien species is generated can be summarized as follows: An Alien Species Risk Analysis Review Panel (ASRARP) is established and tasked to conduct the invasion risk analysis of alien species in the country. This panel, formed of various experts in the field of biological invasion, uses the framework of [35] for alien invasion risk analysis. This framework is grounded on the following five risk assessment criteria: background (of the alien species), likelihood (of the species being introduced to the country, naturalized and invasive), consequences (environmental and socio-economic), management (of the alien species) and reporting (summary of the risk assessment and risk recommendation). Before ASRARP makes a final decision on the risk status of a given species, the opinions of at least two experts, generally one local and one international, are consulted. The list of South Africa's alien species used in the present study emanated from this process and additional expert consultations [21].

2.3. Record of Services of Woody Flora (Native and Alien) in South Africa

We documented through an intensive literature search the different services these species (native and alien) provide to humans in South Africa. First, we used the Web of Science (WoS) to retrieve existing scientific ethnobotanical studies in the region. Second, we searched for each species by using combinations of keywords such as "scientific name of species", "Southern Africa", "Botswana", "Mozambique", "Namibia", "South Africa", "Swaziland", "Lesotho", "Zimbabwe", "uses", "usages", and "benefit". We also made use of Google and Google Scholar for scientific and grey literature using similar keywords to retrieve online resources such as regional and country-specific journals, proceedings, technical reports, herbarium and commercial websites informing on the uses of woody plants in our dataset. The Southern African Plant Invaders Atlas (http://www.agis.agric.za/wip/, accessed on 1 March 2017) was also consulted. In addition, we consulted key books on the regional flora such as *Trees of Southern Africa, Field Guide to Trees of Southern Africa*, and *Guide to Trees Introduced into Southern Africa* [33,36,37]. Additionally, plant uses in South Africa were retrieved from the *Prelude Database for Medicinal Plants in Africa* (http://www.africamuseum.be/collections/external/prelude; accessed on 10 February 2017), a unique database where medicinal plants and uses across the entire African continent since 1847 are documented and frequently updated. Finally, services of plants were updated by consulting the global dataset of plant uses of plants documented on the WEP database (National Plant Germplasm System GRIN-GLOBAL; https://npgsweb.ars-grin.gov/gringlobal/taxon/taxonomysearcheco.aspx, accessed on May 2021) and Diazgranados et al. [38]. All the different services (uses) retrieved from this wide and intensive literature search were grouped into 12 distinct categories of services (Table S1).

2.4. Phylogeny of the Southern Africa's Woody Flora

The phylogenetic tree used in this study is the most comprehensive DNA-based phylogeny ever assembled for both native and alien woody flora of Southern Africa in one of our recent papers [21]. In summary, this phylogeny was based on a matrix of the two DNA barcode regions *matK* (942 bp) and *rbcLa* (552 bp) generated in two recent studies (ref. [34] for native flora and ref. [21] for alien flora; sequences available since 2015 on www.boldsystems.org;). Although four markers are proposed as plant barcodes, the two regions *matK* and *rbcLa* have been shown to be efficient in several ecological studies, e.g., [21,34]. The phylogeny includes 1400 native and alien taxa representing 117 families and 562 genera. The reconstruction of the phylogeny follows the classical widely established Bayesian method (see details in [21]). Importantly, four independent runs of MCMC were performed, each for 100 million generations, sampling every 1000 generations. The MCMC log files for convergence using the effective sample size (ESS) statistics in Tracer v.1.5 [39] were evaluated, and all ESS values >100. Finally, the resulting tree files from the four runs were combined in LogCombiner v.1.7.5 [39], down sampling 1 in 20,000 trees, and discarding the first 25% trees as burn-in. The maximum clade consensus (MCC) phylogeny was generated with TreeAnnotator v.1.7.5 [39]. This MCC phylogeny is used for all phylogenetic analyses in the present study.

2.5. Data Analysis

- Test of phylogenetic signal in services provided by alien woody species

Prior to the analysis, the phylogeny was pruned off the native species. To test whether species used by humans are randomly selected with respect to the services they provide, a matrix of species and service categories was first created for each plant invasion status (alien non-invasive and alien invasive). In this matrix, the 12 categories of services (Table S1) were transformed into binary data, as follows: 1 (if a species provides a given service), and 0 (if not). Then, using the phylogenetic tree of Southern Africa's flora pruned to have only alien species to South Africa, we applied the D statistic [40] on this binary data to assess whether species used for a particular service are phylogenetically more closely related than expected at random (test of phylogenetic signal). D statistic has the advantage of measuring both a phylogenetic signal and its strength. The strength of the signal was interpreted as follows: $D < 0$ means strong signal; $D = 0$ means presence of signal under Brownian Motion model; D between 0 and 0.5 means moderate signal; D between 0.5 and 1 means weak signal; $D = 1$ means no signal; $D > 1$ means over-dispersion. The statistical significance of the observed D value was tested by comparing the observed D value to 0 (expected value for a phylogenetically conserved pattern under a Brownian Motion model) and 1 (random expectation). The p values for significance tests were reported as P_{BM} (giving the result of testing whether D was significantly different from 0) and P_{rand} (giving the result of testing whether D was significantly different from 1). In the scenario of a D value falling between 0 and 1 but being statistically different from 1, this implies that the observed D value shows moderate/weak signal but is non-random. If D value is between 0 and 1 but not statistically different from 1, then the observed value is moderate/weak and not different from random.

- Tests of link between services provided by alien plants and their invasion status

To test if services can be linked to invasion status, we tested whether the diversity of services provided by alien plants (i.e., total number of services for each alien species) correlates with their invasion status. This analysis was carried out by fitting two types of GLM models on "number of services" (response variable) versus "invasion status" (predictor). On one hand, we fitted a Poisson GLM (given the response variable is count data and on the other, we fitted a phyloGLM as implemented in the R library *Phylolm* [41]. The difference between both tests is that the latter corrects for phylogenetic nonindependence of species, allowing us to assess the potential influence of phylogeny on the result reported in the former test.

Finally, we tested whether there was a direct potential link between each service and the invasion status. The test was run by fitting a binomial GLM since invasion status (response variable) was measured as a binary variable (invasive vs. non-invasive following NEMBA).

3. Results

Firstly, we found evidence for non-random selection of alien species intentionally introduced to South Africa for the services they provide, although most phylogenetic signals were weak to moderate (Table 1). Specifically, we found support for phylogenetically non-random selection of alien species for 75% of services, i.e., 9 services out of 12.

Secondly, we found a correlation between the number of services and invasion status, such that alien non-invasive species tend to have more services than the invasive species (Figure 1; $\beta = -0.28 \pm 0.09$, $p = 0.003$). When we corrected for the phylogeny, our finding still confirms this pattern ($\beta = -0.38 \pm 0.39$, $p = 0.04$).

Table 1. Results of phylogenetic signal test in the services provided by alien plant species using D statistic. The *p* values for tests of significance were reported as P_{BM} (giving the result of testing whether D was significantly different from 0) and Prand (giving the result of testing whether D was significantly different from 1). A weak-to-moderate significant signal was detected in 9 out of 12 services.

Categories of Services of Woody Species ($n = 1400$)	Counts of States	Estimated D	P_{rand}	P_{BM}	Interpretation
Service 1 Human Food (edible fruits, edible starchy roots, edible nuts, beverages)	0 = 169 1 = 39	0.5145573	<0.001	0.002	Weak signal, but non-random
Service 2 Livestock (Fodder and forage)	0 = 193 1 = 15	0.6435349	0.006	0.01	Weak signal, but non-random
Service 3 Medicinal (Human and animal treatment-medicinal oils, purgatives, skin infections, ringworms and other ailments)	0 = 178 1 = 30	0.6578688	<0.001	<0.001	Weak signal, but non-random
Service 4 Body and house care (perfume, essential oils for hair and skin, face and Skin Mask, Exfoliants and Wash, Polishes, Soaps, detergents, Shampoos)	0 = 196 1 = 12	0.5787294	0.001	0.047	Weak signal, but non-random
Service 5 Coloring Substances (Tanbarks, Dyes and Inks)	0 = 200 1 = 8	1.022189	0.508	<0.001	No signal, random
Service 6 Insect Attractants Repellents (Butterflies, Bees, Ants, Bugs, Mosquitoes and Worms)	0 = 172 1 = 36	0.5870513	<0.001	0.001	Weak signal, but non-random
Service 7 Hunting Fishing (Fish and Arrow Poison)	0 = 207 1 = 1	−2.105872	0.093	0.766	NA (there is only one state for 1)
Service 8 Soil Management (Soil Stabilization, Sand-binding, Dune Stabilization and Dune Reclamation)	0 = 191 1 = 17	0.2478724	<0.001	0.0217	Moderate signal, but non-random
Service 9 Fuels Biofuels (Firewood, Woodchips, Biofuel and Charcoal)	0 = 179 1 = 29	0.7891897	0.02	<0.001	Weak signal, but non-random
Service 10 Construction and Manufacturing Materials (Poles, Fence Posts, Timber, Shelter, Fibers, Ropes, Fish Nets, Carving, Windbreak, Hedging and Screening)	0 = 97 1 = 111	0.6462119	<0.001	<0.001	Weak signal, but non-random
Service 11 Ornamental (Indoor and Outdoor Ornament, Street Trees and Shade)	0 = 32 1 = 176	0.4971587	<0.001	0.001	Moderate signal, but non-random
Service 12 Cultural Religious (Traditional, Magical, Religious/Spiritual Values)	0 = 204 1 = 4	0.9566721	0.352	0.046	No signal

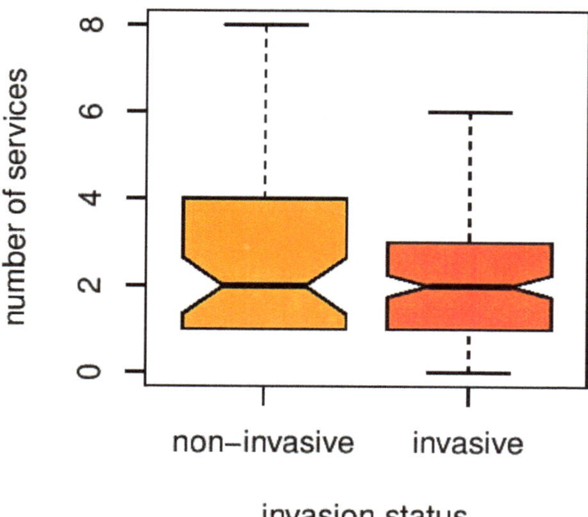

Figure 1. Relationships between the number of known services provided by alien species and their invasion status.

Finally, when we tested for the link between each of the 12 services and invasion status, we found that, among the 12 services recorded, only three services (food, medicine, and fuel) show significant correlations with invasion status, but this correlation is negative (Figure 2), as follows: food ($\beta = -1.20 \pm 0.37$, $p = 0.001$); medicine ($\beta = -1.31 \pm 0.41$, $p = 0.00124$); fuel ($\beta = -0.88 \pm 0.42$, $p = 0.03$), implying that these services tend to be provided by non-invasive species.

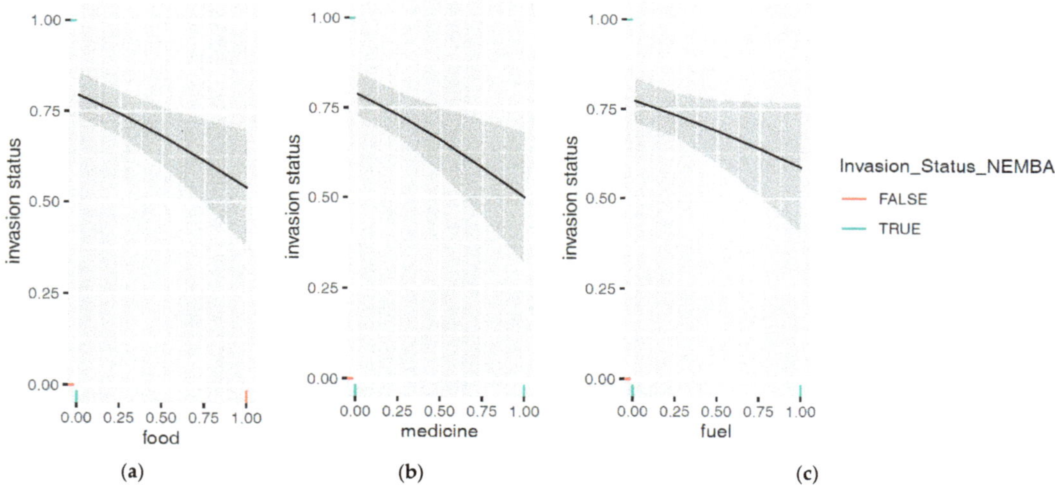

Figure 2. Relationships between invasion status and each service category. Only the three services that show significant correlation with invasion status are presented, which are as follows: (**a**) food, (**b**) medicine, (**c**) fuel. Invasion status is coded as follows: 0 (non-invasive) and 1 (invasive). The following are for service 'food': 0 (a species is not used as food) and 1 (the species is used as food). Same for medicine and fuel.

4. Discussion

The phylogeny used in the present study, and reconstructed using two markers of DNA barcodes for plants (rbcLa + matK), has been used in several studies to test various ecological hypotheses [21,34]. Using this DNA barcode phylogeny, our study indicates that there is a phylogenetic signal in the services provided by intentionally introduced alien woody plant species used by humans in South Africa. From a phylogenetic perspective, this is an indication of non-random selection of alien plants. However, taxonomic non-random plant selection was initially suggested almost four decades ago to explain human–plant interactions, particularly for native plants used in traditional medicine [6,41,42]. This was later supported in several other studies, but mostly for native plants used in traditional medicine [8,43–45]. Nonetheless, only a few studies have tested whether this taxonomic signal translates into a phylogenetic signal (e.g., [46,47]). In addition, the question of whether alien species introduced into a new environment follow the general pattern of non-random selection is not yet widely explored (but see [48]). The present study contributes to filling such a gap, showing that alien woody species in South Africa are not randomly selected; phylogenetically closely related alien species, intentionally introduced into the region, tend to provide similar services. Can services then be used to predict the invasion status of these alien species?

Contradicting evidence has been provided in several studies that investigated the correlates of invasion success, suggesting the context specificity of invasions. Because species' functional roles (ecology and physiology) in ecosystems are linked to the services that they provide to humans [49], and functional traits drive invasion success [21,28], our expectation is that the services should also correlate with invasion success [29]. In addition, if an alien species provides a higher number of services, the chances are greater for that species not only to be sought after, but also to be introduced in a high number and independently multiple times into new environments. This is predicted in the propagule pressure theory [30–32]. Indeed, the propagule pressure theory has been demonstrated in several studies for different taxa in various geographic regions [30,50,51]. In the present study, we found a significant correlation between the number of services (which may indicate propagule pressure) and invasion status, but, contrary to expectation, it is alien non-invasive species that provide more services than invasive species. This pattern is maintained whether we corrected for phylogeny or not, and supports the finding reported in a recent study that naturalized plants provide more services than plants that are not naturalized [29].

This counter-intuitive finding may be expected if our dataset comprises a large proportion of unintentionally introduced alien species (this is not the case). It is also possible that an alien species can be introduced both intentionally and unintentionally into an area. This possibility may a priori complicate the detection of a strong correlation between services and invasion status. However, in our case, we focused only on alien species recorded as intentionally introduced and for which the services these species provide to humans are relatively well documented. As such, even if some of these intentionally introduced species are also transported through unintentional introduction pathways, this would simply increase the propagule pressure of the species and would support our expectations of strong relationships between the number of services (used as proxy for propagule pressure) and invasion status. Furthermore, even if we assume that some of the alien species in our list (Table S1) are unintentionally introduced, the fact that these species are now recorded as providing some services to humans implies that humans may further cultivate these species (for the services they provide), thus contributing to the spread of the species. In such a scenario, our hypothesis of a strong relationship between services and invasion would still hold, since we are not analyzing species' traits, but the services they provide. This scenario would actually make it meaningless to distinguish between intentional and unintentional introduction, since what matters in our approach is the services that species provide (not their ecological traits). In addition, the counter-intuitive finding reported here could possibly be because alien non-invasive species might not yet have enough residence time in their new environment to become invasive. It could also be because the variable "number of services" is not a strong proxy for propagule pressure, as an alien species with a high number of services may not necessarily be introduced multiple times in an area; a species S1 with only one known service may be introduced several times and more often than a species S2 with multiple services if those multiple services are less valuable to communities (i.e., lower use value) than the single service of S1. A further possibility is that the lack of positive relationships between the number of services and invasion status could be because of the differences in species' performances (ecological, physiological) in different environments/habitats in the same country; alien species providing a similar number of services for different human communities may perform differently in different environments or ecosystems. Another possibility is that the NEMBA alien categorization itself may be a source of concern, due to human misjudgment or bias, or decisions that are not ecologically informed, since the NEMBA list was generally criticized for not being science-based (it was allegedly influenced by politics). Potential bias in the list may perhaps lead to the unexpected results that we found. However, the fact that our findings mirror what was recently reported at the global scale (see ref. [29]) means that the NEMBA list may not be a profoundly biased representation of alien invasion status in South Africa. These various scenarios that potentially explain our findings call for future studies that link species' use values to their alien invasion status.

When we tested the link between invasion status and each of the service categories, only three services (food, medicine, and fuel) correlated significantly, in a negative direction, with invasion status, suggesting that these services tend to be provided by non-invasive species. All these findings confirm that alien non-invasive species tend to provide more services to humans than alien invasive species, corroborating a recent finding that naturalized species provide more services to humans at the global scale [29]. These findings prompt the following key question: by harvesting alien plants for human use, do humans limit their ability to invade? Although we did not test this hypothesis, we strongly suspect this possibility, given that alien plants providing more services, or even specific services (food, medicine), tend to be non-invasive (naturalized) or are geographically constrained.

Overall, by aiming to link services to invasion, this study pointed to potential roles played by human choices of specific products (e.g., plants for medicine) in driving species invasion. Our tests reveal unexpected evidence that alien non-invasive species provide more or unique services to humans in comparison to alien invasive species, supporting the recent similar finding reported at the global scale (see ref. [29]). Although a number of scenarios are plausible to explain our finding, we suggest that human utilization/harvest of alien species may constrain their ability to spread and become invasive. This requires further investigations. Other studies in other geographies have shown the following similar finding with that reported in the present study: the uses of alien plants by humans determine their outcome along the introduction–naturalization–invasion continuum [52–54]. Our study also provides additional evidence that DNA barcodes, initially thought of as a taxonomic tool (e.g., [55]), can be used beyond taxonomy and for ecological investigations (see reviews in ref. [56]).

Supplementary Materials: The following are available online at https://www.mdpi.com/1424-2818/13/11/553/s1: Table S1. The raw data collected for the present study. The definitions of Services 1–12 are in Table 1.

Author Contributions: Conceptualization, K.Y.; methodology, K.Y.; software, K.Y.; validation, K.Y.; formal analysis, K.Y. and A.E.A.; investigation, K.Y. and A.E.A.; resources, K.Y.; data curation, A.E.A.; writing—original draft preparation, K.Y.; writing—review and editing, K.Y.; visualization, K.Y.; supervision, K.Y.; project administration, K.Y.; funding acquisition, K.Y. All authors have read and agreed to the published version of the manuscript.

Funding: This research was funded by National Research Foundation, grant number Grant No: 112113.

Institutional Review Board Statement: Not applicable.

Informed Consent Statement: Not applicable.

Data Availability Statement: The data presented in this study are available in Supplementary Materials.

Acknowledgments: We acknowledge the South Africa's National Research Foundation (NRF) Research Development Grants for Y-Rated Researchers (Grant No: 112113). We thank four anonymous reviewers for their contributions to the improvement of an earlier version.

Conflicts of Interest: The authors declare no conflict of interest.

References

1. Poynton, R.J. *Tree Planting in Southern Africa. Volume 3: Other Genera*; Department of Agriculture, Forestry and Fisheries: Pretoria, South Africa, 2009.
2. Bennett, B.M. El Dorado of Forestry: The *Eucalyptus* in India, South Africa and Thailand, 1850–2000. *Int. Rev. Soc. Hist.* **2010**, *55*, 27–50. [CrossRef]
3. Blanchard, R.; Kumschick, S.; Richardson, D.M. Biofuel plants as potential invasive species: Environmental concerns and progress towards objective risk assessment. In *Roadmap for Sustainable Biofuels in Southern Africa: Regulatory Frameworks for Improved Development? Recht und Verfassung in Afrika—Law and Constitution in Africa*; Ruppel, O.C., Dix, H., Eds.; Nomos Verlagsges: Baden-Baden, Germany, 2017; Volume 30, pp. 47–60.
4. Novoa, A.; Le Roux, J.J.; Richardson, D.M.; Wilson, J.R.U. Level of environmental threat posed by horticultural trade in Cactaceae. *Conserv. Biol.* **2017**, *31*, 1066–1075. [CrossRef] [PubMed]
5. Zengeya, T.; Ivey, P.; Woodford, D.J.; Weyl, O.; Novoa, A.; Shackleton, R.; Richardson, D.; Van Wilgen, B. Managing conflict-generating invasive species in South Africa: Challenges and trade-offs. *Bothalia* **2017**, *47*, a2160. [CrossRef]
6. Gaoue, O.G.; Yessoufou, K.; Mankga, L.; Vodouhe, F. Phylogeny reveals non-random medicinal plant organs selection by local people in Benin. *Plants People Planet* **2021**, *3*, 710–720. [CrossRef]
7. Robles Arias, D.M.; Cevallos, D.; Gaoue, O.G.; Fadiman, M.G.; Hindle, T. Non-random medicinal plants selection in the Kichwa community of the Ecuadorian Amazon. *J. Ethnopharmacol.* **2020**, *246*, 112220. [CrossRef]
8. Moerman, D.E. An analysis of the food plants and drug plants of native North America. *J. Ethnopharmacol.* **1996**, *52*, 1–22. [CrossRef]
9. Ford, J.; Gaoue, O.G. Alkaloid-poor plant families, Poaceae and Cyperaceae, are over-utilized for medicine in Hawaiian Pharmacopoeia. *Econ. Bot.* **2017**, *71*, 123–132. [CrossRef]
10. Muleba, I.; Yessoufou, K.; Rampedi, I.T. Testing the non-random hypothesis of medicinal plant selection using the woody flora of the Mpumalanga Province, South Africa. *Environ. Dev. Sustain.* **2021**, *23*, 4162–4173. [CrossRef]
11. Yessoufou, K.; Mearns, K.; Elansary, H.O.; Stoffberg, G.H. Assessing the phylogenetic dimension of Australian Acacia species introduced outside their native ranges. *Bot. Lett.* **2016**, *163*, 33–39.
12. Gaoue, O.G.; Coe, M.A.; Bond, M.; Hart, G.; Seyler, B.C.; McMillen, H. Theories and major hypotheses in ethnobotany. *Econ. Bot.* **2017**, *71*, 269–287. [CrossRef]
13. Essl, F.; Hulme, P.E.; Jeschke, J.M.; Keller, R.; Pyšek, P.; Richardson, D.M.; Saul, W.-C.; Bacher, S.; Dullinger, S.; Estévez, R.A.; et al. Scientific and Normative Foundations for the Valuation of Alien-Species Impacts: Thirteen Core Principles. *BioScience* **2017**, *67*, 166–178. [CrossRef]
14. Richardson, D.M.; Pyšek, P.; Rejmánek, M.; Barbour, M.G.; Panetta, D.F.; West, C.J. Naturalization and invasion of alien plants: Concepts and definitions. *Divers. Distrib.* **2000**, *6*, 93–107. [CrossRef]
15. Pimentel, D.; Zuniga, R.; Morrison, D. Update on the environmental and economic costs associated with alien invasive species in the United States. *Ecol. Econ.* **2005**, *52*, 273–288. [CrossRef]
16. Pyšek, P.; Jarošík, V.; Pergl, J.; Moravcová, L.; Chytrý, M.; Kühn, I. Temperate trees and shrubs as global invaders: The relationship between invasiveness and native distribution depends on biological traits. *Biol. Invasions* **2014**, *16*, 577–589. [CrossRef]
17. Shackleton, R.T.; Witt, A.B.R.; Nunda, W.; Richardson, D.M. *Chromolaena odorata* (Siam weed) in Eastern Africa: Distribution and socio-ecological impacts. *Biol. Invasions* **2017**, *19*, 1285–1298. [CrossRef]
18. Ortega, Y.K.; Pearson, D.E. Weak Vs. Strong Invaders of Natural Plant Communities: Assessing Invasibility and Impact. *Ecol. Soc. Am.* **2005**, *15*, 651–661. [CrossRef]
19. Wilson, J.R.U.; Gaertner, M.; Richardson, D.M.; Van Wilgen, B.W. Contributions to the National Status Report on Biological Invasions in South Africa–Research. *Bothalia* **2017**, *47*, 1–8. [CrossRef]
20. Bezeng, B.S.; Savolainen, V.; Yessoufou, K.; Papadopulos, A.S.T.; Maurin, O.; Van der Bank, M. A phylogenetic approach towards understanding the drivers of plant invasiveness on Robben Island, South Africa. *Bot. J. Linn. Soc.* **2013**, *172*, 142–152. [CrossRef]
21. Bezeng, S.B.; Davies, J.; Yessoufou, K.; Maurin, O.; Van der Bank, M. Data from: Revisiting Darwin's naturalization conundrum: Explaining invasion success of non-native trees and shrubs in Southern Africa. *J. Ecol.* **2015**, *103*, 871–879. [CrossRef]
22. Hirsch, H.; Richardson, D.M.; Le Roux, J.J. Introduction to the special issue: Tree invasions: Towards a better understanding of their complex evolutionary dynamics. *AoB Plants* **2017**, *9*, plx014. [CrossRef]
23. Hui, C.; Richardson, D.M. *Invasion Dynamics*; Oxford University Press: Oxford, UK, 2017.
24. Kolar, C.S.; Lodge, D.M. Progress in invasion biology: Predicting invaders. *Trends Ecol. Evol.* **2001**, *16*, 199–204. [CrossRef]

25. Cadotte, M.W.; Murray, B.R.; Lovett-Doust, J. Evolutionary and ecological influences of plant invader success in the flora of Ontario. *Ecoscience* **2006**, *13*, 388–395. [CrossRef]

26. Pyšek, P.; Richardson, D.M. Traits associated with invasiveness in alien plants: Where do we stand. In *Biological Invasions. Series Ecological Studies*; Nentwig, W., Ed.; Springer: Berlin/Heidelberg, Germany, 2007; Volume 193, pp. 97–126.

27. Wolkovich, E.M.; Davies, T.J.; Schaefer, H.; Cleland, E.E.; Cook, B.I.; Travers, S.E.; Willis, C.G.; Davis, C.C. Temperature-dependent shifts in phenology contribute to the success of exotic species with climate change. *Am. J. Bot.* **2013**, *100*, 1407–1421. [CrossRef]

28. Pyšek, P.; Jarošík, V.; Chytrý, M.; Danihelka, J.; Kühn, I.; Pergl, J.; Tichý, L.; Biesmeijer, J.C.; Ellis, W.N.; Kunin, W.E.; et al. Successful invaders co-opt pollinators of native flora and accumulate insect pollinators with increasing residence time. *Ecol. Soc. Am.* **2011**, *81*, 277–293. [CrossRef]

29. van Kleunen, M.; Xu, X.; Yang, Q.; Maurel, N.; Zhang, Z.; Dawson, W.; Essl, F.; Kreft, H.; Pergl, J.; Pyšek, P.; et al. Economic use of plants is key to their naturalization success. *Nat. Commun.* **2020**, *11*, 3201. [CrossRef] [PubMed]

30. Lockwood, J.L.; Cassey, P.; Blackburn, T. The role of propagule pressure in explaining species invasions. *TRENDS Ecol. Evol.* **2005**, *20*, 223–228. [CrossRef]

31. Blackburn, T.M.; Duncan, R.P. Determinants of establishment success in introduced birds. *Nature* **2001**, *414*, 195–197. [CrossRef]

32. Carlton, J.T. Pattern, process, and prediction in marine invasion ecology. *Biol. Conserv.* **1996**, *78*, 97–106. [CrossRef]

33. Coates-Palgrave, M. *Keith Coates-Palgrave Trees of Southern Africa*, 3rd ed.; Struik Nature: Cape Town, South Africa, 2002.

34. Maurin, O.; Davies, T.J.; Burrows, J.E.; Daru, B.H.; Yessoufou, K.; Muasya, A.M.; Van der Bank, M.; Bond, W.J. Savanna fire and the origins of the 'underground forests' of Africa. *New Phytol.* **2014**, *204*, 201–214. [CrossRef]

35. Kumschick, S.; Wilson, J.R.U.; Llewellyn, C. A framework to support alien species regulation: The Risk Analysis for Alien Taxa (RAAT). *NeoBiota* **2020**, *62*, 213–239. [CrossRef]

36. Van Wyk, B.; Van Wyk, P. *Field Guide to Trees of Southern Africa*, 2nd ed.; Struik Publisher: Cape Town, South Africa, 2013.

37. Glen, H.; Van Wyk, B. *Guide to Trees Introduced into Southern Africa*; Penguin Random House South Africa: Cape Town, South Africa, 2016.

38. Diazgranados, M.; Allkin, B.; Black, N.; Cámara-Leret, R.; Canteiro, C.; Carretero, J.; Eastwood, R.; Hargreaves, S.; Hudson, A.; Milliken, W.; et al. *World Checklist of Useful Plant Species. Produced by the Royal Botanic Gardens, Kew*; Knowledge Network for Biocomplexity: London, UK, 2020.

39. Drummond, A.J.; Rambaut, A. BEAST: Bayesian evolutionary analysis by sampling trees. *BMC Evol. Biol.* **2007**, *7*, 214. [CrossRef]

40. Fritz, S.A.; Purvis, A. Selectivity in mammalian extinction risk and threat types: A new measure of phylogenetic signal strength in binary traits. *Conserv. Biol.* **2010**, *24*, 1042–1051. [CrossRef]

41. Ho, L.S.T. and Ane, C. A linear-time algorithm for Gaussian and non-Gaussian trait evolution models. *Syst. Biol.* **2014**, *63*, 397–408. [PubMed]

42. O'Hara, R.B.; Kotze, D.J. Do not log-transform count data. *Methods Ecol. Evol.* **2010**, *1*, 118–122. [CrossRef]

43. Moerman, D.E. Symbols and selectivity: A statistical analysis of native American medical ethnobotany. *J. Ethnopharmacol.* **1979**, *1*, 111–119. [CrossRef]

44. Moerman, D.E. The medicinal flora of native North America: An analysis. *J. Ethnopharmacol.* **1991**, *31*, 1–42. [CrossRef]

45. Gaoue, O.G. Moutouama, Coe, J.K.; Bond, M.A.; Green, M.O.E.; Sero, N.B.; Bezeng, B.S.; Yessoufou, K. Methodological advances for hypothesis-driven ethnobiology. *Biol. Rev.* **2021**, *96*, 2281–2303. [CrossRef] [PubMed]

46. Yessoufou, K.; Daru, B.H.; Muasya, A.M. Phylogenetic exploration of commonly used medicinal plants in South Africa. *Mol. Ecol. Resour.* **2015**, *15*, 405–413. [CrossRef]

47. Saslis-Lagoudakis, C.H.; Klitgaard, B.B.; Forest, F.; Francis, L.; Savolainen, V.; Williamson, E.M.; Hawkins, J.A. The use of phylogeny to interpret cross-cultural patterns in plant use and guide medicinal plant discovery: An example from pterocarpus (leguminosae). *PLoS ONE* **2011**, *6*, e22275. [CrossRef] [PubMed]

48. Canavan, S.; Richardson, D.M.; Visser, V.; Le Roux, J.J.; Vorontsova, M.S.; Wilson, J.R.U. The global distribution of bamboos: Assessing correlates of introduction and invasion. *AoB PLANTS* **2017**, *9*, plw078. [CrossRef] [PubMed]

49. Reich, P.B. The world-wide 'fast-slow' plant economics spectrum: A traits manifesto. *J. Ecol.* **2014**, *102*, 275–301. [CrossRef]

50. Colautti, R.I.; Grigorovich, I.A.; MacIsaac, H.J. Propagule pressure: A null model for biological invasions. *Biol. Invasions* **2006**, *8*, 1023–1037. [CrossRef]

51. Von Holle, B.; Simberloff, D. Ecological Resistance to Biological Invasion Overwhelmed by Propagule Pressure. *Ecol. Soc. Am.* **2005**, *86*, 3212–3218.

52. Pyšek, P.; Manceur, A.M.; Alba, C.; McGregor, K.F.; Pergl, J.; Stajerová, K.; Chytrý, M.; Danihelka, J.; Kartesz, J.; Klimesova, J.; et al. Naturalization of central European plants in North America: Species traits, habitats, propagule pressure, residence time. *Ecology* **2015**, *96*, 762–774. [CrossRef] [PubMed]

53. Pyšek, P.; Sádlo, J.; Mandák, B.; Jarošik, V. Czech alien flora and the historical pattern of its formation: What came first to Central Europe? *Oecologia* **2003**, *135*, 122–130. [CrossRef] [PubMed]

54. Pergl, J.; Pyšek, P.; Bacher, S.; Essl, F.; Genovesi, P.; Harrower, C.A.; Hulme, P.E.; Jeschke, J.E.; Kenis, M.; Kühn, I.; et al. Troubling travellers: Are ecologically harmful alien species associated with particular introduction pathways? *NeoBiota* **2017**, *32*, 1–20. [CrossRef]

55. Van der Bank, F.H.; Herbert, D.; Greenfield, R.; Yessoufou, K. Revisiting species delimitation within the genus *Oxystele* using DNA barcoding approach. *ZooKeys* **2013**, *365*, 337–354.
56. Bezeng, B.S.; Davies, T.J.; Daru, B.H.; Kabongo, R.M.; Maurin, O.; Yessoufou, K.; van der Bank, H.; van der Bank, M. Ten years of plant DNA barcoding at the African Centre for DNA Barcoding. *Genome* **2017**, *60*, 629–638. [CrossRef] [PubMed]

Review

Using DNA Metabarcoding to Identify Floral Visitation by Pollinators

Abigail Lowe [1,2,*], Laura Jones [1], Lucy Witter [1,3], Simon Creer [2] and Natasha de Vere [4]

[1] Science Department, National Botanic Garden of Wales, Llanarthne SA32 8HG, UK; laura.jones@gardenofwales.org.uk (L.J.); lucy.witter@gardenofwales.org.uk (L.W.)
[2] Molecular Ecology and Evolution Group, School of Natural Sciences, Bangor University, Bangor LL57 2UW, UK; s.creer@bangor.ac.uk
[3] Institute of Biological, Environmental and Rural Sciences, Aberystwyth University, Aberystwyth SY23 3FL, UK
[4] Natural History Museum of Denmark, University of Copenhagen, 1350 Copenhagen K, Denmark; natasha.de.vere@snm.ku.dk
* Correspondence: abigail.lowe@gardenofwales.org.uk

Abstract: The identification of floral visitation by pollinators provides an opportunity to improve our understanding of the fine-scale ecological interactions between plants and pollinators, contributing to biodiversity conservation and promoting ecosystem health. In this review, we outline the various methods which can be used to identify floral visitation, including plant-focused and insect-focused methods. We reviewed the literature covering the ways in which DNA metabarcoding has been used to answer ecological questions relating to plant use by pollinators and discuss the findings of this research. We present detailed methodological considerations for each step of the metabarcoding workflow, from sampling through to amplification, and finally bioinformatic analysis. Detailed guidance is provided to researchers for utilisation of these techniques, emphasising the importance of standardisation of methods and improving the reliability of results. Future opportunities and directions of using molecular methods to analyse plant–pollinator interactions are then discussed.

Keywords: DNA metabarcoding; pollen; pollinators; pollen metabarcoding; plant–pollinator interactions; DNA barcoding; honeybees; bumblebees; hoverflies

Citation: Lowe, A.; Jones, L.; Witter, L.; Creer, S.; de Vere, N. Using DNA Metabarcoding to Identify Floral Visitation by Pollinators. *Diversity* **2022**, *14*, 236. https://doi.org/10.3390/d14040236

Academic Editors: W. John Kress, Morgan Gostel and Michael Wink

Received: 28 January 2022
Accepted: 16 March 2022
Published: 24 March 2022

Publisher's Note: MDPI stays neutral with regard to jurisdictional claims in published maps and institutional affiliations.

1. Background

Understanding the relationship between plants and pollinators is vital for biodiversity conservation, food security, and ecosystem sustainability [1,2]. Worldwide, there are approximately 350,000 animal pollinator species, of which insects contribute a significant proportion [3]. Despite the importance of pollinators, evidence of declines in species richness and abundance are increasing across the globe [4,5]. The most significant drivers of decline are land use change, pesticides, climate change, pests, and pathogens [6–9].

DNA metabarcoding provides a powerful tool for investigating pollinator foraging preferences and should be a standard part of the ecologist's toolkit. The aim of this review is to describe the range of approaches and methods available, along with their opportunities and challenges. We thoroughly explore the ecological questions that can be answered from identifying floral visitation across a range of species and habitats and present a summary of findings from the literature. The entire pollen metabarcoding workflow is described along with considerations and guidance for each step, in the hope of inspiring more researchers to adopt these techniques.

Identifying floral visitation can provide an insight into the resources used by insects and the pollination services they deliver [10]. Whilst the methods described here do not directly detect the process of pollination [11], we use the term pollinators as a general term to refer to flower-visiting insects.

2. Methods for Identifying Floral Visitation by Pollinators

Floral visitation studies may be plant- or insect-focused. Examples of insect-focused methods include observational methods such as mark recapture using paint, plastic tags [12], or harmonic radar [13]. In addition, waggle dances, used by honeybees to communicate the location of resources to the colony [14], can be de-coded to elucidate forage preferences and behaviour [15]. Floral visitation may also be investigated by identifying the pollen collected by the insect. Pollen microscopy has been widely utilised for diet characterisation by identifying pollen grains obtained from body parts of individuals, e.g., mouthparts [16], scopa [17] and entire bodies [18,19], or honey [20,21] and nest provisions [22,23]. However, the identification of pollen grains to species level using light microscopy is difficult and time-consuming [24]. In recent years, automated machine learning systems have been developed to identify pollen from images and are showing great promise [25–27].

Pollen may also be identified by DNA metabarcoding: a process involving large-scale identification of unknown taxa within a mixed sample using DNA barcode markers and high-throughput sequencing [28–30]. The DNA contained in the sample is compared to a reference library composed of DNA sequences of a standard genetic marker. For plants, parts of the genes coding for ribulose bisphosphate carboxylase large subunit (*rbcL*) and maturase K (*matK*) are recommended as standard markers due to their universality across land plants [31]. However, the length of *matK* (around 800 bp) and the requirement for multiple primer combinations to gain taxonomic coverage makes it less suitable for amplicon-based metabarcoding [32]. Instead, additional markers such as the non-coding nuclear internal transcribed region ITS2, the *trnL* intron, and the non-coding intergenic spacer *trnH-psbA* are often used, either alone, or alongside *rbcL* for increased species discrimination [33]. DNA metabarcoding has been used to successfully identify pollen from provisions within nests [34–36], honey [37–39], proboscises [16,40], guts [41,42], and the legs or bodies of insects [43–45] (Table S1).

Shotgun metagenomics is an alternative tool which can be used to identify taxonomic diversity within a mixed sample using untargeted sequencing of genomic fragments mapped to whole genomes or barcode regions [46,47]. By mapping genome-skims to a constructed reference library of plastid genomes, Lang et al. [48] demonstrated quantitative identification of >97% taxa in mixed pollen samples. The advantages of metagenomic methods are the option of PCR-free processes which reduce possible amplification biases, the ability to output long read lengths, and the increased taxonomic resolution compared to targeted sequencing of specific regions [46,49]. The main limitation facing whole-genome studies is that currently, few whole plant genomes are available, resulting in difficulties assembling reference material [46]. A further promising approach is the use of reverse metagenomics to map long reads produced by the MinION to genomic skims, a method which has produced semi-quantitative identification of plant species in mixed pollen loads [49].

Plant-focused methods of identifying floral visitation provide an alternative perspective to insect-focused methods. Interactions between plants and pollinators can be characterised through observing which insects visit plants (plant-focused) [50–53]. Two methods are commonly adopted: timed observations of plants with the frequency of each insect visit recorded [53,54], and transect or plot walks where individuals within a survey area are identified when visiting plants [52,55,56]. For both methods, insects are either identified in the field or captured for later identification. An example of a more novel plant-focused approach to elucidating floral visitation is through the method of obtaining residual insect DNA from flowers [57]. Similarly, the identification of 'microbial signatures' specific to pollinators within nectar can also be used to elucidate visitation [58,59].

3. Plant vs. Pollinator Perspective of Foraging

Recording floral visitation from the perspective of the plant or the insect will yield varying information [60–63], and each method of recording visits from either perspective has its advantages and disadvantages. Plant-focused surveys using visual observations are the most common method of analysing plant–pollinator interactions, providing a quantitative measure of the frequency of interactions between species [55]. A key advantage of visual surveys is that there is an opportunity to supplement observational data with environmental metadata such as the time of interaction [64], weather conditions [65], plant colour [66], and horticultural variety [67], which can be used to explore further questions surrounding foraging behaviour. In addition, the type of resource (pollen, nectar, or resin) collected by pollinators can be identified [55], a vital component of pollinator ecology. It is often possible to identify both the plants and insects to species, providing pollinators are retained for identification through morphology or DNA barcoding [68].

The characterisation of interactions between plants and pollinators using plant-focused observations are usually grouped at the species level [52] due to difficulties tracking individuals [12]. This means that quantitative data (e.g., frequency of visits) can only be gained at the species level and information regarding individual foraging trips is inaccessible. Moreover, the period of observation is often limited both spatially and temporally, resulting in a bias towards abundant pollinator species [63]. As a result, interactions may be missed [62] and those of rare individuals may appear more specialised than in reality [63]. As a result, sampling effort is a major determining factor of the number of relationships which are recorded [69]. Further, the method used to observe interactions (e.g., transects, timed observations) will also lead to biases which should be considered when constructing networks [70]. Increasing the sampling effort by increasing the time spent surveying can increase the likelihood of capturing rare interactions and thus reducing the incidence of specialisation [71]. Identifying floral visitation through molecular analysis of remnant DNA on flowers provides an opportunity to increase the temporal scope of plant-focused surveys, whilst increasing the likelihood of detecting rare interactions in comparison to plant-focused visit surveys [57].

The use of visual and electronic aids to track insects such as paint, plastic tags [12], or harmonic radar [13] provides information on individual foraging to be determined, offering a different perspective compared to plant-focused observations. DNA metabarcoding and pollen microscopy allow for an increased insight into interactions which may be missed through observations [42,61,72–75]. These methods are free from the spatial limitations of observations which come as a result of visual bias, e.g., height [19], as they provide a record of any resources which have been accessed by the individual which may be up to several kilometres away [76]. For example, by analysing pollen loads of bumblebees, Carvell et al. [77] found that the dominant plant in pollen loads was not always the plant the bee had been caught on, demonstrating that observation of floral networks does not reveal all interactions with visitors.

Arstingstall et al. [78] found that when comparing plant–pollinator networks characterised by DNA metabarcoding of pollen to those constructed from observations of foraging bees, networks constructed from molecular analysis had increased species richness and reduced specialisation. By identifying the pollen assemblage carried by insects, it is possible to gain a semi-quantitative measure of frequency of use per individual (discussed in detail within the methodological considerations) [44,79]. The collection of insects for pollen analysis also allows specimens to be retained for identification through traditional morphology or DNA barcoding [68].

Nevertheless, insect-focused methods of identifying floral visitation are not free from biases or limitations. Some insect-focused methods of tracking pollinators can also suffer from spatial limitations such as tag ranges [13]. During observations the time spent foraging can be recorded; however, it is difficult to distinguish the temporal range of pollen found on an insect's body. Further, the identification of pollen from insects does not provide information on whether plants were visited to collect pollen, or incidental pollen transfer through visitation for nectar or resin collection, or, indeed, pollen that has collected on the body of an insect whilst it has been flying. Interactions observed through visual surveys can be undetected using DNA metabarcoding and pollen microscopy, owing to their rarity [74], pollen accessibility [73], or use for nectar with limited or no pollen production [42,80]. These factors reduce the amount of pollen transferred to the insect and therefore identified. However, whilst rare interactions may still be missed through the identification of pollen, they are more likely to be captured than through plant-focused surveys [63].

Whilst both pollen microscopy and DNA metabarcoding yield valuable individual-level information on foraging, identification of plant taxa using DNA eliminates the need for expert palynologists for microscopy. Although also time-consuming and initially expensive [81], molecular processes may be easily scaled up [82,83]. In pollen microscopy, a small sub-sample is fully identified and used to estimate the composition of the total pollen load [63], whereas molecular analysis can sample the entire pollen assemblage on the body of an insect [45,84]. Although there is some congruence between the taxa which are difficult to identify using microscopy and those which are indistinguishable using DNA, e.g., some taxa within the Rosaceae family [38], both methods may detect additional taxa when compared to the other [83,85]. In comparing pollen microscopy and DNA metabarcoding, several authors have found higher taxonomic resolution of plant taxa identified [16,86] and a greater number of species detected [83,85–87] using DNA metabarcoding. For example, when comparing the use of metabarcoding and microscopy to characterise pollen transport networks in moths, Macgregor et al. [16] found that metabarcoding detected more interactions per moth species. This was likely due to the increased discriminatory power of metabarcoding which allows some pollen types to be separated to a lower taxonomic level than through microscopy [16]. Both methods, however, are subject to the stochasticity of detecting rare taxa [87,88].

The method used to identify floral visitation is dependent on the type of question being asked. In order to create highly resolved plant–pollinator interaction networks, it is recommended that a combination of plant- and insect-focused methods are used [16,62,63,89].

4. Using DNA Metabarcoding to Answer Questions about Pollinator Foraging Preferences

The use of DNA metabarcoding to answer ecological questions about pollinator foraging preferences has increased rapidly over recent years alongside key methodological developments (Table S1). A range of taxonomic groups have been studied; however, the research is predominately focused on wild bees (e.g., *Bombus, Megachile, Osmia*), managed bees (e.g., *Apis mellifera, Tetragonula carbonaria*), and hoverflies (Syrphidae). The questions addressed can be broadly grouped into four topics: (1) How does foraging change throughout time and space? (2) How is foraging affected by resource availability? (3) How are resources partitioned between species and individuals in a plant–pollinator network? (4) What is the relationship between plant use and pollinator health?

4.1. How Does Foraging Change throughout Time and Space?

DNA metabarcoding provides a useful method for monitoring plant use across wide spatiotemporal scales, such as multiple countries or regions [90] and, when compared with historical data, time periods over decades or centuries [38,91,92]. The reproducibility of DNA metabarcoding allows for continued sampling of foraging across a species' entire flight period, providing an understanding of plant selection at specific time points. When assessing foraging habits of pollinators throughout the year, it is often found that the amount and diversity of pollen collected is strongly influenced by season, most likely influenced by the phenology of surrounding plants [45,93–95]. In addition to tracking contemporary foraging habits, DNA metabarcoding has been shown to be a useful tool for analysing pollen from historical specimens [91,96,97]. By sequencing plant DNA from pollen obtained from museum specimens, Simanonok et al. [97] successfully identified the plants used by an endangered bumblebee species over 100 years, vastly improving current knowledge of resource use and mechanisms of decline. Similarly, analysing the pollen DNA within UK honey and comparing the plant diversity to samples characterised 65 years prior using microscopy revealed landscape-scale shifts in foraging habits due to changes in agricultural intensification, crop use, and the spread of invasive species [38].

Long-range movements can be tracked by identifying pollen on migrating insects [40,98]. Suchan et al. [98] detected plant species endemic to Africa on butterflies using DNA metabarcoding, significantly improving the understanding of migration patterns which were previously limited when using traditional techniques. As well as increasing the spatial scale of studies, pollen metabarcoding has highlighted the importance of trees and woody species to pollinators, plants with flowers which are often visually restricted and therefore may be missed during observational surveys [37,99]. Whilst most of these spatial assessments of foraging focus on geographic differences, only one study has specifically demonstrated the ability of pollen metabarcoding to elucidate changes in resource use across elevational gradients to better understand how climatic changes in the environment impact foraging of a solitary bee [41].

4.2. How Is Foraging Affected by Resource Availability?

A key area of research in pollinator foraging ecology is understanding why specific plants are used and whether this is driven by preferences relating to characteristics of the plant, e.g., nectar quality [100], or simply a result of resource availability [101]. By conducting floral surveys and comparing the flowering plants available to the plants identified in honey using DNA metabarcoding, de Vere et al. [37] found that honeybees only used 11% of genera available. Park and Nieh [94] also used a metabarcoding method along with herbarium records to illustrate that honeybees used between 2.7 and 10% of flowering species available over three seasons.

Insect visitation can be influenced by the abundance of floral resources in a landscape [102], which is affected both temporally by plant phenology [103] and spatially by habitat type [104]. Timberlake [105] utilised a null model method and DNA metabarcoding of pollen samples collected from bumblebees within farmland to illustrate that floral choice was not always driven by the abundance of plant species, nor their nectar availability. By identifying plants which are visited more than expected compared to their abundance, management recommendations can be given for maintaining appropriate floral provision aimed at the effective conservation of bumblebees on farmland [105]. Likewise, Jones [106] found no significant correlation between the abundance of plant taxa in the landscape and the abundance of plants found in honey samples each month. However, Nürnberger et al. [107] found that the number of plant genera in pollen loads of honeybees identified by metabarcoding was lower when floral availability was reduced. Recent work by Quinlan et al. [108] suggests that whilst honeybees may sometimes preferentially select plants found in high abundance, this is dependent on the time of year and nutritional demand.

DNA metabarcoding can be used to monitor how spatiotemporal changes in resource availability across landscapes affect the diet of wild and managed bees [90,104,109–111]. By assessing honeybee diet across gradients of land use, multiple authors have found that the richness and diversity of pollen collected is not strongly linked to the composition of surrounding landscapes [39,83,111,112]. Instead, seasonality of resources appears to be the greatest driver of diet, irrespective of land use [93,95].

4.3. How Are Resources Partitioned between Species and Individuals in Plant–Pollinator Networks?

The use of DNA-based methods for identifying species interactions allows complex networks to be constructed and analysed [16,89]. Constructing accurate networks is important to help fully understand their structure, as the level of specialisation and generalisation of networks, species, or individuals can affect their robustness against environmental change [113,114].

A number of authors have used molecular approaches to assess resource partitioning within large plant–pollinator networks [16,84,115]. Elliott et al. [116] used DNA metabarcoding to construct an interaction network between honeybees, native bees, and the floral resources used to identify resource overlap. The number of known floral hosts of many species were increased compared to the previous literature based on observational studies, improving the understanding of how wild and introduced bees co-exist in a landscape [116].

The ability to identify an individual's entire pollen assemblage results in the valuable characterisation of interactions at varying hierarchical levels throughout a plant–pollinator community [117]. To date, of the studies that have identified resource partitioning within plant–pollinator networks using DNA metabarcoding, all have found that generalised networks or species are made up of specialised individuals [44,84,115,118]. This presents a promising area of research to further investigate the levels of specialisation and generalisation exhibited by pollinators.

4.4. What Is the Relationship between Plant Use and Pollinator Health?

Floral resources vary in the quality of their nectar and pollen rewards [100], and consequently, the diversity of resources used has been found to impact pollinator fitness [119]. Insights into the nutritional ecology of pollinators can be unearthed using DNA metabarcoding, by quantifying the relationship between plant taxa found and either the protein, carbohydrate, lipid, and amino acid content of pollen [111,120,121] or the physiological glycogen, lipid, and protein levels of insects themselves [122].

As well as affecting the nutritional quality of provisions, the plant species visited by pollinators may also influence the bacteria present in the nest [123]. DNA metabarcoding allows plant–microbe relationships to be explored, increasing the understanding of plant–pollinator interactions throughout an insect's lifecycle. The relationship between the diversity of pollen species collected and the diversity of the microbiome appears complex. However, both positive and negative associations have been found between particular pollen types and bacteria [124–126]. For example, Voulgari-Kokota et al. [126] found that the presence of Acinetobacteria in pollen provisions of solitary bees was positively associated with the presence of some taxa such as European goldenrod (*Solidago virgaurea*), oxeye daisy (*Leucanthemum vulgare*), and yarrow (*Achillea millefolium*), but negatively associated with spear thistle (*Cirsium vulgare*), red poppy (*Papaver rhoeas*), and sycamore (*Acer pseudoplatanus*).

The identification of pollen in nests has also been used to investigate the relationship between mass-flowering crops and the prevalence of parasites in nests of mason bees (*Osmia* spp.), finding that increased abundance of resources may help to reduce transmission by diluting parasite transmission through reducing visitation frequency per flower [127].

5. Key Methodological Considerations for Using DNA Approaches and Their Challenges
5.1. Study Design and Sampling

Careful considerations are required for every stage of the molecular approach, from the initial stages of study design to the downstream bioinformatic analysis (Table 1). Firstly, the nature of the study system must be considered in order to understand the information which will be produced. For example, sampling pollen from a single bee which is actively foraging will yield different results to pollen collected through pollen traps or honey, as the latter methods represent the foraging efforts of multiple bees over numerous trips [37]. In addition, morphological features such as body size and pilosity (hairiness) of insects can influence the number and diversity of pollen retained [128]. Pollen may be transferred from plants visited solely for nectar [55], and some plants do not produce nectar at all [129]. In addition, nectar can itself be contaminated with pollen as a result of plant visitors [130]. Therefore, molecular analysis of pollen generates information on which plants have been visited for both pollen and nectar collection. Another important consideration is that the presence of pollen on insects does not assume pollination has occurred [10], and therefore the identification of pollen represents floral visitation only. It is also important to consider that when identifying plant material within nest provisions, contamination may occur from multiple sources of plant DNA such as pollen provisions or leaf or soil material used to build nests [86].

Capturing methods such as on transect walks or during observations will also influence the number and diversity of insects caught and therefore the resulting sampling universe. The flight times of insects and phenology of plants must also be considered due to their influence on foraging. For example, sampling one species across its entire flight period will provide a more complete picture of resources used compared to studies undertaken within a shorter time period, which have limited information on the total resources used. Further, the time of day at which pollinators are sampled will affect the resultant species collected [131].

The nature of pollen sampling from insect bodies results in a risk of cross-contamination occurring in the field; therefore, samples should be collected using a combination of nets and sterile tubes, with nets changed regularly and sterilised between surveys [44]. Airborne pollen may also contaminate samples [61], leading some authors to use thresholds to exclude rare taxa (reviewed in [132]) or removing all wind-pollinated species from analysis [133]. However, it should be noted that rare taxa may include real interactions, and some pollinators are known to visit wind-pollinated plants [134,135]. Further work to quantify the prevalence of residual pollen left on plants by insect visitors would be useful to infer thresholds for removal [78].

The method of preserving samples may also affect the success of the study [136]. Whilst successful sequencing of pollen from historical specimens is possible [91], samples should be preserved quickly to avoid degradation of DNA. Most pollen metabarcoding studies have preserved samples by freezing at −20 °C; however, recent work by Quaresma et al. [137] suggests that the use of silica gel for preserving pollen should not be overlooked, particularly when samples are collected by citizen scientists.

Table 1. Key considerations required for each step of the pollen metabarcoding workflow.

Step	Description of Method	Consideration	Recommendations
Sampling	Plant DNA can be captured through a number of sampling methods:	Source of pollen influences information obtained	Collect insects in sterile pots and replace nets if any pollen transfer is suspected
		Morphological features of insects, such as body size and pilosity (hairiness), can influence the amount of pollen retained	
	1. Pollen obtained from individuals collected from light traps, on transects, or within observational plots	Capture methods influence the number and diversity of insects caught	
	2. Pollen obtained from within nest provisions	Contamination may occur	
	3. Pollen obtained from honey samples	Sampling period limits the knowledge which can be gained	
Sample preservation	Avoidance of DNA degradation	Preservation method may affect downstream success	Store pollen samples at −20 °C or dry using silica gel to limit degradation of DNA
DNA extraction	Extraction of DNA from pollen	Quantity of DNA obtained is affected by extraction method	Membrane-based commercial kits offer a fast and simple way of yielding DNA, although they are costly
		Success of DNA extraction may depend on pollen type and source	Additional purification step is required for honey samples, e.g., Zymo OneStep PCR Inhibitor Removal Kit
		Contamination may occur	Stringent cleaning procedures are required using 10% bleach solution before and after each process Use of filter tips Use of negative controls
Amplification	PCR amplification of extracted DNA using primers which target specific region of interest	Choice of marker will influence which taxa are recovered and their taxonomic resolution	We recommend a multi-locus approach using *rbcL* [138,139] and ITS2 [140,141] Primer recommendations in Table S2, Supporting Information
		Contamination may occur	Stringent cleaning procedures are required using 10% bleach solution before and after each process Use of filter tips Use of positive and negative controls
		Biases may be introduced through primer specificity	Complete three rounds of PCR per sample and pool

Table 1. *Cont.*

Step	Description of Method	Consideration	Recommendations
Multiplexing and library preparation	Addition of nucleotide sequences to primers to allow for pooling of samples and compatibility with sequencing platforms	Each method has a trade-off between multiple factors including overall cost, risk of contamination and PCR efficiency Tag-jumping can occur causing misidentification	Index strategy used should be based on research question and experimental set-up A two-step PCR approach allows for cost-effective indexing
Sequencing	Identification of nucleotide sequences	Sequencing strategy is dependent on choice of marker	Illumina MiSeq (2 × 300 bp) allows sequencing of *rbcL* and ITS2
Reference library	Comparison of DNA sequences to a reference library for identification	Identifications made through DNA metabarcoding will only be as good as the reference library	Create a reference library which is appropriate to the question being asked and ensure that it is complete and well curated
Bioinformatic analysis	Automated processes used to curate sequences for analysis including quality control	Species may be incorrectly assigned during automated processes	Requires manual verification steps by someone with knowledge of relevant plant taxa
		Metabarcoding data are considered to be semi-quantitative	Treat proportion of sequences as relative read abundance for analysis

5.2. DNA Extraction

Numerous DNA isolation methods exist which can influence the quality of the DNA template [142,143]. Membrane-based isolation techniques are most commonly used for pollen metabarcoding studies, providing a fast and simple way of yielding DNA, although they are costly [142]. Regardless of the technique used, standard principles are followed: first the pollen cell wall (exine) is lysed to enable access to genomic material whilst preventing DNA degradation. Methods for pollen exine rupture can be chemical or mechanical, e.g., bead beating (the most common method) [143]. This lysis step is followed by degradation of the cell membrane, removal of contaminants, and finally precipitation of DNA from protein. Prior to amplification, additional purification steps may be required to remove PCR inhibitors, a common step when using honey as a source of pollen [38].

5.3. Amplification

The choice of barcode marker is regarded as one of the most important considerations of DNA barcoding studies and its applications, ultimately affecting the number of taxa recovered and the level of species discrimination obtained [32]. DNA barcode markers require high universality so that a large proportion of species in a sample are amplified, but also low intra-specific and high inter-specific variation for effective species discrimination [33]. Short markers allow for amplification of degraded DNA; however, these come with a caveat of reduced taxonomic resolution [144].

There is no single marker which meets the ideal requirements for a plant barcode [31,32]. For pollen metabarcoding, five regions are commonly used: *rbcL*, ITS2, *matK*, *trnL*, and *trnH-psbA* (Table S1). A multi-locus approach is recommended to ensure the greatest number of taxa are identified [24,38,144]. The length of *matK* (800 bp), restricts its use in metabarcoding due to limitations in read length on standard sequencing platforms [32]. Therefore, it is recommended that *rbcL* and ITS2 are used for pollen metabarcoding, due to their ability to identify taxa at varying taxonomic levels along with additional taxa unique to one marker which provides accurate identification of plant species within mixed pollen samples [38,45,78].

Contamination may also occur in the laboratory; therefore, stringent cleaning procedures are required to minimise these risks. The use of controls (negative in extraction, positive and negative in PCR amplification) helps in the identification of sources of contamination and should be sequenced with samples [38,145]. If sequences occur in negative controls, the number of reads of each taxon should be removed from all samples [81].

5.4. Multiplexing and Library Preparation

The ability to scale up metabarcoding studies relies on the use of sample-specific labels in the form of unique sequences of nucleotides which are attached to amplicons. These unique identifiers allow hundreds or thousands of samples to be pooled for sequencing (multiplexing), significantly increasing the capacity of one sequencing run. Methods for indexing of samples occur either during the initial PCR amplification through nucleotide additions to amplicons or through a secondary PCR amplification along with adapters to allow successful sequencing (library indices) (reviewed in [146]). Each of the methods comes with trade-offs between many factors, mainly the risk of cross-contamination, efficiency of PCR amplification, and overall cost [146]. The two-step PCR approach is most widely used in pollen metabarcoding studies (Table S1), allowing a cost-effective approach to sample labelling whilst allowing effective detection of cross-contamination, but comes with the caveat of increased risk of biases due to an additional amplification stage [146].

5.5. Sequencing

Following amplification of DNA, the sequencing strategy used is dependent on a variety of factors including the choice of marker, with most studies thus far utilising the Illumina MiSeq platform. Although concerns are raised over the maximum read length of Illumina platforms (2 × 300 bp) [29,98], multiple studies have demonstrated successful sequencing of longer markers such as *rbcL* (around 500 bp) along with additional adapters and primers [45,80]. Newer sequencing technologies such as the MinION (Oxford Nanopore Technologies) and SMRT platform (PACBIO, Pacific Biosciences) produce longer read lengths, but they generate less reads than Illumina [29]. The development of ultra-deep short read sequencing technologies such as Illumina NovaSeq provide an opportunity to increase sequencing depth and improve the detection rate of taxa. The requirement for high quality and quantity of input DNA may be a limiting factor for some applications of these technologies [49].

5.6. Reference Library

The accuracy of DNA barcoding is reliant on a comprehensive reference library [32,147]. The creation of large-scale, complete DNA barcode reference libraries for a national flora has been achieved in the UK [32,139] and Canada [148] using a multi-locus approach, allowing reliable species identification in subsequent pollen metabarcoding studies [37,38]. The curation of reference libraries from chloroplast genomes and nuclear ribosomal DNA sequences can also provide coverage of standard barcodes; however, these methods are more costly [149]. If a complete regional reference database is not available [150], then authors are encouraged to compile custom, relevant reference libraries using the sequences available in GenBank (https://www.ncbi.nlm.nih.gov/genbank/, accessed on 27 January 2022). Curation of these libraries is required, however, to identify and remove incorrect sequences [78,112,116,151]. It is critically important to understand the coverage of the reference library being used compared to the plant taxa that could be detected [32].

5.7. Bioinformatic Analysis

The quantity of data produced from DNA metabarcoding studies requires automated processes for curation of sequences, including steps for quality control. The main purpose of this process is to remove any additional nucleotide sequences (index tags, adapter tags, and primers) and to separate each sample for subsequent analysis (demultiplexing). The reduction of the need for expert taxonomists to identify pollen grains is often cited as one of the major advantages of molecular methods over pollen microscopy [91]. However, few authors have highlighted the importance of having good knowledge of the taxonomic group in question (i.e., plants in pollen metabarcoding), including their distribution and phenology for accurate species identification [37,38,83,152]. Misidentifications may occur during the bioinformatic process due to low interspecific variance [32] or incorrectly identified sequences in GenBank [153]. In order to mitigate misidentifications, deployment of a manual verification step in the assignment process, underpinned by botanical expertise, will reduce incorrect species assignments.

5.8. Towards Standardisation of Methods

Although each step of the pollen metabarcoding process has a range of different approaches, only certain elements of the entire pollen metabarcoding workflow have been reviewed [132,143,146], leaving a large proportion of the study design to the authors' discretion. Without a standardisation of approaches, comparison of results across multiple studies must be interpreted with caution. Until each stage has been critically reviewed and a robust, standardised approach is established, we encourage researchers to carefully assess the considerations outlined in Table 1 for guidance prior to conducting a pollen metabarcoding study. Further, we call upon authors to be transparent in reporting every aspect of their molecular methods to ensure studies are reproducible, utilising supporting information where word limits are restricting.

5.9. How Quantitative Is DNA Metabarcoding?

Finally, there is continued debate over whether DNA metabarcoding may characterise pollen samples in a quantitative manner, with mixed results across studies [74,85,86,154,155]. Quantification has been found to be affected by a combination of marker and primer used, pollen type, mixture characteristics, and PCR conditions [88,156–158]. It is likely that relationships between the proportion of DNA reads and pollen counts are more likely for the most abundant taxa within a sample [83,159]. Similar to microscopy, rare taxa are difficult to detect using pollen metabarcoding [87]. Whilst this is a limitation, studies examining insect floral resource use often place greater focus on those plants detected at higher abundance. For this reason, along with the potential biases which can occur, DNA metabarcoding should be considered as semi-quantitative and relative read abundance used for downstream analysis [160]. We do not recommend the use of presence/absence approaches due to rare taxa being overstated and abundant taxa devalued [160].

6. Opportunities and Future Directions

The use of DNA metabarcoding as a tool to investigate pollinator foraging has allowed increased insight into the interactions between plants and pollinators; however, it is still a developing field. Most studies focus on the identification of pollen; however, other plant material may be used to identify relationships between insects and plants. For example, recently, the characterisation of resin within the nests of solitary bees through DNA metabarcoding has been suggested as a promising approach to identify which plants are important for nest building [161]. DNA metabarcoding is also not free from limitations. Overall, the greatest limitation is the cost and reproducibility of the molecular techniques [162], which determine which methods are used. Whilst the interpretation of data remains semi-quantitative, future work may lead to the ability to accurately measure pollen abundance, significantly improving the application of this technique [157,158]. Quantification may be improved by using PCR-free approaches which also provide a greater representation of the genome [46]. Recent work by Bell et al. [46] has demonstrated that whole-genome shotgun sequencing of pollen DNA is a reliable method for identification of pollen species mixtures. However, coverage of eukaryotic organisms in reference libraries remains low, and assembly of whole genomes is currently more expensive than metabarcoding per sample [46]. It is likely that DNA metabarcoding will remain the standard technique until genome-level coverage improves. Until then, genome-skimming techniques may hold promise to identify beyond the species level, e.g., to population or individual, if the nuclear genome is retained [163].

7. Final Remarks

This review describes the range of approaches available to investigate floral visitation by pollinators using DNA metabarcoding. We demonstrate how the ability to yield valuable individual-to-community-level information on foraging over large spatiotemporal scales allows for a breadth of ecological questions to be explored, for the benefit of both the conservation of pollinators and the maintenance of the ecosystem services they provide.

DNA metabarcoding has become a standard tool for the characterisation of complex plant–pollinator interactions, allowing for an improved understanding of threatened global biodiversity.

Supplementary Materials: The following supporting information can be downloaded at https://www.mdpi.com/article/10.3390/d14040236/s1, Table S1: Details of studies which use plant DNA metabarcoding to identify floral visitation by pollinators or developed methods to support. Table S2: Recommended primer sequences used to amplify the *rbcL* and ITS2 barcode regions.

Author Contributions: Conceptualisation, A.L., N.d.V. and S.C.; methodology, A.L. and N.d.V.; validation, A.L., L.J., L.W., S.C. and N.d.V.; investigation, A.L.; data curation, A.L.; writing—original draft preparation, A.L. and N.d.V.; writing—review and editing, A.L., L.J., L.W., S.C. and N.d.V.; supervision, N.d.V. and S.C.; funding acquisition, N.d.V. All authors have read and agreed to the published version of the manuscript.

Funding: A.L., L.J. and N.d.V. have received funding through the Welsh Government Rural Communities—Rural Development Programme 2014–2020, which is funded by the European Agricultural Fund for Rural Development and the Welsh Government. A.L. and L.W. were supported by a Knowledge Economy Skills Scholarship (KESS2), part funded by the Welsh Government's European Social Fund (ESF).

Institutional Review Board Statement: Not applicable.

Data Availability Statement: All relevant data are provided in the Supplementary Materials.

Conflicts of Interest: The authors declare no conflict of interest.

References

1. Klein, A.-M.; Vaissiere, B.E.; Cane, J.H.; Steffan-Dewenter, I.; Cunningham, S.A.; Kremen, C.; Tscharntke, T. Importance of pollinators in changing landscapes for world crops. *Proc. R. Soc. B Biol. Sci.* **2007**, *274*, 303–313. [CrossRef]
2. Potts, S.G.; Imperatriz-Fonseca, V.; Ngo, H.T.; Aizen, M.A.; Biesmeijer, J.C.; Breeze, T.D.; Dicks, L.V.; Garibaldi, L.A.; Hill, R.; Settele, J.; et al. Safeguarding pollinators and their values to human well-being. *Nature* **2016**, *540*, 220–229. [CrossRef] [PubMed]
3. Ollerton, J. Pollinator Diversity: Distribution, Ecological Function, and Conservation. *Annu. Rev. Ecol. Evol. Syst.* **2017**, *48*, 353–376. [CrossRef]
4. Powney, G.D.; Carvell, C.; Edwards, M.; Morris, R.K.A.; Roy, H.E.; Woodcock, B.A.; Isaac, N.J.B. Widespread losses of pollinating insects in Britain. *Nat. Commun.* **2019**, *10*, 1018. [CrossRef]
5. Wepprich, T.; Adrion, J.R.; Ries, L.; Wiedmann, J.; Haddad, N.M. Butterfly abundance declines over 20 years of systematic monitoring in Ohio, USA. *PLoS ONE* **2019**, *14*, e0216270. [CrossRef]
6. Dicks, L.V.; Breeze, T.D.; Ngo, H.T.; Senapathi, D.; An, J.; Aizen, M.A.; Basu, P.; Buchori, D.; Galetto, L.; Garibaldi, L.A.; et al. A global-scale expert assessment of drivers and risks associated with pollinator decline. *Nat. Ecol. Evol.* **2021**, *5*, 1453–1461. [CrossRef]
7. Hristov, P.; Shumkova, R.; Palova, N.; Neov, B. Factors associated with honey bee colony losses: A mini-review. *Vet. Sci.* **2020**, *7*, 166. [CrossRef]
8. Neov, B.; Georgieva, A.; Shumkova, R.; Radoslavov, G.; Hristov, P. Biotic and abiotic factors associated with colonies mortalities of managed honey bee (*Apis mellifera*). *Diversity* **2019**, *11*, 237. [CrossRef]
9. Goulson, D.; Nicholls, E.; Botias, C.; Rotheray, E.L. Bee declines driven by combined stress from parasites, pesticides, and lack of flowers. *Science* **2015**, *347*, 1255957. [CrossRef]
10. Ballantyne, G.; Baldock, K.C.R.; Willmer, P.G. Constructing more informative plant-pollinator networks: Visitation and pollen deposition networks in a heathland plant community. *Proc. R. Soc. B Biol. Sci.* **2015**, *282*, 20151130. [CrossRef]
11. King, C.; Ballantyne, G.; Willmer, P.G. Why flower visitation is a poor proxy for pollination: Measuring single-visit pollen deposition, with implications for pollination networks and conservation. *Methods Ecol. Evol.* **2013**, *4*, 811–818. [CrossRef]
12. Heinrich, B. The foraging specializations of individual bumblebees. *Ecol. Monogr.* **1976**, *46*, 105–128. [CrossRef]
13. Osborne, J.L.; Clark, S.J.; Morris, R.J.; Williams, I.H.; Riley, J.R.; Smith, A.D.; Reynolds, D.R.; Edwards, A.S. A landscape-scale study of bumble bee foraging range and constancy, using harmonic radar. *J. Appl. Ecol.* **1999**, *36*, 519–533. [CrossRef]
14. Seeley, T. *The Wisdom of the Hive*; Harvard University Press: London, UK, 1995.
15. Balfour, N.J.; Fensome, K.A.; Samuelson, E.E.W.; Ratnieks, F.L.W. Following the dance: Ground survey of flowers and flower-visiting insects in a summer foraging hotspot identified via honey bee waggle dance decoding. *Agric. Ecosyst. Environ.* **2015**, *213*, 265–271. [CrossRef]
16. Macgregor, C.J.; Kitson, J.J.N.; Fox, R.; Hahn, C.; Lunt, D.H.; Pocock, M.J.O.; Evans, D.M. Construction, validation, and application of nocturnal pollen transport networks in an agro-ecosystem: A comparison using light microscopy and DNA metabarcoding. *Ecol. Entomol.* **2019**, *44*, 17–29. [CrossRef]

17. Wood, T.J.; Holland, J.M.; Goulson, D. Providing foraging resources for solitary bees on farmland: Current schemes for pollinators benefit a limited suite of species. *J. Appl. Ecol.* **2017**, *54*, 323–333. [CrossRef]

18. Eckhardt, M.; Haider, M.; Dorn, S.; Müller, A. Pollen mixing in pollen generalist solitary bees: A possible strategy to complement or mitigate unfavourable pollen properties? *J. Anim. Ecol.* **2014**, *83*, 588–597. [CrossRef]

19. Wood, T.J.; Holland, J.M.; Goulson, D. Diet characterisation of solitary bees on farmland: Dietary specialisation predicts rarity. *Biodivers. Conserv.* **2016**, *25*, 2655–2671. [CrossRef]

20. Ebenezer, I.O.; Olugbenga, M.T. Pollen characterisation of honey samples from North Central Nigeria. *J. Biol. Sci.* **2010**, *10*, 43–47. [CrossRef]

21. Ponnuchamy, R.; Bonhomme, V.; Prasad, S.; Das, L.; Patel, P.; Gaucherel, C.; Pragasam, A.; Anupama, K. Honey pollen: Using melissopalynology to understand foraging preferences of bees in tropical south India. *PLoS ONE* **2014**, *9*, e101618. [CrossRef]

22. Lawson, S.P.; Ciaccio, K.N.; Rehan, S.M. Maternal manipulation of pollen provisions affects worker production in a small carpenter bee. *Behav. Ecol. Sociobiol.* **2016**, *70*, 1891–1900. [CrossRef]

23. Williams, N.M.; Kremen, C. Resource distributions among habitats determine solitary bee offspring production in a mosaic landscape. *Ecol. Appl.* **2007**, *17*, 910–921. [CrossRef] [PubMed]

24. Bell, K.L.; de Vere, N.; Keller, A.; Richardson, R.T.; Gous, A.; Burgess, K.S.; Brosi, B.J. Pollen DNA barcoding: Current applications and future prospects. *Genome* **2016**, *59*, 629–640. [CrossRef] [PubMed]

25. Holt, K.A.; Bennett, K.D. Principles and methods for automated palynology. *New Phytol.* **2014**, *203*, 735–742. [CrossRef] [PubMed]

26. Polling, M.; Li, C.; Cao, L.; Verbeek, F.; de Weger, L.A.; Belmonte, J.; De Linares, C.; Willemse, J.; de Boer, H.; Gravendeel, B. Neural networks for increased accuracy of allergenic pollen monitoring. *Sci. Rep.* **2021**, *11*, 11357. [CrossRef] [PubMed]

27. Sevillano, V.; Aznarte, J.L. Improving classification of pollen grain images of the POLEN23E dataset through three different applications of deep learning convolutional neural networks. *PLoS ONE* **2018**, *13*, e0201807. [CrossRef]

28. Cristescu, M.E. From barcoding single individuals to metabarcoding biological communities: Towards an integrative approach to the study of global biodiversity. *Trends Ecol. Evol.* **2014**, *29*, 566–571. [CrossRef]

29. Evans, D.M.; Kitson, J.J. Molecular ecology as a tool for understanding pollination and other plant–insect interactions. *Curr. Opin. Insect Sci.* **2020**, *38*, 26–33. [CrossRef]

30. Leidenfrost, R.M.; Bänsch, S.; Prudnikow, L.; Brenig, B.; Westphal, C.; Wünschiers, R. Analyzing the Dietary Diary of Bumble Bee. *Front. Plant Sci.* **2020**, *11*, 287. [CrossRef]

31. CBOL Plant Working Group A DNA barcode for land plants. *Proc. Natl. Acad. Sci. USA.* **2009**, *106*, 12794–12797. [CrossRef]

32. Jones, L.; Twyford, A.D.; Ford, C.R.; Rich, T.C.G.; Davies, H.; Forrest, L.L.; Hart, M.L.; McHaffie, H.; Brown, M.R.; Hollingsworth, P.M.; et al. Barcode UK: A complete DNA barcoding resource for the flowering plants and conifers of the United Kingdom. *Mol. Ecol. Resour.* **2021**, 1755–0998.13388. [CrossRef]

33. Hollingsworth, P.M.; Graham, S.W.; Little, D.P. Choosing and using a plant DNA barcode. *PLoS ONE* **2011**, *6*, e19254. [CrossRef]

34. Eeraerts, M.; Piot, N.; Pisman, M.; Claus, G.; Meeus, I.; Smagghe, G. Landscapes with high amounts of mass-flowering fruit crops reduce the reproduction of two solitary bees. *Basic Appl. Ecol.* **2021**, *56*, 122–131. [CrossRef]

35. Gresty, C.E.A.; Clare, E.; Devey, D.S.; Cowan, R.S.; Csiba, L.; Malakasi, P.; Lewis, O.T.; Willis, K.J. Flower preferences and pollen transport networks for cavity-nesting solitary bees: Implications for the design of agri-environment schemes. *Ecol. Evol.* **2018**, *8*, 7574–7587. [CrossRef] [PubMed]

36. Vaudo, A.D.; Biddinger, D.J.; Sickel, W.; Keller, A.; López-Uribe, M.M. Introduced bees (*Osmia cornifrons*) collect pollen from both coevolved and novel host-plant species within their family-level phylogenetic preferences. *R. Soc. Open Sci.* **2020**, *7*, 200225. [CrossRef]

37. De Vere, N.; Jones, L.E.; Gilmore, T.; Moscrop, J.; Lowe, A.; Smith, D.; Hegarty, M.J.; Creer, S.; Ford, C.R. Using DNA metabarcoding to investigate honey bee foraging reveals limited flower use despite high floral availability. *Sci. Rep.* **2017**, *7*, 42838. [CrossRef]

38. Jones, L.; Brennan, G.L.; Lowe, A.; Creer, S.; Ford, C.R.; de Vere, N. Shifts in honeybee foraging reveal historical changes in floral resources. *Commun. Biol.* **2021**, *4*, 37. [CrossRef]

39. Lucek, K.; Galli, A.; Gurten, S.; Hohmann, N.; Maccagni, A.; Patsiou, T.; Willi, Y. Metabarcoding of honey to assess differences in plant-pollinator interactions between urban and non-urban sites. *Apidologie* **2019**, *50*, 317–329. [CrossRef]

40. Chang, H.; Guo, J.; Fu, X.; Liu, Y.; Wyckhuys, K.A.G.; Hou, Y.; Wu, K. Molecular-assisted pollen grain analysis reveals spatiotemporal origin of long-distance migrants of a noctuid moth. *Int. J. Mol. Sci.* **2018**, *19*, 567. [CrossRef]

41. Mayr, A.V.; Keller, A.; Peters, M.K.; Grimmer, G.; Krischke, B.; Geyer, M.; Schmitt, T.; Steffan-Dewenter, I. Cryptic species and hidden ecological interactions of halictine bees along an elevational gradient. *Ecol. Evol.* **2021**, *11*, 7700–7712. [CrossRef]

42. Wilson, E.E.; Sidhu, C.S.; Levan, K.E.; Holway, D.A. Pollen foraging behaviour of solitary Hawaiian bees revealed through molecular pollen analysis. *Mol. Ecol.* **2010**, *19*, 4823–4829. [CrossRef] [PubMed]

43. Fahimee, J.; Badrulisham, A.S.; Zulidzham, M.S.; Reward, N.F.; Muzammil, N.; Jajuli, R.; Md-Zain, B.M.; Yaakop, S. Metabarcoding in diet assessment of Heterotrigona itama based on trnL marker towards domestication program. *Insects* **2021**, *12*, 205. [CrossRef] [PubMed]

44. Lucas, A.; Bodger, O.; Brosi, B.J.; Ford, C.R.; Forman, D.W.; Greig, C.; Hegarty, M.; Neyland, P.J.; de Vere, N. Generalisation and specialisation in hoverfly (Syrphidae) grassland pollen transport networks revealed by DNA metabarcoding. *J. Anim. Ecol.* **2018**, *87*, 1008–1021. [CrossRef] [PubMed]

45. Lowe, A.; Jones, L.; Brennan, G.L.; Creer, S.; de Vere, N. Seasonal progression and differences in major floral resource use by bees and hoverflies in a diverse horticultural and agricultural landscape revealed by DNA metabarcoding. *J. Appl. Ecol.* **2022**. [CrossRef]

46. Bell, K.L.; Petit, R.A.; Cutler, A.; Dobbs, E.K.; Macpherson, J.M.; Read, T.D.; Burgess, K.S.; Brosi, B.J. Comparing whole-genome shotgun sequencing and DNA metabarcoding approaches for species identification and quantification of pollen species mixtures. *Ecol. Evol.* **2021**, *11*, 16082–16098. [CrossRef] [PubMed]

47. Creer, S.; Deiner, K.; Frey, S.; Porazinska, D.; Taberlet, P.; Thomas, W.K.; Potter, C.; Bik, H.M. The ecologist's field guide to sequence-based identification of biodiversity. *Methods Ecol. Evol.* **2016**, *7*, 1008–1018.

48. Lang, D.; Tang, M.; Hu, J.; Zhou, X. Genome-skimming provides accurate quantification for pollen mixtures. *Mol. Ecol. Resour.* **2019**, *19*, 1433–1446. [CrossRef]

49. Peel, N.; Dicks, L.V.; Clark, M.D.; Heavens, D.; Percival-Alwyn, L.; Cooper, C.; Davies, R.G.; Leggett, R.M.; Yu, D.W. Semi-quantitative characterisation of mixed pollen samples using MinION sequencing and Reverse Metagenomics (RevMet). *Methods Ecol. Evol.* **2019**, *10*, 1690–1701. [CrossRef]

50. Goulson, D.; Lye, G.C.; Darvill, B. Diet breadth, coexistence and rarity in bumblebees. *Biodivers. Conserv.* **2008**, *17*, 3269–3288. [CrossRef]

51. Hanley, M.E.; Awbi, A.J.; Franco, M. Going native? Flower use by bumblebees in English urban gardens. *Ann. Bot.* **2014**, *113*, 799–806. [CrossRef]

52. Klečka, J.; Hadrava, J.; Biella, P.; Akter, A. Flower visitation by hoverflies (Diptera: Syrphidae) in a temperate plant-pollinator network. *PeerJ* **2018**, *6*, e6025. [CrossRef] [PubMed]

53. Rollings, R.; Goulson, D. Quantifying the attractiveness of garden flowers for pollinators. *J. Insect Conserv.* **2019**, *23*, 803–817. [CrossRef]

54. Salisbury, A.; Armitage, J.; Bostock, H.; Perry, J.; Tatchell, M.; Thompson, K. Enhancing gardens as habitats for flower-visiting aerial insects (pollinators): Should we plant native or exotic species? *J. Appl. Ecol.* **2015**, *52*, 1156–1164. [CrossRef]

55. Goulson, D.; Hanley, M.E.; Darvill, B.; Ellis, J.S.; Knight, M.E. Causes of rarity in bumblebees. *Biol. Conserv.* **2005**, *122*, 1–8. [CrossRef]

56. Memmott, J. The structure of a plant-pollinator food web. *Ecol. Lett.* **1999**, *2*, 276–280. [CrossRef]

57. Thomsen, P.F.; Sigsgaard, E.E. Environmental DNA metabarcoding of wild flowers reveals diverse communities of terrestrial arthropods. *Ecol. Evol.* **2019**, *9*, 1665–1679. [CrossRef]

58. Aizenberg-Gershtein, Y.; Izhaki, I.; Halpern, M. Do Honeybees Shape the Bacterial Community Composition in Floral Nectar? *PLoS ONE* **2013**, *8*, e67556. [CrossRef]

59. Ushio, M.; Yamasaki, E.; Takasu, H.; Nagano, A.J.; Fujinaga, S.; Honjo, M.N.; Ikemoto, M.; Sakai, S.; Kudoh, H. Microbial communities on flower surfaces act as signatures of pollinator visitation. *Sci. Rep.* **2015**, *5*, 8695. [CrossRef]

60. Cuff, J.P.; Windsor, F.M.; Tercel, M.P.T.G.; Kitson, J.J.N.; Evans, D.M. Overcoming the pitfalls of merging dietary metabarcoding into ecological networks. *Methods Ecol. Evol.* **2021**, *13*, 545–559. [CrossRef]

61. Pornon, A.; Andalo, C.; Burrus, M.; Escaravage, N. DNA metabarcoding data unveils invisible pollination networks. *Sci. Rep.* **2017**, *7*, 16828. [CrossRef]

62. Olesen, J.M.; Bascompte, J.; Dupont, Y.L.; Elberling, H.; Rasmussen, C.; Jordano, P. Missing and forbidden links in mutualistic networks. *Proc. R. Soc. B Biol. Sci.* **2011**, *278*, 725–732. [CrossRef]

63. Bosch, J.; Martín González, A.M.; Rodrigo, A.; Navarro, D. Plant-pollinator networks: Adding the pollinator's perspective. *Ecol. Lett.* **2009**, *12*, 409–419. [CrossRef] [PubMed]

64. Comba, L.; Corbet, S.A.; Hunt, L.; Warren, B. Flowers, Nectar and Insect Visits: Evaluating British Plant Species for Pollinator-friendly Gardens. *Ann. Bot.* **1999**, *83*, 369–383. [CrossRef]

65. Peat, J.; Goulson, D. Effects of experience and weather on foraging rate and pollen versus nectar collection in the bumblebee, Bombus terrestris. *Behav. Ecol. Sociobiol.* **2005**, *58*, 152–156. [CrossRef]

66. Garbuzov, M.; Ratnieks, F.L.W. Using the British national collection of asters to compare the attractiveness of 228 varieties to flower-visiting insects. *Environ. Entomol.* **2015**, *44*, 638–646. [CrossRef]

67. Garbuzov, M.; Ratnieks, F.L.W. Quantifying variation among garden plants in attractiveness to bees and other flower-visiting insects. *Funct. Ecol.* **2014**, *28*, 364–374. [CrossRef]

68. Hebert, P.D.N.; Cywinska, A.; Ball, S.L.; deWaard, J.R. Biological identifications through DNA barcodes. *Proc. R. Soc. B Biol. Sci.* **2003**, *270*, 313–321. [CrossRef]

69. Chacoff, N.P.; Vázquez, D.P.; Lomáscolo, S.B.; Stevani, E.L.; Dorado, J.; Padrón, B. Evaluating sampling completeness in a desert plant-pollinator network. *J. Anim. Ecol.* **2012**, *81*, 190–200. [CrossRef]

70. Gibson, R.H.; Knott, B.; Eberlein, T.; Memmott, J. Sampling method influences the structure of plant-pollinator networks. *Oikos* **2011**, *120*, 822–831. [CrossRef]

71. Petanidou, T.; Kallimanis, A.S.; Tzanopoulos, J.; Sgardelis, S.P.; Pantis, J.D. Long-term observation of a pollination network: Fluctuation in species and interactions, relative invariance of network structure and implications for estimates of specialization. *Ecol. Lett.* **2008**, *11*, 564–575. [CrossRef]

72. De Manincor, N.; Hautekèete, N.; Mazoyer, C.; Moreau, P.; Piquot, Y.; Schatz, B.; Schmitt, E.; Zélazny, M.; Massol, F. How biased is our perception of plant-pollinator networks? A comparison of visit- and pollen-based representations of the same networks. *Acta Oecologica* **2020**, *105*, 103551. [CrossRef]

73. Galliot, J.N.; Brunel, D.; Bérard, A.; Chauveau, A.; Blanchetête, A.; Lanore, L.; Farruggia, A. Investigating a flower-insect forager network in a mountain grassland community using pollen DNA barcoding. *J. Insect Conserv.* **2017**, *21*, 827–837. [CrossRef]

74. Pornon, A.; Escaravage, N.; Burrus, M.; Holota, H.; Khimoun, A.; Mariette, J.; Pellizzari, C.; Iribar, A.; Etienne, R.; Taberlet, P.; et al. Using metabarcoding to reveal and quantify plant-pollinator interactions. *Sci. Rep.* **2016**, *6*, 27282. [CrossRef] [PubMed]

75. Zhao, Y.H.; Lázaro, A.; Ren, Z.X.; Zhou, W.; Li, H.D.; Tao, Z.B.; Xu, K.; Wu, Z.K.; Wolfe, L.M.; Li, D.Z.; et al. The topological differences between visitation and pollen transport networks: A comparison in species rich communities of the Himalaya–Hengduan Mountains. *Oikos* **2019**, *128*, 551–562. [CrossRef]

76. Beekman, M.; Ratnieks, F.L.W. Long-range foraging by the honey-bee, *Apis mellifera* L. *Funct. Ecol.* **2000**, *14*, 490–496. [CrossRef]

77. Carvell, C.; Westrich, P.; Meek, W.R.; Pywell, R.F.; Nowakowski, M. Assessing the value of annual and perennial forage mixtures for bumblebees by direct observation and pollen analysis. *Apidologie* **2006**, *37*, 326–340. [CrossRef]

78. Arstingstall, K.A.; DeBano, S.J.; Li, X.; Wooster, D.E.; Rowland, M.M.; Burrows, S.; Frost, K. Capabilities and limitations of using DNA metabarcoding to study plant-pollinator interactions. *Mol. Ecol.* **2021**, *30*, 5266–5297. [CrossRef]

79. Tur, C.; Vigalondo, B.; Trøjelsgaard, K.; Olesen, J.M.; Traveset, A. Downscaling pollen-transport networks to the level of individuals. *J. Anim. Ecol.* **2014**, *83*, 306–317. [CrossRef]

80. Potter, C.; De Vere, N.; Jones, L.E.; Ford, C.R.; Hegarty, M.J.; Hodder, K.H.; Diaz, A.; Franklin, E.L. Pollen metabarcoding reveals broad and species-specific resource use by urban bees. *PeerJ* **2019**, *7*, e5999. [CrossRef]

81. Bell, K.L.; Fowler, J.; Burgess, K.S.; Dobbs, E.K.; Gruenewald, D.; Lawley, B.; Morozumi, C.; Brosi, B.J. Applying Pollen DNA Metabarcoding to the Study of Plant–Pollinator Interactions. *Appl. Plant Sci.* **2017**, *5*, 1600124. [CrossRef]

82. Sickel, W.; Ankenbrand, M.J.; Grimmer, G.; Holzschuh, A.; Härtel, S.; Lanzen, J.; Steffan-Dewenter, I.; Keller, A. Increased efficiency in identifying mixed pollen samples by meta-barcoding with a dual-indexing approach. *BMC Ecol.* **2015**, *15*, 20. [CrossRef]

83. Smart, M.D.; Cornman, R.S.; Iwanowicz, D.D.; McDermott-Kubeczko, M.; Pettis, J.S.; Spivak, M.S.; Otto, C.R.V. A comparison of honey bee-collected pollen from working agricultural lands using light microscopy and its metabarcoding. *Environ. Entomol.* **2017**, *46*, 38–49. [CrossRef] [PubMed]

84. Lucas, A.; Bodger, O.; Brosi, B.J.; Ford, C.R.; Forman, D.W.; Greig, C.; Hegarty, M.; Jones, L.; Neyland, P.J.; de Vere, N. Floral resource partitioning by individuals within generalised hoverfly pollination networks revealed by DNA metabarcoding. *Sci. Rep.* **2018**, *8*, 5133. [CrossRef] [PubMed]

85. Richardson, R.T.; Lin, C.-H.; Sponsler, D.B.; Quijia, J.O.; Goodell, K.; Johnson, R.M. Application of ITS2 Metabarcoding to Determine the Provenance of Pollen Collected by Honey Bees in an Agroecosystem. *Appl. Plant Sci.* **2015**, *3*, 1400066. [CrossRef]

86. Keller, A.; Danner, N.; Grimmer, G.; Ankenbrand, M.; von der Ohe, K.; von der Ohe, W.; Rost, S.; Hartel, S.; Steffan-Dewenter, I. Evaluating multiplexed next-generation sequencing as a method in palynology for mixed pollen samples. *Plant Biol.* **2015**, *17*, 558–566. [CrossRef] [PubMed]

87. Hawkins, J.; De Vere, N.; Griffith, A.; Ford, C.R.; Allainguillaume, J.; Hegarty, M.J.; Baillie, L.; Adams-Groom, B. Using DNA metabarcoding to identify the floral composition of honey: A new tool for investigating honey bee foraging preferences. *PLoS ONE* **2015**, *10*, e0134735. [CrossRef]

88. Richardson, R.T.; Lin, C.; Quijia, J.O.; Riusech, N.S.; Goodell, K.; Johnson, R.M. Rank-based characterization of pollen assemblages collected by honey bees using a multi-locus metabarcoding approach. *Appl. Plant Sci.* **2015**, *3*, 1500043. [CrossRef]

89. Evans, D.M.; Kitson, J.J.N.; Lunt, D.H.; Straw, N.A.; Pocock, M.J.O. Merging DNA metabarcoding and ecological network analysis to understand and build resilient terrestrial ecosystems. *Funct. Ecol.* **2016**, *30*, 1904–1916. [CrossRef]

90. Lu, H.; Dou, F.; Hao, Y.; Li, Y.; Zhang, K.; Zhang, H.; Zhou, Z.; Zhu, C.; Huang, D.; Luo, A. Metabarcoding Analysis of Pollen Species Foraged by Osmia excavata Alfken (Hymenoptera: Megachilidae) in China. *Front. Ecol. Evol.* **2021**, *9*, 730549. [CrossRef]

91. Gous, A.; Swanevelder, D.Z.H.; Eardley, C.D.; Willows-Munro, S. Plant–pollinator interactions over time: Pollen metabarcoding from bees in a historic collection. *Evol. Appl.* **2019**, *12*, 187–197. [CrossRef]

92. Polling, M.; ter Schure, A.T.M.; van Geel, B.; van Bokhoven, T.; Boessenkool, S.; MacKay, G.; Langeveld, B.W.; Ariza, M.; van der Plicht, H.; Protopopov, A.V.; et al. Multiproxy analysis of permafrost preserved faeces provides an unprecedented insight into the diets and habitats of extinct and extant megafauna. *Quat. Sci. Rev.* **2021**, *267*, 107084. [CrossRef]

93. Danner, N.; Keller, A.; Härtel, S.; Steffan-Dewenter, I. Honey bee foraging ecology: Season but not landscape diversity shapes the amount and diversity of collected pollen. *PLoS ONE* **2017**, *12*, e0183716. [CrossRef]

94. Park, B.; Nieh, J.C. Seasonal trends in honey bee pollen foraging revealed through DNA barcoding of bee-collected pollen. *Insectes Soc.* **2017**, *64*, 425–437. [CrossRef]

95. Wilson, R.S.; Keller, A.; Shapcott, A.; Leonhardt, S.D.; Sickel, W.; Hardwick, J.L.; Heard, T.A.; Kaluza, B.F.; Wallace, H.M. Many small rather than few large sources identified in long-term bee pollen diets in agroecosystems. *Agric. Ecosyst. Environ.* **2021**, *310*, 107296. [CrossRef]

96. Gous, A.; Eardley, C.D.; Johnson, S.D.; Swanevelder, D.Z.H.; Willows-Munro, S. Floral hosts of leaf-cutter bees (Megachilidae) in a biodiversity hotspot revealed by pollen DNA metabarcoding of historic specimens. *PLoS ONE* **2021**, *16*, e0244973. [CrossRef]

97. Simanonok, M.P.; Otto, C.R.V.; Cornman, R.S.; Iwanowicz, D.D.; Strange, J.P.; Smith, T.A. A century of pollen foraging by the endangered rusty patched bumble bee (*Bombus affinis*): Inferences from molecular sequencing of museum specimens. *Biodivers. Conserv.* **2021**, *30*, 123–137. [CrossRef]

98. Suchan, T.; Talavera, G.; Sáez, L.; Ronikier, M.; Vila, R. Pollen metabarcoding as a tool for tracking long-distance insect migrations. *Mol. Ecol. Resour.* **2018**, *19*, 149–162. [CrossRef]

99. Kratschmer, S.; Petrović, B.; Curto, M.; Meimberg, H.; Pachinger, B. Pollen availability for the Horned mason bee (*Osmia cornuta*) in regions of different land use and landscape structures. *Ecol. Entomol.* **2020**, *45*, 525–537. [CrossRef]

100. Hicks, D.M.; Ouvrard, P.; Baldock, K.C.R.; Baude, M.; Goddard, M.A.; Kunin, W.E.; Mitschunas, N.; Memmott, J.; Morse, H.; Nikolitsi, M.; et al. Food for pollinators: Quantifying the nectar and pollen resources of urban flower meadows. *PLoS ONE* **2016**, *11*, e0158117. [CrossRef]

101. Hegland, S.J.; Boeke, L. Relationships between the density and diversity of floral resources and flower visitor activity in a temperate grassland community. *Ecol. Entomol.* **2006**, *31*, 532–538. [CrossRef]

102. Fowler, R.E.; Rotheray, E.L.; Goulson, D. Floral abundance and resource quality influence pollinator choice. *Insect Conserv. Divers.* **2016**, *9*, 481–494. [CrossRef]

103. Timberlake, T.P.; Vaughan, I.P.; Memmott, J. Phenology of farmland floral resources reveals seasonal gaps in nectar availability for bumblebees. *J. Appl. Ecol.* **2019**, *56*, 1585–1596. [CrossRef]

104. Richardson, R.T.; Eaton, T.D.; Lin, C.H.; Cherry, G.; Johnson, R.M.; Sponsler, D.B. Application of plant metabarcoding to identify diverse honeybee pollen forage along an urban–agricultural gradient. *Mol. Ecol.* **2021**, *30*, 310–323. [CrossRef] [PubMed]

105. Timberlake, T. Mind the Gap: The Importance of Flowering Phenology in Pollinator Conservation. Ph.D. Thesis, University of Bristol, Bristol, UK, November 2019.

106. Jones, L. Investigating the Foraging Preferences of the Honeybee, *Apis mellifera* L., Using DNA Metabarcoding. Ph.D. Thesis, Bangor University, Bangor, UK, April 2020.

107. Nürnberger, F.; Keller, A.; Härtel, S.; Steffan-Dewenter, I. Honey bee waggle dance communication increases diversity of pollen diets in intensively managed agricultural landscapes. *Mol. Ecol.* **2019**, *28*, 3602–3611. [CrossRef]

108. Quinlan, G.; Milbrath, M.; Otto, C.; Smart, A.; Iwanowicz, D.; Cornman, R.S.; Isaacs, R. Honey bee foraged pollen reveals temporal changes in pollen protein content and changes in forager choice for abundant versus high protein flowers. *Agric. Ecosyst. Environ.* **2021**, *322*, 107645. [CrossRef]

109. Bontšutšnaja, A.; Karise, R.; Mänd, M.; Smagghe, G. Bumble bee foraged pollen analyses in spring time in southern estonia shows abundant food sources. *Insects* **2021**, *12*, 922. [CrossRef]

110. Casanelles-Abella, J.; Müller, S.; Keller, A.; Aleixo, C.; Alós Orti, M.; Chiron, F.; Deguines, N.; Hallikma, T.; Laanisto, L.; Pinho, P.; et al. How wild bees find a way in European cities: Pollen metabarcoding unravels multiple feeding strategies and their effects on distribution patterns in four wild bee species. *J. Appl. Ecol.* **2021**, *59*, 457–470. [CrossRef]

111. Simanonok, M.P.; Otto, C.R.V.; Iwanowicz, D.D.; Cornman, R.S. Honey bee-collected pollen richness and protein content across an agricultural land-use gradient. *Apidologie* **2021**, *52*, 1291–1304. [CrossRef]

112. Tommasi, N.; Biella, P.; Guzzetti, L.; Lasway, J.V.; Njovu, H.K.; Tapparo, A.; Agostinetto, G.; Peters, M.K.; Steffan-Dewenter, I.; Labra, M.; et al. Impact of land use intensification and local features on plants and pollinators in Sub-Saharan smallholder farms. *Agric. Ecosyst. Environ.* **2021**, *319*, 107560. [CrossRef]

113. Biesmeijer, J.C.; Roberts, S.P.M.; Reemer, M.; Ohlemuller, R.; Edwards, M.; Peeters, T.; Schaffers, A.P.; Potts, S.G.; Kleukers, R.; Thoimas, C.D.; et al. Parallel Declines in Pollinators and Insect-Pollinated Plants in Britain and the Netherlands. *Science* **2006**, *313*, 351–354. [CrossRef]

114. Memmott, J.; Waser, N.M.; Price, M.V. Tolerance of pollination networks to species extinctions. *Proc. R. Soc. B Biol. Sci.* **2004**, *271*, 2605–2611. [CrossRef]

115. Pornon, A.; Baksay, S.; Escaravage, N.; Burrus, M.; Andalo, C. Pollinator specialization increases with a decrease in a mass-flowering plant in networks inferred from DNA metabarcoding. *Ecol. Evol.* **2019**, *9*, 13650–13662. [CrossRef] [PubMed]

116. Elliott, B.; Wilson, R.; Shapcott, A.; Keller, A.; Newis, R.; Cannizzaro, C.; Burwell, C.; Smith, T.; Leonhardt, S.D.; Kämper, W.; et al. Pollen diets and niche overlap of honey bees and native bees in protected areas. *Basic Appl. Ecol.* **2021**, *50*, 169–180. [CrossRef]

117. Brosi, B.J. Pollinator specialization: From the individual to the community. *New Phytol.* **2016**, *210*, 1190–1194. [CrossRef]

118. Klečka, J.; Mikát, M.; Koloušková, P.; Hadrava, J.; Straka, J. Individual-level specialisation and interspecific resource partitioning in bees revealed by pollen DNA metabarcoding. *bioRxiv* **2021**. [CrossRef]

119. Kaluza, B.F.; Wallace, H.M.; Heard, T.A.; Minden, V.; Klein, A.; Leonhardt, S.D. Social bees are fitter in more biodiverse environments. *Sci. Rep.* **2018**, *8*, 12353. [CrossRef]

120. Donkersley, P.; Rhodes, G.; Pickup, R.W.; Jones, K.C.; Power, E.F.; Wright, G.A.; Wilson, K. Nutritional composition of honey bee food stores vary with floral composition. *Oecologia* **2017**, *185*, 749–761. [CrossRef]

121. Trinkl, M.; Kaluza, B.F.; Wallace, H.; Heard, T.A.; Keller, A.; Leonhardt, S.D. Floral species richness correlates with changes in the nutritional quality of larval diets in a stingless bee. *Insects* **2020**, *11*, 125. [CrossRef]

122. Mogren, C.L.; Benítez, M.S.; McCarter, K.; Boyer, F.; Lundgren, J.G. Diverging landscape impacts on macronutrient status despite overlapping diets in managed (*Apis mellifera*) and native (*Melissodes desponsa*) bees. *Conserv. Physiol.* **2020**, *8*, coaa109. [CrossRef]

123. Dew, R.M.; McFrederick, Q.S.; Rehan, S.M. Diverse diets with consistent core microbiome in wild bee pollen provisions. *Insects* **2020**, *11*, 499. [CrossRef]

124. McFrederick, Q.S.; Rehan, S.M. Characterization of pollen and bacterial community composition in brood provisions of a small carpenter bee. *Mol. Ecol.* **2016**, *25*, 2302–2311. [CrossRef] [PubMed]

125. McFrederick, Q.S.; Rehan, S.M. Wild Bee Pollen Usage and Microbial Communities Co-vary Across Landscapes. *Microb. Ecol.* **2019**, *77*, 513–522. [CrossRef] [PubMed]

126. Voulgari-Kokota, A.; Ankenbrand, M.J.; Grimmer, G.; Steffan-Dewenter, I.; Keller, A. Linking pollen foraging of megachilid bees to their nest bacterial microbiota. *Ecol. Evol.* **2019**, *9*, 10788–10800. [CrossRef] [PubMed]

127. Piot, N.; Eeraerts, M.; Pisman, M.; Claus, G.; Meeus, I.; Smagghe, G. More is less: Mass-flowering fruit tree crops dilute parasite transmission between bees. *Int. J. Parasitol.* **2021**, *51*, 777–785. [CrossRef]

128. Cullen, N.; Xia, J.; Wei, N.; Kaczorowski, R.; Arceo-Gómez, G.; O'Neill, E.; Hayes, R.; Ashman, T.-L. Diversity and composition of pollen loads carried by pollinators are primarily driven by insect traits, not floral community characteristics. *Oecologia* **2021**, *196*, 131–143. [CrossRef]

129. Stout, J.C.; Kells, A.R.; Goulson, D. Pollination of the invasive exotic shrub Lupinus arboreus (Fabaceae) by introduced bees in Tasmania. *Biol. Conserv.* **2002**, *106*, 425–434. [CrossRef]

130. Willmer, P.G. The effects of insect visitors on nectar constituents in temperate plants. *Oecologia* **1980**, *47*, 270–277. [CrossRef]

131. Macgregor, C.J.; Pocock, M.J.O.; Fox, R.; Evans, D.M. Pollination by nocturnal Lepidoptera, and the effects of light pollution: A review. *Ecol. Entomol.* **2015**, *40*, 187–198. [CrossRef]

132. Tommasi, N.; Ferrari, A.; Labra, M.; Galimberti, A.; Biella, P. Harnessing the Power of Metabarcoding in the Ecological Interpretation of Plant-Pollinator DNA Data: Strategies and Consequences of Filtering Approaches. *Diversity* **2021**, *13*, 437. [CrossRef]

133. Tanaka, K.; Nozaki, A.; Nakadai, H.; Shiwa, Y.; Shimizu-Kadota, M. Using pollen DNA metabarcoding to profile nectar sources of urban beekeeping in Kōtō-ku, Tokyo. *BMC Res. Notes* **2020**, *13*, 515. [CrossRef]

134. Bertrand, C.; Eckerter, P.W.; Ammann, L.; Entling, M.H.; Gobet, E.; Herzog, F.; Mestre, L.; Tinner, W.; Albrecht, M. Seasonal shifts and complementary use of pollen sources by two bees, a lacewing and a ladybeetle species in European agricultural landscapes. *J. Appl. Ecol.* **2019**, *56*, 2431–2442. [CrossRef]

135. Rotheray, G.E.; Gilbert, F. *The Natural History of Hoverflies*; Forrest Text: Ceredigion, UK, 2011.

136. Liu, M.; Clarke, L.J.; Baker, S.C.; Jordan, G.J.; Burridge, C.P. A practical guide to DNA metabarcoding for entomological ecologists. *Ecol. Entomol.* **2020**, *45*, 373–385. [CrossRef]

137. Quaresma, A.; Brodschneider, R.; Gratzer, K.; Gray, A.; Keller, A.; Kilpinen, O.; Rufino, J.; van der Steen, J.; Vejsnæs, F.; Pinto, M.A. Preservation methods of honey bee-collected pollen are not a source of bias in ITS^2 metabarcoding. *Environ. Monit. Assess.* **2021**, *193*, 785. [CrossRef] [PubMed]

138. Kress, W.J.; Erickson, D.L. A Two-Locus Global DNA Barcode for Land Plants: The Coding rbcL Gene Complements the Non-Coding trnH-psbA Spacer Region. *PLoS ONE* **2007**, *2*, e508. [CrossRef]

139. De Vere, N.; Rich, T.C.G.; Ford, C.R.; Trinder, S.A.; Long, C.; Moore, C.W.; Satterthwaite, D.; Davies, H.; Allainguillaume, J.; Ronca, S.; et al. DNA barcoding the native flowering plants and conifers of Wales. *PLoS ONE* **2012**, *7*, e37945. [CrossRef]

140. Chiou, S.J.; Yen, J.H.; Fang, C.L.; Chen, H.L.; Lin, T.Y. Authentication of medicinal herbs using PCR-amplified ITS2 with specific primers. *Planta Med.* **2007**, *73*, 1421–1426. [CrossRef] [PubMed]

141. Moorhouse-Gann, R.J.; Dunn, J.C.; De Vere, N.; Goder, M.; Cole, N.; Hipperson, H.; Symondson, W.O.C. New universal ITS2 primers for high-resolution herbivory analyses using DNA metabarcoding in both tropical and temperate zones. *Sci. Rep.* **2018**, *8*, 8542. [CrossRef]

142. Abdel-Latif, A.; Osman, G. Comparison of three genomic DNA extraction methods to obtain high DNA quality from maize. *Plant Methods* **2017**, *13*, 1. [CrossRef]

143. Swenson, S.J.; Gemeinholzer, B. Testing the effect of pollen exine rupture on metabarcoding with Illumina sequencing. *PLoS ONE* **2021**, *16*, e0245611. [CrossRef]

144. Richardson, R.T.; Curtis, H.R.; Matcham, E.G.; Lin, C.H.; Suresh, S.; Sponsler, D.B.; Hearon, L.E.; Johnson, R.M. Quantitative multi-locus metabarcoding and waggle dance interpretation reveal honey bee spring foraging patterns in Midwest agroecosystems. *Mol. Ecol.* **2019**, *28*, 686–697. [CrossRef]

145. Brennan, G.L.; Potter, C.; de Vere, N.; Griffith, G.W.; Skjøth, C.A.; Osborne, N.J.; Wheeler, B.W.; McInnes, R.N.; Clewlow, Y.; Barber, A.; et al. Temperate airborne grass pollen defined by spatio-temporal shifts in community composition. *Nat. Ecol. Evol.* **2019**, *3*, 750–754. [CrossRef] [PubMed]

146. Bohmann, K.; Elbrecht, V.; Carøe, C.; Bista, I.; Leese, F.; Bunce, M.; Yu, D.W.; Seymour, M.; Dumbrell, A.J.; Creer, S. Strategies for sample labelling and library preparation in DNA metabarcoding studies. *Mol. Ecol. Resour.* **2021**. [CrossRef]

147. Geiger, M.; Moriniere, J.; Hausmann, A.; Haszprunar, G.; Wägele, W.; Hebert, P.; Rulik, B. Testing the Global Malaise Trap Program—How well does the current barcode reference library identify flying insects in Germany? *Biodivers. Data J.* **2016**, *4*, e10671. [CrossRef]

148. Kuzmina, M.L.; Braukmann, T.W.A.; Fazekas, A.J.; Graham, S.W.; Dewaard, S.L.; Rodrigues, A.; Bennett, B.A.; Dickinson, T.A.; Saarela, J.M.; Catling, P.M.; et al. Using Herbarium-Derived DNAs to Assemble a Large-Scale DNA Barcode Library for the Vascular Plants of Canada. *Appl. Plant Sci.* **2017**, *5*, 1700079. [CrossRef]

149. Alsos, I.G.; Lavergne, S.; Merkel, M.K.F.; Boleda, M.; Lammers, Y.; Alberti, A.; Pouchon, C.; Denoeud, F.; Pitelkova, I.; Puşcaş, M.; et al. The treasure vault can be opened: Large-scale genome skimming works well using herbarium and silica gel dried material. *Plants* **2020**, *9*, 432. [CrossRef]
150. Kress, W.J. Plant DNA barcodes: Applications today and in the future. *J. Syst. Evol.* **2017**, *55*, 291–307. [CrossRef]
151. Biella, P.; Tommasi, N.; Akter, A.; Guzzetti, L.; Klečka, J.; Sandionigi, A.; Labra, M.; Galimberti, A. Foraging strategies are maintained despite workforce reduction: A multidisciplinary survey on the pollen collected by a social pollinator. *PLoS ONE* **2019**, *14*, e0224037. [CrossRef]
152. Cornman, R.S.; Otto, C.R.V.; Iwanowicz, D.; Pettis, J.S. Taxonomic characterization of honey bee (Apis mellifera) pollen foraging based on non-overlapping paired-end sequencing of nuclear ribosomal loci. *PLoS ONE* **2015**, *10*, e0145365. [CrossRef]
153. Harris, D.J. Can you bank on GenBank? *Trends Ecol. Evol.* **2003**, *18*, 317–319. [CrossRef]
154. Bell, K.L.; Burgess, K.S.; Botsch, J.C.; Dobbs, E.K.; Read, T.D.; Brosi, B.J. Quantitative and qualitative assessment of pollen DNA metabarcoding using constructed species mixtures. *Mol. Ecol.* **2019**, *28*, 431–455. [CrossRef]
155. Polling, M.; Sin, M.; de Weger, L.A.; Speksnijder, A.G.C.L.; Koenders, M.J.F.; de Boer, H.; Gravendeel, B. DNA metabarcoding using nrITS2 provides highly qualitative and quantitative results for airborne pollen monitoring. *Sci. Total Environ.* **2022**, *806*, 150468. [CrossRef] [PubMed]
156. Baksay, S.; Pornon, A.; Burrus, M.; Mariette, J.; Andalo, C.; Escaravage, N. Experimental quantification of pollen with DNA metabarcoding using ITS1 and trnL. *Sci. Rep.* **2020**, *10*, 4202. [CrossRef] [PubMed]
157. Lamb, P.D.; Hunter, E.; Pinnegar, J.K.; Creer, S.; Davies, R.G.; Taylor, M.I. How quantitative is metabarcoding: A meta-analytical approach. *Mol. Ecol.* **2019**, *28*, 420–430. [CrossRef] [PubMed]
158. Piñol, J.; Senar, M.A.; Symondson, W.O.C. The choice of universal primers and the characteristics of the species mixture determine when DNA metabarcoding can be quantitative. *Mol. Ecol.* **2019**, *28*, 407–419. [CrossRef] [PubMed]
159. Bänsch, S.; Tscharntke, T.; Ratnieks, F.L.W.; Härtel, S.; Westphal, C. Foraging of honey bees in agricultural landscapes with changing patterns of flower resources. *Agric. Ecosyst. Environ.* **2020**, *291*, 106792. [CrossRef]
160. Deagle, B.E.; Thomas, A.C.; McInnes, J.C.; Clarke, L.J.; Vesterinen, E.J.; Clare, E.L.; Kartzinel, T.R.; Eveson, J.P. Counting with DNA in metabarcoding studies: How should we convert sequence reads to dietary data? *Mol. Ecol.* **2019**, *28*, 391–406. [CrossRef]
161. Chui, S.X.; Keller, A.; Leonhardt, S.D. Functional resin use in solitary bees. *Ecol. Entomol.* **2021**, *47*, 115–136. [CrossRef]
162. Deiner, K.; Bik, H.M.; Mächler, E.; Seymour, M.; Lacoursière-Roussel, A.; Altermatt, F.; Creer, S.; Bista, I.; Lodge, D.M.; de Vere, N.; et al. Environmental DNA metabarcoding: Transforming how we survey animal and plant communities. *Mol. Ecol.* **2017**, *26*, 5872–5895. [CrossRef]
163. Bohmann, K.; Mirarab, S.; Bafna, V.; Gilbert, M.T.P. Beyond DNA barcoding: The unrealized potential of genome skim data in sample identification. *Mol. Ecol.* **2020**, *29*, 2521–2534. [CrossRef]

Review

The Expanding Role of DNA Barcodes: Indispensable Tools for Ecology, Evolution, and Conservation

Morgan R. Gostel [1,2,*] and **W. John Kress** [2,3,4]

1 Botanical Research Institute of Texas (BRIT), Fort Worth, TX 76107, USA
2 National Museum of Natural History, Smithsonian Institution, Washington, DC 20560, USA; kressj@si.edu
3 Department of Biological Sciences, Dartmouth College, Hanover, NH 03755, USA
4 The Arnold Arboretum, Harvard University, Boston, MA 02130, USA
* Correspondence: mgostel@brit.org; Tel.: +1-(817)-463-4199

Abstract: DNA barcoding has transformed the fields of ecology, evolution, and conservation by providing a rapid and effective tool for species identification. The growth of DNA barcodes as a resource for biologists has followed advances in computational and sequencing technology that have enabled high-throughput barcoding applications. The global DNA barcode database is expanding to represent the diversity of species on Earth thanks to efforts by international consortia and expanding biological collections. Today, DNA barcoding is instrumental in advancing our understanding of how species evolve, how they interact, and how we can slow down their extirpation and extinction. This review focuses on current applications of DNA barcode sequences to address fundamental lines of research, as well as new and expanding applications of which DNA barcoding will play a central role.

Keywords: high-throughput sequencing; species interactions; metabarcoding; symbioses

Citation: Gostel, M.R.; Kress, W.J. The Expanding Role of DNA Barcodes: Indispensable Tools for Ecology, Evolution, and Conservation. *Diversity* **2022**, *14*, 213. https://doi.org/10.3390/d14030213

Academic Editor: Mario A. Pagnotta

Received: 19 February 2022
Accepted: 7 March 2022
Published: 13 March 2022

Publisher's Note: MDPI stays neutral with regard to jurisdictional claims in published maps and institutional affiliations.

1. Introduction

The fields of ecology, evolution, and conservation are being transformed by novel resources and techniques in the biological sciences. One of these, DNA barcoding, has now realized its potential for the research community. Since the concept of DNA barcodes was first introduced in 2003 [1], tens of millions of barcode sequences have been made publicly available in reference databases for comparative research applications across the Tree of Life (Table 1). The growth of DNA barcode data in public repositories has been driven by several factors, including advances in sequencing technology, novel database management and other computational software, and the expansion of national and international consortia that support DNA barcode sequencing. Recent reviews have highlighted the growth of DNA barcode applications for phylogenetics and taxonomy (e.g., [2]). Other overviews suggest that DNA barcoding is a resilient field that will continue to grow as sequence databases are enriched, throughput expands, and automation provides an ever-expanding user-community with increased accessibility to DNA barcodes, as reported by [3]. This review highlights the advances and applications in DNA barcode sequencing that have been leveraged for novel research in ecology, evolution, and conservation.

1.1. Accurate and Reliable Identification of Species in Taxonomy, Ecology, Evolution, and Conservation

Hypothesis testing is at the heart of the biological sciences and is the standard for how we understand the complexity of the natural world. For most biodiversity research, the reliability and repeatability of hypothesis testing is dependent on accurate identifications of the species under investigation. Faulty identifications can result in faulty hypotheses. A fundamental challenge for any biologist, therefore, is to determine in a reliable and repeatable fashion the correct identification of any given biological sample. "DNA barcodes," i.e., standardized short sequences of DNA between 400 and 800 base pairs long, which in theory can be easily isolated and characterized for all species on the planet, were originally conceived to facilitate this task [1]. By combining the strengths of molecular biology, sequencing technology, and bioinformatics, DNA barcodes offer a quick and accurate means to recognize previously known, described, and named species and to retrieve information about them.

Table 1. Diversity and number of barcode sequences available in the Barcode of Life Data System (BOLD) database, taxon labels follow the BOLD format.

Taxon		Barcode Sequences Available [1]
Animals		11,607,692
	Acanthocephala	2302
	Acoelomorpha	20
	Annelida	112,010
	Arthropoda	11,486,730
	Brachiopoda	326
	Bryozoa	4529
	Chaetognatha	1775
	Chordata	877,866
	Cnidaria	32,680
	Ctenophora	649
	Echinodermata	326
	Entoprocta	76
	Gastrotricha	1351
	Gnathostomulida	24
	Hemichordata	263
	Kinorhyncha	720
	Mollusca	258,885
	Nematoda	36,513
	Nematomorpha	408
	Nemertea	6443
	Onychophora	1394
	Phoronida	172
	Placozoa	20
	Platyhelminthes	41,262
	Porifera	9668
	Priapulida	151
	Rhombozoa	48
	Rotifera	13,758
	Sipuncula	1367
	Tardigrada	3175
	Xenacoelomorpha	18
Fungi		178,231
	Ascomycota	99,779
	Basidiomycota	71,120
	Chytridiomycota	293
	Glomeromycota	3529
	Myxomycota	235
	Zygomycota	3275
Plants		572,154
	Bryophyta	22,675
	Chlorophyta	18,286
	Lycopodiophyta	1338
	Magnoliophyta	454,329
	Pinophyta	7661
	Pteridophyta	11,671
	Rhodophyta	56,194
Protists		10,463
	Chlorarachniophyta	67
	Ciliophora	819
	Heterokontophyta	7238
	Pyrrophycophyta	2339
Total		12,368,540

[1] Data accessed from https://www.boldsystems.org/index.php/TaxBrowser_Home, accessed on 26 January 2022.

For plants, DNA barcoding has truly become a universal tool for hypothesis testing by expanding the ability to identify a species at all stages of its life history (i.e., fruits, seeds, seedlings, mature individuals both fertile and sterile) from damaged or preserved specimens, as well as environmental samples with multiple species. Accordingly, DNA barcodes have been applied to address fundamental questions in ecology, evolution, and conservation biology, such as: how are species assembled in communities; what is the extent and specificity of multispecies interactions in well-studied and previously poorly known environments; and where are the most evolutionarily rich habitats for priority conservation and natural area protection in this age of habitat degradation [4,5]. With regard to the applied users of taxonomy, DNA barcodes also serve as a means to identify regulated species, invasive species, and endangered species.

1.2. Generating, Applying, and Sharing DNA Barcodes

1.2.1. Sequencing Technology

Advances in sequencing technology have radically transformed the potential for DNA barcoding over the last decade by significantly reducing costs and time [6]. The current state-of-the-art sequencing platforms can rapidly sequence tens to hundreds of millions of short-length DNA fragments (50–300 base pairs with Illumina) or tens to hundreds of thousands of long DNA fragments (10,000–30,000 base pairs on PacBio® and Oxford Nanopore). The scale of targeted sequencing projects has expanded such that a single researcher can generate barcode sequences from hundreds or thousands of extracted DNA samples in a matter of hours [7–9]. The expanding scale of sequencing presents a great opportunity for the barcoding community, as it allows for rapid generation of a universal DNA barcode library across the Tree of Life. This is critical, as high-throughput sequencing leads to a better curated database of barcode sequences from known species, but also a greater representation of sequences from unidentified species (e.g., dark taxa, [10]). The universal library of DNA barcodes from known species is being populated at an increasing pace, but the global scientific community still lacks reference barcode data for a majority of species across all major lineages (Figure 1).

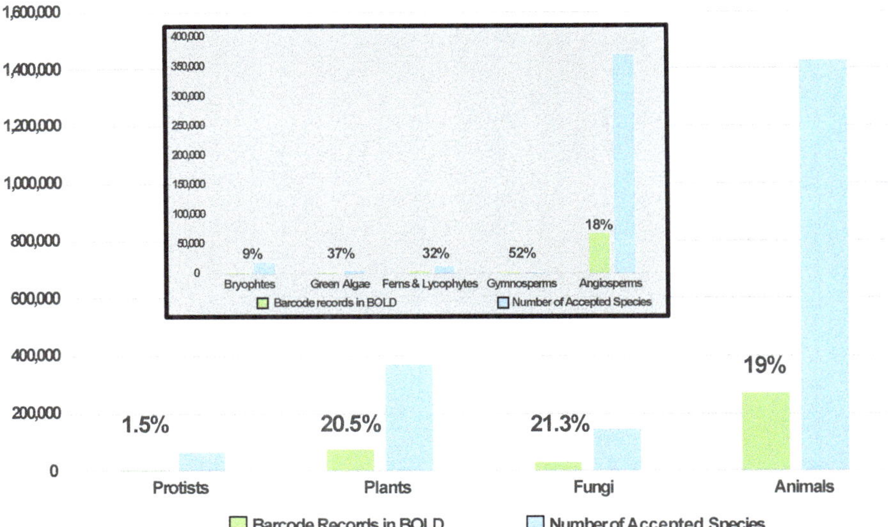

Figure 1. Numbers of species and DNA barcodes across the Tree of Life. The number of species in each of the four major groups of organisms on Earth (blue bars) according to the Catalog of Life are given along with the number of published barcode sequences in BOLD (green bars). Inset shows the major green plant clades (blue bars) with the number of barcode sequences in BOLD (green bars) adjacent to the number of accepted species (according to [11]). The estimated percentage of all species with DNA barcode sequences for that group is provided above the bars in this plot.

As the sequencing technology landscape continues to expand (also see Section 3 below), so does the traditional view of DNA barcodes. Longer sequence reads have led some researchers to consider longer barcode sequences with potentially greater discriminatory power for taxonomic identification. A number of recent studies have presented "super-barcodes" [12,13] or "ultra-barcodes" [14,15] as approaches that leverage whole organelle genomes (e.g., the chloroplast) or a combination of organellar and ribosomal DNA to provide significantly longer sequence data for barcoding. The super- or ultra-barcoding approach has been most commonly used for plants, which present a number of challenges to traditional DNA barcoding. Another alternative for traditional DNA barcoding leverages high-throughput sequencing technology to "skim" the genome (e.g., genome skimming, low-coverage sequence reads from a whole genome) as a universal barcode [16]. This approach circumvents the need for PCR, which can be problematic for preserved specimens with degraded DNA and also provides a method for less ambiguous reference databases for taxonomic identification [17]. Regardless of where the standard for DNA barcode technology is headed, barcode sequence databases will benefit from a growing number of sequences generated for known species.

1.2.2. Novel Computational Resources and Software

The Barcode of Life Data System (BOLD, https://www.boldsystems.org/, accessed on 26 January 2022 [18]) has been the core bioinformatics resource dedicated to hosting DNA barcode sequence data since it was launched in 2007. In addition, many computational resources and software have been developed to accommodate the expanding role of DNA barcodes. Some of these packages (e.g., MDOP [19]) help researchers to organize DNA barcoding data before uploading to databases, such as BOLD and NCBI's Genbank, and still others are designed to assess the quality of data that have already been made publicly available (e.g., BAGS [20] and MACER [21]). The quality of DNA barcode data can be impacted by a number of factors, including poor sequence annotation, a lack of physical specimen vouchers, poor sequence quality, and incorrect consensus sequence building. The last of these factors is especially problematic for DNA barcoding methods based on high-throughput sequence reads. Fortunately, several recent software packages have been developed to address challenges with consensus sequence building, such as PIPEBAR, OverlapPER [22] and NGSpeciesID [23].

Taxonomic assignment is key for downstream applications of DNA barcode sequences and the accuracies of approaches, which assign sequences from unknown taxa to a recognized barcode sequence, are critical [24]. Despite the development of several tools to accurately assign sequences to taxa represented in barcode sequence databases, comparison across software has demonstrated that it remains challenging to accurately assign sequences to taxa at or below the level of genus [25]. Taxonomic assignment methods are being developed and refined rapidly, with several options published in just the last four years. Among these are the QIIME2 feature classifier [26], IDTAXA [27], MeTaxa2 [28], and Basta [29]. Although the methodology to perform taxonomic assessment is quickly evolving, older methods are accurate, still perform well, and continue to be used, such as Kraken2 [30], Protax [31], and the longstanding BLAST tool [32]. Beyond these methods, other options are optimized for clade-based metabarcoding reference databases (e.g., Fungi: funbarRF [33]) or have been developed as part of custom pipelines that have more specific user needs (e.g., the Anacapa Toolkit [34]). Ultimately, the ability of any computational method to accurately match a sequence from an unknown species is dependent upon well-curated, annotated, and comprehensive reference sequence databases. Focus should remain on populating DNA barcode reference databases with high-quality sequence data from accurately identified and vouchered collections.

1.2.3. National and International Sequencing Consortia

The effort to contribute DNA barcode sequence data is coordinated worldwide through both national and international organizations. Coordination of international barcoding activities began in 2004 with the Consortium for the Barcode of Life, followed by the International Barcode of Life Project (iBOL, https://ibol.org/, accessed on 26 January 2022) in 2008. National efforts have also been launched in Austria (ABOL), Finland (FinBOL), Germany (GBOL), the Netherlands (NBOL), Norway (NorBOL), and Switzerland (Swiss-BOL) to name a few. Most recently, BIOSCAN [35], an international project organized by iBOL, was initiated and includes 1000 researchers in over 30 countries with the objective of generating DNA barcodes to discover species, to understand species interactions, and to monitor species in a global biological surveillance system. Once achieved, the collective goals of these organizations will result in a DNA barcode library for nearly all species on Earth.

In the nearly two decades since DNA barcodes were first proposed, other ambitious and sweeping networks have emerged that also reflect the fundamental goal of the DNA barcoding community: to leverage organismal DNA to understand life on Earth. One of these, the Global Genome Biodiversity Network (GGBN, https://www.ggbn.org/ggbn_portal/, accessed on 26 January 2022 [36]) represents a network of well-curated tissue collections that seeks to develop standards, share collection information, and facilitate biodiversity genomics research. More recently, the Earth BioGenome Project (EBP; https://www.earthbiogenome.org/, accessed on 26 January 2022) was launched [37] as a "moonshot" [38] for biology that aims to sequence whole genomes of all eukaryotic species on Earth in ten years. Although not specifically aimed at DNA barcode loci, EBP will indirectly provide a wealth of sequence data for the major DNA barcode loci of plants, animals, and fungi. DNA barcoding, which was originally considered to be at one end of the sequence spectrum, is now converging with entire genomes [39]. These global efforts, which have been described as "networks of networks," build connections among more localized, often national endeavors.

The organization of DNA barcoding projects has often followed geopolitical boundaries and the most common denominator for large sequence programs reflects local, regional, or national funding structures. Some examples of these at a regional and national levels include the African Centre for DNA Barcoding (https://www.acdb.co.za/, accessed on 26 January 2022 [40]), the Canadian Centre for DNA Barcoding (https://ccdb.ca/, accessed on 26 January 2022), and the China Plant BOL (Barcode of Life) Group [41]. In a similar way, the United Kingdom's Darwin Tree of Life Project (https://www.darwintreeoflife.org/, accessed on 26 January 2022 [42]) takes a geopolitical approach toward their goal to sequence the whole genomes of all eukaryotic species in Britain and Ireland. These focused, localized research networks contribute to international goals that help support the shared priority of advancing a global understanding of biodiversity and facilitate the use of DNA barcodes and other genetic tools for broader ecological, evolutionary, and conservation purposes.

1.2.4. Building the Plant DNA Barcode Library

With more than half a million plant DNA barcode sequences available today in the Barcode of Life Data Systems (BOLD, Figure 1), continuing to populate the global library is a major effort of botanists. In addition to the national and multinational projects described above, building the plant DNA barcode library can be enhanced by taking advantage of a number of diverse efforts, such as forest monitoring plots, individual lineage-based taxonomic studies, and regional floristic efforts. Forest monitoring plots, such as the Smithsonian's Forest Global Earth Observatories (ForestGEO) and the National Science Foundation's Long Term Ecological Research (LTER) sites, are rich resources because they have well-verified identifications, vouchered collections, and individually tagged trees that can be revisited by botanists if necessary [43–46]. Even if no specific monitoring plots have been established, many studies have generated DNA barcode libraries for specific habitats [47], plant communities [48], or regional taxa [49–51] and are thereby expanding the global plant genetic library. Individual taxonomists are also generating DNA barcodes for specific groups of plants as either standard markers (e.g., [52–55]) or as an offshoot of their basic molecular phylogenetic investigations aimed at understanding evolutionary relationships. Preserved museum specimens can also be used to generate DNA barcodes [56]. It is significant that one recent study has encouraged a large-scale effort to sequence DNA barcodes from all types of specimens [57]. All of these DNA sequences add to the library of standard DNA barcode markers even if they do not carry the official GenBank DNA barcode designation.

Other efforts to generate DNA barcodes for entire regional floras are in some cases complete or just getting underway. One of the most impressive is the library that has been built for identifying the vascular plants of Canada [58], which includes sequence records (*rbcL*, *matK*, and ITS2) for 96% of the 5108 species known from that country. Another success story for plant DNA barcodes is the China Plant Barcode of Life [41]. This sixteen-year project has now generated and made available for use 120,000 DNA barcodes for 16,000 species, representing a significant sampling of the entire flora of China. Other examples are the recently completed DNA barcode library for the plants of the UK [59], and work that has started on the flora of the Arabian peninsula [60].

1.3. The Purpose and Structure of This Review

Today, more than ever, DNA barcodes are being used to advance our understanding of how species evolve, how they interact, and how we can slow down their extirpation and extinction (e.g., [61–63]). As sequencing technologies have improved and sequencing costs have declined, the use of DNA barcoding is skyrocketing and some of the most exciting prospects for using this new taxonomic tool are being realized. A number of comprehensive reviews of the application of plant DNA barcodes to the fields of ecology, evolution, and conservation have been provided in the past [5,64–68]. This review and the Special Issue of *Diversity* of which it is a part focus on current areas of research as well as new applications of DNA barcodes that are the direct result of the accumulation of barcode reference sequences, including past trials, experiments, and applications of this twenty-first century biological tool (Figure 2).

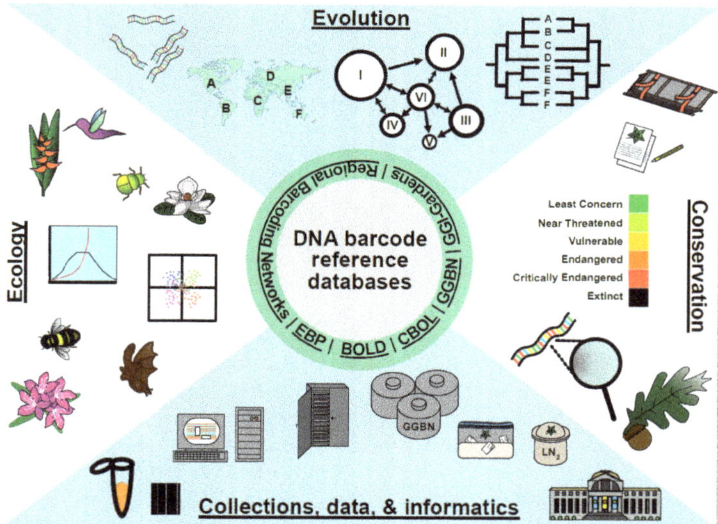

Figure 2. A graphical representation of DNA barcoding today. DNA barcode applications in ecology (left), evolution (top), and conservation (right) are supported by a foundation of collections, metadata, and informatics (bottom). These applications are facilitated by increasingly large DNA barcode reference databases (center circle) that are reciprocally built from and contribute to the major biological disciplines. National and international initiatives that support the growth of DNA barcode reference databases are core resources (green circle).

2. Current Applications

2.1. Improving Taxonomy and Species Identification

2.1.1. Defining Species Boundaries

Taxonomists have been using morphological features for the identification of both plants and animals since before the time of Carl von Linné. Yet, even after centuries of taxonomic work, perhaps only 20 percent of the species on Earth have been formally named [69]. Much work remains to be done. DNA barcoding provides a relatively new and significant tool to aid in the determination of species boundaries and discovery of new taxa. Entomologists have been pioneers in incorporating DNA barcode technologies for species discovery in the tropics, where the majority of biodiversity is found (e.g., [70–73]). Although the discriminatory power of barcode markers for plants is less than for insects, botanists have also used DNA barcodes as a taxonomic resource. Early studies, which have mostly focused on trees in tropical forest monitoring plots (e.g., [62,74,75]), demonstrated the difficulties of using DNA barcodes in plants (also see [76] for a recent study on African trees). However, the same studies also pointed out the advantages of being able to accurately identify sterile and juvenile specimens that lack morphological features required for identification. Costion et al. [77] applied a three-locus DNA barcode to estimate tree species diversity in a taxonomically poorly known tropical rain forest plot in Queensland, Australia, and concluded that DNA barcodes were a significant aid in rapid biodiversity assessment and determination of cryptic tree populations. A similar study in a central African rain forest plot recognized the high discriminatory power of barcode markers at the genus-level (95–100%), but somewhat lower species-level success (71–88%) in identification, especially in species-rich clades [78] or those with high rates of molecular evolution.

One of the major issues faced by plant taxonomists and ecologists attempting to use DNA barcodes in diverse forests, especially in the tropics, is that many species are new to science, therefore lack Latin binomials, and/or are members of poorly circumscribed species complexes that are difficult to identify even with traditional morphological data. Inventories and assessments of plant diversity in these habitats can be greatly enhanced by building DNA barcode libraries of these taxa [79]. Standardizing the DNA barcode markers and bioinformatics tools being used in different forest inventory projects (e.g., RAINFOR, http://www.rainfor.org/, accessed on 26 January 2022; the Amazon Tree Diversity Network [80], CForBio, http://www.cfbiodiv.org/, accessed on 26 January 2022; and ForestGEO [43]) will provide more confidence in identifications and maybe even allow rapid discovery and description of unknown taxa in these species-rich forests [79].

In addition to discovering new species, the introduction of integrative taxonomy has encouraged closer collaboration among biologists with different backgrounds, and in turn has promoted the use of DNA barcoding as a new tool in a broad taxonomic toolkit [81]. For very poorly documented regions or "understudied and hyperdiverse" taxa, DNA barcoding can be a key part of integrative workflows for species description and identification [82].

2.1.2. Regional Biodiversity Assessments

DNA barcode studies both benefit from and serve a key role in support of local and regional biodiversity assessments, including floristics. In many biodiverse regions, where species diversity is poorly known, collections-based exploration and inventory studies are vital for alpha taxonomy and conservation. Modern approaches to field expeditions employ a variety of strategies to collect and document species, which often include the collection of various data to inform biodiversity studies. These data incorporate traditional natural history specimens, photographs, ecological notes, and, more recently, vouchers intended for genetic and/or genomics research [83]. The collection of genetic vouchers and sequencing of DNA barcodes in standard species inventories help to build the global barcode reference database [84] as mentioned above, and often result in surprising discoveries of cryptic diversity (e.g., [85–87]).

2.2. Quantify Species Diversity

2.2.1. Species Richness and Phylogenetic Diversity

Fundamental to biodiversity research is the quantification of organismal diversity. Different approaches to this task may provide different interpretations by ecologists, evolutionary biologists, and conservation biologists regarding the role that biodiversity plays in ecosystem function, niche allocation, and species preservation. Phylogenetic diversity was proposed as a metric that quantifies diversity by summing the branch lengths of a given phylogenetic tree [88,89] and is arguably a more descriptive measure of biodiversity than alternative indices such as simple species richness and abundance [90]. DNA barcoding provides an efficient and rapid resource for generating phylogenies to measure phylogenetic diversity, particularly when combined with metabarcoding [91].

It should be noted however that despite the utility of DNA barcoding approaches in diversity assessment, limitations exist. Winter et al. [92] described some of the limitations of phylogenetic diversity insofar as the metric is applied to conservation applications. And although phylogenetic diversity has been lauded as an indicator of species interactions and ecosystem functions [93–95], caution has been urged against using this measure alone to conserve functional trait diversity in ecosystems. The growth of DNA barcode databases and new sequencing methods are facilitating the ability to analyze and understand phylogenetic diversity, but if these data are to be used as predictors for conservation and estimates of ecosystem function, they need to be carefully evaluated in combination with detailed trait databases. Among the earliest uses of DNA barcoding to quantify biodiversity were investigations of community assembly and function in long-term forest monitoring plots in Panama.

2.2.2. BCI as an Exemplar Tropical Field Site for DNA Barcoding

More than a decade ago the first community phylogeny based on DNA barcode sequence data was published for the trees in a forest dynamics plot on Barro Colorado Island (BCI) in Panama [62]. This publication set off a storm of new investigations that were able to add a well-supported evolutionary perspective to understanding species diversity and assembly in plant communities (e.g., [96–100]). The DNA barcode phylogeny generated for the approximately 300 species of trees on BCI also served as a template for a number of investigations of functional traits. The evolutionary context of such characteristics as soil associations [101], leaf toughness [102], wood nitrogen concentration and life-history strategies [103], foliar spectral traits [104], and anti-herbivore defense traits [105] was found to vary in each of these functional traits across the tree species in the BCI plot. Although some have concluded that phylogenetic indictors are not always tied to ecological determinants of community assembly [106], both phylogenetic- and trait-based approaches have greatly enhanced the understanding of community structure and function on BCI.

Belowground interactions among species have also been investigated at BCI using the DNA barcode library for trees. Jones et al. [107] mapped the belowground distribution of all trees and lianas greater than one centimeter in diameter using their genetic DNA barcode signature. Comparing underground species distributions with aboveground distributions showed that species interactions and spatial overlap was greater belowground than expected based on aboveground stem densities. Although this study raised several questions about methodology and analysis, it concluded that the potential for using DNA barcodes in this type of investigation was high.

The DNA barcode library for trees on BCI has now been expanded to include many of the shrubs and lianas as well as some epiphytes that occur in the forest on the island. Efforts to build DNA barcode libraries and apply DNA barcode methodologies to other groups of organisms (e.g., insects [108]) are underway. This rich genetic resource will greatly enhance future studies of ecological interactions and evolutionary signal in this tropical forest community in Panama.

2.3. Determining Community Structure and Species Interactions

2.3.1. Community Evolution and Assembly

DNA barcoding has played a significant role in expanding collaboration between systematists, who focus on species identification and evolutionary relationships, and ecologists, who investigate species interactions and patterns of associations [109]. As noted above for work conducted on Barro Colorado Island in Panama, plant DNA barcoding has been a boon to community ecologists seeking to understand the factors, such as species diversity pools and functional traits, that control the assembly of species into ecological communities [100,106]. Estimating a third component that may determine species assembly, namely evolutionary history, has always been hampered by the lack of well-resolved phylogenetic hypotheses on species relationships in communities. Determining if species in a community are more closely related than by chance (phylogenetic clustering), more distantly related than by chance (phylogenetic overdispersion), or randomly distributed across the plant tree of life is now readily ascertained by building a DNA barcode-based phylogenetic tree. The assumption follows that species in a community that are phylogenetically clustered are more likely to have similar ecological niches (i.e., phylogenetic niche conservation) and have been assembled via abiotic filtering. The contrasting assumption is that phylogenetic overdispersion in a community is the result of biotic interactions among sympatric species. Based on these assumptions the impact of evolutionary history on community structure has been investigated using DNA barcodes across stages of forest succession [99], among habitats within a forest type [62,110,111], among forests across habitat gradients [112], and among communities across an entire country [113,114] and across the globe [45,115]. The generation of such community phylogenies has great promise for further testing the basic assumptions and rules governing species assemblies in plant communities (see [45]).

2.3.2. Herbivory and Food Webs

The accurate and repeatable identifications of species is imperative if we are to fully understand the ecology and evolution of interactions among partners in natural and human-altered environments. This requirement is especially true for specialized interactions, including mutualisms and antagonisms. The application of DNA barcodes as species-level markers has revolutionized our ability to track species interactions and the community networks they form, in boreal, temperate, and tropical habitats.

Food web interactions have been greatly clarified with the application of DNA barcodes. Smith et al. [116] using the CO1 DNA barcode marker were able to verify the food web structure of the spruce budworm and its numerous parasitoids to understand the population dynamics of this major pest of trees in boreal forests. The utility of DNA barcodes to identify the diversity of host plants for herbivorous beetles have been demonstrated in both neotropical [62,117] and Asian tropical forests [118]. These early studies used a limited number of molecular markers and were only able to identity the hosts at the generic or familial level.

The most comprehensive analyses between herbivorous beetles and their host plants have been conducted by García-Robledo and colleagues [72,73,119]. The host-specific relationships between rolled-leaf beetles in the genera *Cephaloleia* and *Chelobasis* (Chrysomelidae) and plants in the order Zingiberales have been well-studied by ecologists [120], but the application of DNA barcodes to both the beetles and the hosts have provided a much more detailed and quantitative measure of these interactions [74]. One of the advantages of using an easily extracted DNA barcode is that the beetles can be identified at any of their life stages and not only as adults as in most previous investigations using morphological features [119]. Once the basic network of food web interactions is established using DNA barcodes, comparisons can be made across habitats, elevations, and temperature gradients. Most recently Palmer et al. [121] have extended this methodology to the interactions among katydids and their host plants in a wet forest habitat in Panama. They found that, in general, these insects consumed a broad range of flowering plants and were rarely specialists on only a few species. It has been shown in numerous cases (e.g., Hebert et al. [70]) that

DNA barcodes can detect the presence of cryptic species, especially in insects. This power of DNA barcoding has greatly improved the understanding of species boundaries in the rolled-leaf beetles, allowing for more precise mapping of the insect–host networks. The detection of these cryptic species clearly demonstrated that the elevational distributions and thermal tolerances of the beetles was much narrower than previously thought, which will have an impact on the food web networks as climate change differentially impacts both host and herbivore migrations [73].

DNA barcodes have also altered our view of why tropical biomes are so diverse. It has long been held that specialized ecological interactions, which are common in tropical forests, will lead to reproductive isolation and speciation, and hence greater biotic diversity in the tropics. One such specialization is that between tropical flowers and the nectar-robbing floral mites that are caried from plant to plant on the bills of hummingbird pollinators [122]. This specialization allows floral mites to easily find mates and reproduce, because many conspecifics accumulate in the flowers of only a few species of plants. This "mating rendezvous hypothesis" [123] accounted for the host specialization in these mites. However, using DNA barcode markers to identify the mites, rather than morphological identification, has now shown that most floral mites are generalists and not specialists [124]. The mating rendezvous hypothesis is no longer supported, at least for mite diversification.

This detailed understanding of herbivore–host interactions using DNA barcodes has also been applied to large mammalian herbivores. In a semiarid African savannah, Kartzinel et al. [125] determined the extent that sympatric mammalian browsers and grazers could partition their diets. After building a library of plant DNA barcodes for the local flora, they quantified the diet breadth, composition, and overlap for seven co-occurring mammal species, ranging in size from dik-diks to elephants using DNA metabarcoding. Earlier conclusions on competition and coexistence in these habitats based on low-resolution analyses were shown to be misleading, according to the more high-resolution taxonomic data provided by the metabarcoding results. This work in Africa has now been extended to demonstrate that the abundance and diversity of food plants is negatively associated with their mammalian herbivores, apparently to avoid consumption [126]. The same type of DNA barcoding protocol has also been adapted to tracking and identifying the vectors of bird-dispersed fruits and seeds in the field [127] in order to build a quantifiable network of frugivores and seed dispersal interactions.

2.3.3. Symbiotic Relationships and Plant-Pollinator Interactions

Symbioses, perhaps the most characteristic of "species interactions," entail very close relationships between two or more species living together, and DNA barcodes have facilitated researchers studying such close interactions [128]. In some groups (e.g., fungi, [129]), DNA barcoding has revolutionized the field, especially where symbiotic partners are very closely associated and interactions often exist at a cellular level (e.g., in lichens [130]). The use of DNA barcodes to understand symbioses is common in all major clades, including arthropods [131], vertebrates [132], green plants [133], and fungi [134]. An especially powerful tool for symbiosis-based research is metabarcoding [135], which allows for pooled sequencing from closely associated, symbiotic organisms that are otherwise difficult to isolate. The application of DNA barcodes to more closely track and untangle symbiotic relationships is still in its infancy (see below Section 3.1).

The interactions between plants and pollinators is a symbiotic mutualism that is critical for the survival of both partners. An understanding of the dynamics of these interactions is a priority for plant and insect ecologists to conserve biodiversity and to protect the agricultural crop supply chain. DNA barcodes have been explored for more than two decades as a means to identify plants from the insects that have visited them as pollinators [136]. Given the nature of pollination dynamics, samples removed either from plants or their pollinators can include a mixed community of pollen and, therefore, metabarcoding approaches provide a unique tool to identify the diversity contained in these mixed samples [137,138].

Clare et al. [139] were among the first to apply metabarcoding to study plant–pollinator interactions, extending the concepts earlier proposed by Valentini et al. [68] and Soininen et al. [140]. A key threshold for advancing these methods is a comprehensive DNA barcode sequence reference database. For example, the first national DNA barcode sequence reference database of Wales [141] has provided a benchmark for DNA metabarcoding studies of plant–pollinator interactions and this has recently grown into a comprehensive database for all of the United Kingdom [59]. Together, these databases have proven powerful for reconstructing bee foraging behavior [142–144]. These and other studies [145–147] have built a strong foundation for using DNA metabarcoding to study plant–pollinator interactions.

2.4. Protecting Species

2.4.1. Forensics and Monitoring Traffic in Endangered Plants

It is abundantly clear to all biologists that biodiversity is under severe threat across the globe due to natural resource overutilization and exploitation, major habitat degradation, and climate change caused by humans. Biodiversity conservation is imperative. DNA barcoding, as a tool primarily for species identification, can be used in three general ways to further biodiversity conservation: (1) to accurately monitor and thereby protect endangered species subject to illegal commercial trade (i.e., point-of-origin tracing [148,149]), (2) to track biological invasions, and (3) to provide data that will assist in the estimation of phylogenetic diversity for setting conservation priorities [150].

Although DNA barcode-based discrimination at the species-level is not possible in all groups of organisms, DNA barcodes have been utilized for forensic identification of algae [151], plants [152,153], invertebrates [154], and vertebrates [155]. A significant driving force in developing DNA barcode technology for plants has been the need for an accurate and inexpensive tool for the identification of illegal timber products, especially those listed in the Convention on International Trade of Endangered Species (CITES). For example, in tests of the commercially important mahogany family (Meliaceae), most of the standard DNA barcode markers fell short of expectations for discriminating species, although the nuclear ribosomal internal transcribed spacer (ITS) was able to identify some species in this family [156]. A higher level of discrimination using standard markers was demonstrated among commercially important and threatened species of trees collected at timber processing plants in the tropical dry forests of India [157]. This same success was demonstrated in timber species found in *Araucaria* rain forests of the southern Atlantic coast of Brazil [158], which contains many threatened species of trees, especially in the family Lauraceae. In Madagascar, a recognized biodiversity hotspot, Hassold et al. [159] used DNA barcodes in an effort to monitor illegal timber trade, especially in species of rosewood (*Dalbergia* in the Fabaceae). They demonstrated the limitations of the standard genetic markers in identifying closely related species of this genus, although some success was achieved. In addition to timber trees, DNA barcode libraries have been developed for other taxonomic groups of threatened and endangered plant taxa listed in CITES, e.g., orchids [160]. Currently no more effective tool than DNA barcoding exists for accurate identification of products sold in public markets [161–163] or as illegally harvested species intended for trade intercepted at ports [164,165]. As global DNA barcode reference libraries grow, so too does the capacity to enforce conservation laws and to monitor illegal trade in endangered plants.

Traditional medicines, teas, and herbal supplements are another important component of the commercial need for accurate plant species identifications by regulators and quality control specialists. It is estimated that medicinal plants account for billions of US dollars in annual revenues in the United States alone [166]. From the initial use of plant DNA barcodes, applications to monitor this market have been in development [167]. However, many of these trials to use DNA barcodes to identify commercial medicines and herbal supplements have shown limited success in discriminating among species. Some of the major obstacles have been the lack of comprehensive DNA barcode libraries required to make accurate comparisons among species of herbal teas and supplements, and the absence of standardized taxonomy and common names listed in the herbal catalogs and pharmacopeias (e.g., Stoeckle et al. [168]; de Boer et al. [169]). Building the required DNA barcode libraries (see below) and unifying the taxonomy in the literature on traditional medicines are challenges for the future.

2.4.2. Tracking Biological Invasions

The field of conservation biology has also benefited from the accuracy of DNA barcoding methods to trace biological invasions. It has been estimated that the control of invasive species costs more than $27 billion annually in the United States alone [170]. Fast detection can significantly reduce the cost of controlling biological invasions, and DNA barcodes and metabarcoding in particular have been demonstrated to provide the earliest invasive species detection methods available [171]. For example, one of the most widespread threats to marine ecosystems is the invasive zebra mussel, *Dreissena polymorpha*, and recent studies [172], using metabarcoding (or environmental DNA), have proven this method to be cost effective for early detection of this species in marine environments. Studies that quantify regional biodiversity using DNA barcodes have also proven effective for identifying biological invasions [173], where higher than expected phylogenetic diversity may result from the occurrence of non-native or invasive species.

In some groups of plants, invasive and weedy species are remarkably difficult to visually distinguish from non-invasive, endemic species, and several studies suggest DNA barcoding will facilitate proper identification and management by non-specialists. For example, current DNA barcodes in many plant taxa are unable to distinguish taxa at or below the genus-level, but new paradigms in barcode sequencing provide greater distinction of closely related species. Wang et al. [174] have advocated the use of super- or ultra-barcodes (e.g., whole chloroplast genomes) to monitor and detect flaxleaf fleabane, *Conyza bonariensis*, because, unlike traditional plant DNA barcodes, these super-barcodes are able to distinguish among closely related species in this diverse and difficult to identify genus.

2.4.3. Conservation Assessment

The taxonomic impediment [175] is also a significant problem in assessing species diversity and making accurate species determinations for conservation monitoring. This case is especially true in tropical biomes, where biodiversity is poorly known and a greater number of species lack verified scientific names. Species identification by non-taxonomists can be extremely difficult, especially when using non-fertile specimens often only labeled as "morphospecies" [176]. In such cases, DNA barcoding offers a solution for more uniform and accurate identifications, recognizing that some logistical hurdles may still impede the widespread use of DNA barcodes in this fashion [177].

In the relatively poorly known tropical forests of northern Queensland, Australia, it has been demonstrated that plant DNA barcodes can play a key role in estimating species richness and thereby determining conservation priorities [77]. Similarly, in the fragmented rain forest habitats in South Eastern Queensland, Shapcott et al. [61,169] generated plant DNA barcodes for 86% of the flora (770 species in 111 families) and calculated phylogenetic diversity (PD; see Owen [178]) measures for each of the 18 subregions in the area. For these forests, which have lately received renewed conservation attention and are taxonomically rich at the generic-level and less so at the species-level, species richness may not be the most appropriate measure for setting conservation priorities. The phylogenetic diversity estimates calculated from the DNA barcode data were used to prioritize subregions for conservation action and it was concluded that the local floristic patterns were consistent with both ancient ecological refugia and recent lineage range expansions [179].

Even though the Earth may be undergoing its sixth major extinction with extinction rates over one thousand times that of "normal" periods [69], observing an extinction event is rare. For plants the extinction of only 571 species over the last several hundred years has been carefully documented [180]. On the island of Palau in Micronesia, plant DNA barcodes were used to verify that a narrow range endemic tree described in the 1980s known from only two mature individuals was *Timonius salsedoi* Fosberg and Sachet in the family Rubiaceae [181]. Subsequently, after a typhoon hit the area, a survey of the island revealed that both trees had perished. Although previously assessed as Critically Endangered by IUCN criteria, it is suspected that this species is now extinct [181].

DNA barcodes have significant potential as a tool for understanding and enhancing conservation efforts. Using standardized and comparable genetic information for species across broad geographic regions can have a substantial impact on basic biodiversity research (e.g., Mi et al. [112]; Erickson et al. [45]; Pei et al. [98]), as well as conservation monitoring and priority assessments in threatened habitats, in local communities, and across large geographic regions (e.g., Shapcott et al. [61]).

3. Looking Forward: The Expanding Technological Spectrum of DNA Barcodes

3.1. Metabarcoding

DNA metabarcoding [135] has emerged as a powerful technique to rapidly characterize species composition, species interactions, and—when combined with trait databases—functional aspects of biological diversity in communities. This method leverages high-throughput sequencing technology to sequence and/or extract DNA barcodes from pooled community or environmental samples. These samples represent DNA isolated from multiple species or other taxa that have been collected in bulk and targeted sequencing is performed on libraries enriched with (typically) DNA barcode amplicons [182]. Metabarcoding is an emergent field that leverages expansive DNA barcode sequence databases and the increasingly high-throughput capacity of DNA sequencing technology.

This technique allows ecologists to explore species interactions through a new lens and is illuminating species distribution and occurrence from ecosystems and habitats that have remained all but invisible. Metabarcoding is able to provide high-resolution inventories from the hidden worlds of below-ground microbial diversity [183], freshwater [184] and marine [185] benthic communities, and the movement and dispersal of airborne fungi [186] and plants [187]; however, this method is dependent upon well-curated reference collections and databases [188].

Beyond enhanced characterization of species communities, metabarcoding has been used to explore species interactions in a variety of contexts. Some of the earliest applications of DNA metabarcoding involved the analysis of vertebrate diets [189] and this method remains a powerful tool for understanding herbivory and predation (see [125,190–192]). More recently, metabarcoding has been used to reconstruct plant–pollinator networks [146,147] and identify economically important taxa [155] or those relevant to human health [193,194].

DNA metabarcoding was developed using short-read high-throughput sequencing platforms and while these are still the norm, they pose some limitations for the technique [8], especially for longer DNA barcode loci (e.g., *matK* for plants). As the technological standard moves toward long-read sequencing approaches, new sequencing platforms and software [195] are being developed. Some recent programs (e.g., Sahlin et al. [23]) have already been used to successfully extract DNA barcode sequences from mixed samples in previously published long-read data.

3.2. Super- and Ultrabarcoding

Much of the expanding role of barcodes in the past decade has been driven by the rapid growth of high-throughput sequencing technology. As opposed to traditional DNA barcodes, which target individual loci or a set of short loci with universal primers, "super-barcodes" and "ultra-barcodes" have been proposed as alternatives that compare information from entire organellar genomes and/or other long regions [12,13]. For plants, whole chloroplast sequencing has been common for over a decade [196]. Super- and ultra-barcoding provide some unique advantages over traditional barcoding. For example, in some large clades (e.g., the green plant tree of life), traditional DNA barcode loci are not present in all taxa [197] and universal PCR primers often don't exist for some taxa in a given clade (see [8]). In these cases, ultra-barcoding provides a simple solution to challenges with traditional DNA barcodes, in which the entire chloroplast genome can serve as one single, long barcode locus or in combination with other loci (e.g., nuclear ribosomal DNA, [14]). Moreover, some traditional DNA barcode loci (e.g., *matK*, ca. 1000 bp) are simply too long for amplicon-based approaches using short-read sequencing platforms. Lastly, chloroplast genomes are abundant and typically easy to sequence even from recalcitrant (i.e., old and/or preserved) tissues and it's increasingly common to assemble whole organellar genomes from off-target reads even in targeted/capture-based sequencing applications [198].

As sequence databases grow, the concept of super- or ultra-barcodes is certain to follow. Rather than viewing alternative barcoding strategies as either/or choices, novel DNA barcoding strategies are complementary to locus-based markers, and each contributes to a growing, cumulative database of well-curated data for molecular species identification.

3.3. Macrogenetics

Computational science, international collaboration, and data accessibility are facilitating massive, integrative research across the biological sciences. Driven by the era of "big data" and increasingly interoperable datasets, new and emerging fields of research are making it possible to pursue "big questions" like never before. These expanding opportunities have led to the emergence of new fields of study and one of these, "macrogenetics" [199], has been facilitated by the growth of publicly available genetic and genomic datasets. The concept for this field is intended to echo that of "macroecology" and emphasizes the integration of large-scale datasets in genetics with other large, interoperable databases [200], such as the Global Biodiversity Information Facility (GBIF, [201]), WorldClim [202], DRYAD [203], the International Nucleotide Sequence Database Collaboration (INSDC [204]), and BOLD. DNA barcodes provide a vital source of information that can facilitate the emerging field of macrogenetics and indeed, the development of BOLD is credited as one of the key advances that underlies macrogenetics.

As a new and emerging field, macrogenetics is presented as the intersection of several biological foundations, united by large-scale genetic resources and including rich ecological data, collections science and museomics, biogeography, phylogeography, and evolutionary biology [200]. The promise of this new field is to synthesize big data across biological disciplines using genetic data to facilitate priorities for ecology, evolution, and conservation at global scale. Undoubtedly, the expanding role of DNA barcodes will play a central role in the development of macrogenetics. It is an exciting time to study ecology, evolution, and conservation.

4. Conclusions

In nearly two decades since DNA barcodes were first proposed, a remarkable increase has taken place in the representation, use, and integration of DNA barcodes across the biological sciences. Although sequence variation in traditional DNA barcodes is often insufficient for species-level discrimination in many large clades, the advances in computational and sequencing technology are changing the concept of DNA barcodes, from just a few loci to large, genome-scale sequences from organelles or genome-skim data. As technology expands and genome sequence representation increases across the Tree of Life, we envision a future in which the concept of DNA barcodes extends to a much larger interpretation of genome space. DNA barcoding continues to evolve with methodological and technological advances in conjunction with the increasing accessibility to high-throughput sequencing and the growing database of whole genome sequences fostered through international consortia, such as the Earth BioGenome Project [37,38]. A diversity of genetic tools is especially needed in clades, such as green plants, with highly complex genomes that require significant resources to assemble [205]. Until there is a corresponding breakthrough in computational capacity for the comparative analysis of large and highly complex genomes, DNA barcode sequences will play a vital role for species identification in community ecology, evolutionary biology, and conservation. DNA barcodes are a powerful resource and the databases that maintain them continue to grow as they complement and benefit from the rapidly expanding frontiers of computational science and high-throughput sequencing technology.

Author Contributions: Both authors contributed equally to this publication. All authors have read and agreed to the published version of the manuscript.

Funding: This research received no external funding.

Institutional Review Board Statement: Not applicable.

Data Availability Statement: Not applicable.

Acknowledgments: The authors acknowledge the diversity of contributions made by the hundreds of scientists from around the world to the field of DNA barcoding over the last decades. We are especially grateful to the Smithsonian Institution Barcode Network, the Global Genome Initiative, and the Consortium for the Barcode of Life which have supported our own research.

Conflicts of Interest: The authors declare no conflict of interest.

References

1. Hebert, P.D.N.; Cywinska, A.; Ball, S.L.; DeWaard, J.R. Biological Identifications through DNA Barcodes. *Proc. R. Soc. B Biol. Sci.* **2003**, *270*, 313–321. [CrossRef] [PubMed]
2. DeSalle, R.; Goldstein, P. Review and Interpretation of Trends in DNA Barcoding. *Front. Ecol. Evol.* **2019**, *7*, 302. [CrossRef]
3. Grant, D.M.; Brodnicke, O.B.; Evankow, A.M.; Ferreira, A.O.; Fontes, J.T.; Hansen, A.K.; Jensen, M.R.; Gba, T.; Kalaycı, E.; Leeper, A.; et al. The Future of DNA Barcoding: Reflections from Early Career Researchers. *Diversity* **2021**, *13*, 313. [CrossRef]
4. Kress, W.J.; Erickson, D.L. *DNA Barcodes: Methods and Protocols*; Humana Press: Totowa, NJ, USA, 2012; volume 858, pp. 3–8.
5. Kress, W.J. Plant DNA Barcodes: Applications Today and in the Future. *Front. Plant Syst. Evol.* **2017**, *55*, 291–307. [CrossRef]
6. Wilkinson, M.J.; Szabo, C.; Ford, C.S.; Yarom, Y.; Croxford, A.E.; Camp, A.; Gooding, P. Replacing Sanger with Next Generation Sequencing to Improve Coverage and Quality of Reference DNA Barcodes for Plants. *Sci. Rep.* **2017**, *7*, 46040. [CrossRef]
7. Cruaud, P.; Rasplus, J.-Y.; Rodriguez, L.J.; Cruaud, A. High-Throughput Sequencing of Multiple Amplicons for Barcoding and Integrative Taxonomy. *Sci. Rep.* **2017**, *7*, 41948. [CrossRef] [PubMed]
8. Gostel, M.R.; Zúñiga, J.D.; Kress, W.J.; Funk, V.A.; Puente-Lelievre, C. Microfluidic Enrichment Barcoding (MEBarcoding): A New Method for High Throughput Plant DNA Barcoding. *Sci. Rep.* **2020**, *10*, 8701. [CrossRef]
9. Liu, Y.; Xu, C.; Sun, Y.; Chen, X.; Dong, W.; Yang, X.; Zhou, S. Method for Quick DNA Barcode Reference Library Construction. *Ecol. Evol.* **2021**, *11*, 11627–11638. [CrossRef]
10. Page, R.D.M. DNA Barcoding and Taxonomy: Dark Taxa and Dark Texts. *Philos. Trans. R. Soc. B* **2016**, *371*. [CrossRef]
11. Nic Lughadha, E.M.; Govaerts, R.H.A.; Belyaeva, I.; Black, N.; Lindon, H.; Allkin, R.; Magill, R.E.; Nicolson, N. Counting Counts: Revised Estimates of Numbers of Accepted Species of Flowering Plants, Seed Plants, Vascular Plants and Land Plants with a Review of Other Recent. *Phytotaxa* **2016**, *272*, 82–88. [CrossRef]
12. Kane, N.; Ecology, Q.C.-M. Botany without Borders: Barcoding in Focus. *Mol. Ecol.* **2008**, *17*, 5175–5176. [CrossRef] [PubMed]
13. Li, X.; Yang, Y.; Henry, R.J.; Rossetto, M.; Wang, Y.; Chen, S. Plant DNA Barcoding: From Gene to Genome. *Biol. Rev.* **2015**, *90*, 157–166. [CrossRef] [PubMed]
14. Kane, N.; Sveinsson, S.; Dempewolf, H.; Yang, J.Y.; Zhang, D.; Engels, J.M.M.; Cronk, Q. Ultra-Barcoding in Cacao (Theobroma Spp.; Malvaceae) Using Whole Chloroplast Genomes and Nuclear Ribosomal DNA. *Am. J. Bot.* **2012**, *99*, 320–329. [CrossRef] [PubMed]
15. Ji, Y.; Liu, C.; Yang, J.; Jin, L.; Yang, Z.; Yang, J.B. Ultra-Barcoding Discovers a Cryptic Species in Paris Yunnanensis (Melanthiaceae), a Medicinally Important Plant. *Front. Plant Sci.* **2020**, *11*, 411. [CrossRef]
16. Coissac, E.; Hollingsworth, P.M.; Lavergne, S.; Taberlet, P. From Barcodes to Genomes: Extending the Concept of DNA Barcoding. *Mol. Ecol.* **2016**, *25*, 1423–1428. [CrossRef]
17. Bohmann, K.; Mirarab, S.; Bafna, V.; Gilbert, M.T.P. Beyond DNA Barcoding: The Unrealized Potential of Genome Skim Data in Sample Identification. *Mol. Ecol.* **2020**, 2521–2534. [CrossRef]
18. Ratnasingham, S.; Hebert, P.D.N. BOLD: The Barcode of Life Data System: Barcoding. *Mol. Ecol. Notes* **2007**, *7*, 355–364. [CrossRef]
19. Young, R.G.; Yu, J.; Cote, M.J.; Hanner, R.H. The Molecular Data Organization for Publication (MDOP) R Package to Aid the Upload of Data to Shared Databases. *Biodivers. Data J.* **2020**, *8*, e50630. [CrossRef]
20. Fontes, J.T.; Vieira, P.E.; Ekrem, T.; Soares, P.; Costa, F.O. BAGS: An Automated Barcode, Audit & Grade System for DNA Barcode Reference Libraries. *Mol. Ecol. Resour.* **2021**, *21*, 573–583. [CrossRef]
21. Young, R.; Gill, R.; Gillis, D.; Hanner, R. Molecular Acquisition, Cleaning and Evaluation in R (MACER)—A Tool to Assemble Molecular Marker Datasets from BOLD and GenBank. *Biodivers. Data J.* **2021**, *9*. [CrossRef]
22. Oliveira, R.R.M.; Nunes, G.L.; de Lima, T.G.L.; Oliveira, G.; Alves, R. PIPEBAR and OverlapPER: Tools for a Fast and Accurate DNA Barcoding Analysis and Paired-End Assembly. *BMC Bioinform.* **2018**, *19*, 297. [CrossRef] [PubMed]

23. Sahlin, K.; Lim, M.C.W.; Prost, S. NGSpeciesID: DNA Barcode and Amplicon Consensus Generation from Long-Read Sequencing Data. *Ecol. Evol.* **2021**, *11*, 1392–1398. [CrossRef]
24. Bik, H.M. Just Keep It Simple? Benchmarking the Accuracy of Taxonomy Assignment Software in Metabarcoding Studies. *Mol. Ecol. Resour.* **2021**, *21*, 2187–2189. [CrossRef]
25. Hleap, J.S.; Littlefair, J.E.; Steinke, D.; Hebert, P.D.N.; Cristescu, M.E. Assessment of Current Taxonomic Assignment Strategies for Metabarcoding Eukaryotes. *Mol. Ecol. Resour.* **2021**, *21*, 2190–2203. [CrossRef]
26. Bokulich, N.A.; Kaehler, B.D.; Rideout, J.R.; Dillon, M.; Bolyen, E.; Knight, R.; Huttley, G.A.; Gregory Caporaso, J. Optimizing Taxonomic Classification of Marker-Gene Amplicon Sequences with QIIME 2's Q2-Feature-Classifier Plugin. *Microbiome* **2018**, *6*, 90. [CrossRef] [PubMed]
27. Murali, A.; Bhargava, A.; Wright, E.S. IDTAXA: A Novel Approach for Accurate Taxonomic Classification of Microbiome Sequences. *Microbiome* **2018**, *6*, 140. [CrossRef] [PubMed]
28. Bengtsson-Palme, J.; Richardson, R.T.; Meola, M.; Wurzbacher, C.; Tremblay, É.D.; Thorell, K.; Kanger, K.; Eriksson, K.M.; Bilodeau, G.J.; Johnson, R.M.; et al. MetaxA2 Database Builder: Enabling Taxonomic Identification from Metagenomic or Metabarcoding Data Using Any Genetic Marker. *Bioinformatics* **2018**, *34*, 4027–4033. [CrossRef]
29. Kahlke, T.; Ralph, P. BASTA–Taxonomic Classification of Sequences and Sequence Bins Using Last Common Ancestor Estimations. *Methods Ecol. Evol.* **2018**, *10*, 100–103. [CrossRef]
30. Wood, D.E.; Salzberg, S.L. Kraken: Ultrafast Metagenomic Sequence Classification Using Exact Alignments. *Genome Biol.* **2014**, *15*, R46. [CrossRef]
31. Somervuo, P.; Koskela, S.; Pennanen, J.; Nilsson, R.H.; Ovaskainen, O. Unbiased Probabilistic Taxonomic Classification for DNA Barcoding. *Bioinformatics* **2016**, *32*, 2920–2927. [CrossRef]
32. Altschul, S.F.; Gish, W.; Miller, E.W.; Lipman, D.J. Basic Local Alignment Search Tool. *J. Mol. Biol.* **1990**, *215*, 403–410. [CrossRef]
33. Meher, P.K.; Sahu, T.K.; Gahoi, S.; Tomar, R.; Rao, A.R. FunbarRF: DNA Barcode-Based Fungal Species Prediction Using Multiclass Random Forest Supervised Learning Model. *BMC Genet.* **2019**, *20*, 2. [CrossRef] [PubMed]
34. Curd, E.E.; Gold, Z.; Kandlikar, G.S.; Gomer, J.; Ogden, M.; O'Connell, T.; Pipes, L.; Schweizer, T.M.; Rabichow, L.; Lin, M.; et al. Anacapa Toolkit: An Environmental DNA Toolkit for Processing Multilocus Metabarcode Datasets. *Methods Ecol. Evol.* **2019**, *10*, 1469–1475. [CrossRef]
35. Hobern, D. Bioscan: Dna Barcoding to Accelerate Taxonomy and Biogeography for Conservation and Sustainability. *Genome* **2021**, *64*, 161–164. [CrossRef]
36. Seberg, O.; Droege, G.; Barker, K.; Coddington, J.A.; Funk, V.; Gostel, M.; Petersen, G.; Smith, P.P. Global Genome Biodiversity Network: Saving a Blueprint of the Tree of Life—A Botanical Perspective. *Ann. Bot.* **2016**, *118*, 393–399. [CrossRef]
37. Lewin, H.A.; Robinson, G.E.; Kress, W.J.; Baker, W.J.; Coddington, J.; Crandall, K.A.; Durbin, R.; Edwards, S.V.; Forest, E.; Thomas, M.; et al. Earth BioGenome Project: Sequencing Life for the Future of Life. *Proc. Natl. Acad Sci. USA* **2018**, *115*, 4325–4333. [CrossRef]
38. Lewin, H.; Richards, S.; Aiden, E.L.; Allende, M.L.; Archibald, J.M.; Bálint, M.; Barker, K.B.; Baumgartner, B.; Belov, K.; Bertorelle, G.; et al. The Earth BioGenome Project 2020: Starting the Clock. *Proc. Natl. Acad Sci. USA* **2022**, *119*, e2115635118. [CrossRef] [PubMed]
39. Kress, W.J.; Erickson, D.L. DNA Barcodes: Genes, Genomics, and Bioinformatics. *Proc. Natl. Acad. Sci. USA* **2008**, *105*, 2761–2762. [CrossRef]
40. Bezeng, B.S.; Davies, T.J.; Daru, B.H.; Kabongo, R.M.; Maurin, O.; Yessoufou, K.; Van Der Bank, H.; Van Der Bank, M. Ten Years of Barcoding at the African Centre for DNA Barcoding. *Genome* **2017**, *60*, 629–638. [CrossRef] [PubMed]
41. Yang, C.Q.; Lv, Q.; Zhang, A.B. Sixteen Years of DNA Barcoding in China: What Has Been Done? What Can Be Done? *Front. Ecol. Evol.* **2020**, *8*, 1–13. [CrossRef]
42. The Darwin Tree of Life Project Consortium. Sequence Locally, Think Globally: The Darwin Tree of Life Project. *Proc. Natl. Acad Sci. USA* **2022**, *119*, e2115642118. [CrossRef]
43. Anderson-Teixeira, K.J.; Davies, S.J.; Bennett, A.C.; Gonzalez-Akre, E.B.; Muller-Landau, H.C.; Joseph Wright, S.; Abu Salim, K.; Almeyda Zambrano, A.M.; Alonso, A.; Baltzer, J.L.; et al. CTFS-ForestGEO: A Worldwide Network Monitoring Forests in an Era of Global Change. *Glob. Chang. Biol.* **2015**, *21*, 528–549. [CrossRef] [PubMed]
44. Kress, W.J.; Lopez, I.C.; Erickson, D.L. Generating Plant DNA Barcodes for Trees in Long-Term Forest Dynamics Plots. *Methods Mol. Biol.* **2012**, *858*, 441–458. [CrossRef] [PubMed]
45. Erickson, D.L.; Jones, F.A.; Swenson, N.G.; Pei, N.; Bourg, N.; Chen, W.; Davies, S.J.; Ge, X.; Hao, Z.; Howe, R.W.; et al. Comparative Evolutionary Diversity and Phylogenetic Structure across Multiple Forest Dynamics Plots: A Mega-Phylogeny Approach. *Front. Genet.* **2014**, *5*, 358. [CrossRef]
46. Liu, J.; Liu, J.; Shan, Y.-X.; Ge, X.-J.; Burgess, K.S. The Use of DNA Barcodes to Estimate Phylogenetic Diversity in Forest Communities of Southern China. *Ecol. Evol.* **2019**, *9*, 5372–5379. [CrossRef] [PubMed]
47. Li, S.; Qian, X.; Zheng, Z.; Shi, M.; Chang, X.; Li, X.; Liu, J.; Tu, T.; Zhang, D. DNA Barcoding the Flowering Plants from the Tropical Coral Islands of Xisha (China). *Ecol. Evol.* **2018**, *8*, 10587–10593. [CrossRef] [PubMed]
48. Gill, B.A.; Musili, P.M.; Kurukura, S.; Hassan, A.A.; Goheen, J.R.; Kress, W.J.; Kuzmina, M.; Pringle, R.M.; Kartzinel, T.R. Plant DNA-Barcode Library and Community Phylogeny for a Semi-Arid East African Savanna. *Mol. Ecol. Resour.* **2019**, *19*, 838–846. [CrossRef]

49. Bringloe, T.T.; Sjøtun, K.; Saunders, G.W. A DNA Barcode Survey of Marine Macroalgae from Bergen (Norway). *Mar. Biol. Res.* **2019**, *15*, 580–589. [CrossRef]

50. Hernández-León, S.; Little, D.P.; Acevedo-Sandoval, O.; Gernandt, D.S.; Rodríguez-Laguna, R.; Saucedo-García, M.; Arce-Cervantes, O.; Razo-Zárate, R.; Espitia-López, J. Plant Core DNA Barcode Performance at a Local Scale: Identification of the Conifers of the State of Hidalgo, Mexico. *Syst. Biodivers.* **2018**, *16*, 791–806. [CrossRef]

51. Nitta, J.H.; Ebihara, A.; Smith, A.R. A Taxonomic and Molecular Survey of the Pteridophytes of the Nectandra Cloud Forest Reserve, Costa Rica. *PLoS ONE* **2020**, *15*, e0241231. [CrossRef]

52. Chen, S.; Yao, H.; Han, J.; Liu, C.; Song, J.; Shi, L.; Zhu, Y.; Ma, X.; Gao, T.; Pang, X.; et al. Validation of the ITS2 Region as a Novel DNA Barcode for Identifying Medicinal Plant Species. *PLoS ONE* **2010**, *5*, e8613. [CrossRef] [PubMed]

53. Chen, J.; Zhao, J.; Erickson, D.L.; Xia, N.; Kress, W.J. Testing DNA Barcodes in Closely Related Species of Curcuma (Zingiberaceae) from Myanmar and China. *Mol. Ecol. Resour.* **2014**, *15*, 337–348. [CrossRef]

54. Wang, D.Y.; Wang, Q.; Wang, Y.L.; Xiang, X.G.; Huang, L.Q.; Jin, X.H. Evaluation of DNA Barcodes in Codonopsis (Campanu-laceae) and in Some Large Angiosperm Plant Genera. *PLoS ONE* **2017**, *12*, e0170286. [CrossRef]

55. Nieves, K.G.; van Ee, B.W. DNA Barcoding of Adiantum (Pteridaceae: Vittarioideae) in Puerto Rico. *Caribb. J. Sci.* **2021**, *51*, 287–313. [CrossRef]

56. Dal Forno, M.; Lawrey, J.D.; Moncada, B.; Bungartz, F.; Grube, M.; Schuettpelz, E.; Lücking, R. DNA Barcoding of Fresh and Historical Collections of Lichen-Forming Basidiomycetes in the Genera Cora and Corella (Agaricales: Hygrophoraceae): A Success Story? *Diversity* **2022**, *14*.

57. Sánchez-C, D.; Richardson, J.E.; Hart, M.; Serrano, J.; Cárdenas, D.; Gonzalez, M.A.; Cortés-B, R. A Plea to DNA Barcode Type Specimens: An Example from Micropholis (Sapotaceae). *Taxon* **2021**, *71*, 154–167. [CrossRef]

58. Braukmann, T.W.A.; Kuzmina, M.L.; Sills, J.; Zakharov, E.V.; Hebert, P.D.N. Testing the Efficacy of DNA Barcodes for Identifying the Vascular Plants of Canada. *PLoS ONE* **2017**, *12*, e0169515. [CrossRef]

59. Jones, L.; Twyford, A.D.; Ford, C.R.; Rich, T.C.G.; Davies, H.; Forrest, L.L.; Hart, M.L.; McHaffie, H.; Brown, M.R.; Hollingsworth, P.M.; et al. Barcode UK: A Complete DNA Barcoding Resource for the Flowering Plants and Conifers of the United Kingdom. *Mol. Ecol. Resour.* **2021**, *21*, 2050–2062. [CrossRef] [PubMed]

60. Mosa, K.A.; Gairola, S.; Jamdade, R.; El-Keblawy, A.; Al Shaer, K.I.; Al Harthi, E.K.; Shabana, H.A.; Mahmoud, T. The Promise of Molecular and Genomic Techniques for Biodiversity Research and DNA Barcoding of the Arabian Peninsula Flora. *Front. Plant Sci.* **2019**, *9*, 1929. [CrossRef] [PubMed]

61. Shapcott, A.; Forster, P.I.; Guymer, G.P.; McDonald, W.J.F.; Faith, D.P.; Erickson, D.; Kress, W.J. Mapping Biodiversity and Setting Conservation Priorities for SE Queensland's Rainforests Using DNA Barcoding. *PLoS ONE* **2015**, *10*, e0122164. [CrossRef] [PubMed]

62. Kress, W.J.; Erickson, D.L.; Jones, F.A.; Swenson, N.G.; Perez, R.; Sanjur, O.; Bermingham, E. Plant DNA Barcodes and a Community Phylogeny of a Tropical Forest Dynamics Plot in Panama. *Proc. Natl. Acad Sci. USA* **2009**, *106*, 18621–18626. [CrossRef] [PubMed]

63. Jurado-Rivera, J.A.; Vogler, A.P.; Reid, C.A.M.; Petitpierre, E.; Gómez-Zurita, J. DNA Barcoding Insect-Host Plant Associations. *Proc. R. Soc. B Biol. Sci.* **2009**, *276*, 639–648. [CrossRef] [PubMed]

64. Hollingsworth, P.M.; Graham, S.W.; Little, D.P. Choosing and Using a Plant DNA Barcode. *PLoS ONE* **2011**, *6*, e19254. [CrossRef]

65. Hollingsworth, P.M.; Li, D.Z.; Van Der Bank, M.; Twyford, A.D. Telling Plant Species Apart with DNA: From Barcodes to Genomes. *Philos. Trans. R. Soc. B Biol. Sci.* **2016**, *371*. [CrossRef] [PubMed]

66. Kress, W.J.; Wurdack, K.J.; Zimmer, E.A.; Weigt, L.A.; Janzen, D.H. Use of DNA Barcodes to Identify Flowering Plants. *Proc. Natl. Acad. Sci. USA* **2005**, *102*, 8369–8374. [CrossRef] [PubMed]

67. Kress, W.J.; García-Robledo, C.; Uriarte, M.; Erickson, D.L. DNA Barcodes for Ecology, Evolution, and Conservation. *Trends Ecol. Evol.* **2014**, *30*, 25–35. [CrossRef] [PubMed]

68. Valentini, A.; Miquel, C.; Nawaz, M.A.; Bellemain, E.; Coissac, E.; Pompanon, F.; Gielly, L.; Cruaud, C.; Nascetti, G.; Wincker, P.; et al. New Perspectives in Diet Analysis Based on DNA Barcoding and Parallel Pyrosequencing: The TrnL Approach. *Mol. Ecol. Resour.* **2009**, *9*, 51–60. [CrossRef] [PubMed]

69. Wilson, E.O. *Half Earth: Our Planet's Fight for Life*; WW Norton and Company: New York, NY, USA, 2016.

70. Hebert, P.D.N.; Penton, E.H.; Burns, J.M.; Janzen, D.H.; Hallwachs, W. Ten Species in One: DNA Barcoding Reveals Cryptic Species in the Neotropical Skipper Butterfly Astraptes Fulgerator. *Proc. Natl. Acad Sci. USA* **2004**, *104*, 41. [CrossRef]

71. Janzen, D.H.; Hallwachs, W.; Blandin, P.; Burns, J.M.; Cadiou, J.M.; Chacon, I.; Dapkey, T.; Deans, A.R.; Epstein, M.E.; Espinoza, B.; et al. Integration of DNA Barcoding into an Ongoing Inventory of Complex Tropical Biodiversity. *Mol. Ecol. Resour.* **2009**, *9*, 1–26. [CrossRef]

72. García-Robledo, C.; Kuprewicz, E.K.; Staines, C.L.; Kress, W.J.; Erwin, T.L. Using a Comprehensive DNA Barcode Library to Detect Novel Egg and Larval Host Plant Associations in a Cephaloleia Rolled-Leaf Beetle (Coleoptera: Chrysomelidae). *Sci. Rep.* **2013**, *8*, 13398.

73. García-Robledo, C.; Kuprewicz, E.K.; Staines, C.L.; Erwin, T.L.; Kress, W.J. Limited Tolerance by Insects to High Temperatures across Tropical Elevational Gradients and the Implications of Global Warming for Extinction. *Proc. Natl. Acad Sci. USA* **2016**, *113*. [CrossRef] [PubMed]

74. Gonzalez, M.A.; Baraloto, C.; Engel, J.; Mori, S.A.; Pétronelli, P.; Riéra, B.; Roger, A.; Thébaud, C.; Chave, J. Identification of Amazonian Trees with DNA Barcodes. *PLoS ONE* **2009**, *4*, e7483. [CrossRef] [PubMed]

75. Dexter, K.G.; Pennington, T.D.; Cunningham, C.W. Using DNA to Assess Errors in Tropical Tree Identifications: How Often Are Ecologists Wrong and When Does It Matter? *Ecol. Monogr.* **2010**, *80*, 267–286. [CrossRef]

76. Kenfack, D.; Abiem, I.; Chapman, H. The Efficiency of DNA Barcoding in the Identification of Afromontane Forest Tree Species. *Diversity* **2022**, *14*.

77. Costion, C.; Ford, A.; Cross, H.; Crayn, D.; Harrington, M.; Lowe, A. Plant Dna Barcodes Can Accurately Estimate Species Richness in Poorly Known Floras. *PLoS ONE* **2011**, *6*, e26841. [CrossRef]

78. Parmentier, I.; Duminil, J.; Kuzmina, M.; Philippe, M.; Thomas, D.W.; Kenfack, D.; Chuyong, G.B.; Cruaud, C.; Hardy, O.J. How Effective Are DNA Barcodes in the Identification of African Rainforest Trees? *PLoS ONE* **2013**, *8*, e54921. [CrossRef]

79. Dick, C.W.; Webb, C.O. Plant DNA Barcodes, Taxonomic Management, and Species Discovery in Tropical Forests. *DNA Barcodes* **2012**, *858*, 379–393. [CrossRef]

80. Ter Steege, H.; Pitman, N.C.A.; Sabatier, D.; Baraloto, C.; Salomão, R.P.; Guevara, J.E.; Phillips, O.L.; Castilho, C.V.; Magnusson, W.E.; Molino, J.F.; et al. Hyperdominance in the Amazonian Tree Flora. *Science* **2013**, *342*, 1243092. [CrossRef]

81. Pante, E.; Schoelinck, C.; Puillandre, N. From Integrative Taxonomy to Species Description: One Step Beyond. *Syst. Biol.* **2015**, *64*, 152–160. [CrossRef]

82. Meierotto, S.; Sharkey, M.J.; Janzen, D.H.; Hallwachs, W.; Hebert, P.D.N.; Chapman, E.G.; Smith, M.A. A Revolutionary Protocol to Describe Understudied Hyperdiverse Taxa and Overcome the Taxonomic Impediment. *Dtsch. Entomol. Zeitschrift* **2019**, *66*, 119–145. [CrossRef]

83. Funk, V.A.; Gostel, M.; Devine, A.; Kelloff, C.L.; Wurdack, K.; Tuccinardi, C.; Radosavljevic, A.; Peters, M.; Coddington, J. Guidelines for Collecting Vouchers and Tissues Intended for Genomic Work (Smithsonian Institution): Botany Best Practices. *Biodivers. Data J.* **2017**, *5*, e11625. [CrossRef] [PubMed]

84. Weigand, H.; Beermann, A.J.; Čiampor, F.; Costa, F.O.; Csabai, Z.; Duarte, S.; Geiger, M.F.; Grabowski, M.; Rimet, F.; Rulik, B.; et al. DNA Barcode Reference Libraries for the Monitoring of Aquatic Biota in Europe: Gap-Analysis and Recommendations for Future Work. *Sci. Total Environ.* **2019**, *678*, 499–524. [CrossRef] [PubMed]

85. Litman, J.; Chittaro, Y.; Birrer, S.; Praz, C.; Wermeille, E.; Fluri, M.; Stalling, T.; Schmid, S.; Wyler, S.; Gonseth, Y. A DNA Barcode Reference Library for Swiss Butterflies and Forester Moths as a Tool for Species Identification, Systematics and Conservation. *PLoS ONE* **2018**, *13*, e0208639. [CrossRef] [PubMed]

86. Bezeng, B.S.; Van Der Bank, H.F. DNA Barcoding of Southern African Crustaceans Reveals a Mix of Invasive Species and Potential Cryptic Diversity. *PLoS ONE* **2019**, *14*, e0222047. [CrossRef]

87. Del-Prado, R.; Buaruang, K.; Lumbsch, H.T.; Crespo, A.; Divakar, P.K. DNA Sequence-Based Identification and Barcoding of a Morphologically Highly Plastic Lichen Forming Fungal Genus (Parmotrema, Parmeliaceae) from the Tropics. *Bryologist* **2019**, *122*, 281–291. [CrossRef]

88. Faith, D.P. Conservation Evaluation and Phylogenetic Diversity. *Biol. Conserv.* **1992**, *61*, 1–10. [CrossRef]

89. Faith, D.P. Phylogenetic Pattern and the Quantification of Organismal Biodiversity. *Philos. Trans. R. Soc. Lond. B. Biol. Sci.* **1994**, *345*, 45–58. [CrossRef]

90. Miller, J.T.; Jolley-Rogers, G.; Mishler, B.D.; Thornhill, A. Phylogenetic Diversity Is a Better Measure of Biodiversity than Taxon Counting. *J. Syst. Evol.* **2018**, *56*, 663–667. [CrossRef]

91. Steinke, D.; de Waard, S.L.; Sones, J.E.; Ivanova, N.V.; Prosser, S.W.J.; Perez, K.; Braukmann, T.W.A.; Milton, M.; Zakharov, E.V.; de Waard, J.R.; et al. Message in a Bottle–Metabarcoding Enables Biodiversity Comparisons Across Ecoregions. *bioRxiv* **2021**. [CrossRef]

92. Winter, M.; Devictor, V.; Schweiger, O. Phylogenetic Diversity and Nature Conservation: Where Are We? *Trends Ecol. Evol.* **2013**, *28*, 199–204. [CrossRef]

93. Flynn, D.F.B.; Mirotchnick, N.; Jain, M.; Palmer, M.I.; Naeem, S. Functional and Phylogenetic Diversity as Predictors of Biodiversity-Ecosystem-Function Relationships. *Ecology* **2011**, *92*, 1573–1581. [CrossRef] [PubMed]

94. Mazel, F.; Mooers, A.; Riva, G.V.D.; Pennell, M.W. Conserving Phylogenetic Diversity Can Be a Poor Strategy for Conserving Functional Diversity. *Biol. Sci.* **2017**, *66*, 1019–1027. [CrossRef] [PubMed]

95. Mazel, F.; Pennell, M.W.; Cadotte, M.W.; Diaz, S.; Riva, V.D.; Grenyer, R.; Leprieur, F.; Mooers, A.O.; Mouillot, D.; Tucker, C.M.; et al. Prioritizing Phylogenetic Diversity Captures Functional Diversity Unreliably. *Nat. Commun.* **2018**, *9*, 2888. [CrossRef] [PubMed]

96. Gonzalez, M.A.; Roger, A.; Courtois, E.A.; Jabot, F.; Norden, N.; Paine, C.E.T.; Baraloto, C.; Thébaud, C.; Chave, J. Shifts in Species and Phylogenetic Diversity between Sapling and Tree Communities Indicate Negative Density Dependence in a Lowland Rain Forest. *J. Ecol.* **2010**, *98*, 137–146. [CrossRef]

97. Kress, W.J.; Erickson, D.L.; Swenson, N.G.; Thompson, J.; Uriarte, M.; Zimmerman, J.K. Advances in the Use of DNA Barcodes to Build a Community Phylogeny for Tropical Trees in a Puerto Rican Forest Dynamics Plot. *PLoS ONE* **2010**, *5*, e15409. [CrossRef]

98. Pei, N.; Erickson, D.; Chen, B.; Ge, X.; Mi, X.; Swenson, M.G.; Zhang, J.-L.; Jones, F.A.; Huang, C.L.; Ye, W.; et al. Closely-Related Taxa Influence Woody Species Discrimination via DNA Barcoding: Evidence from Global Forest Dynamics Plots. *Sci. Rep.* **2015**, *5*, 15127. [CrossRef]

99. Whitfeld, T.J.S.; Kress, W.J.; Erickson, D.L.; Weiblen, G.D. Change in Community Phylogenetic Structure during Tropical Forest Succession: Evidence from New Guinea. *Ecography* **2012**, *35*, 821–830. [CrossRef]
100. Swenson, N.G. Phylogenetic Analyses of Ecological Communities Using DNA Barcode Data. *Methods Mol. Biol.* **2012**, *858*, 409–419. [CrossRef]
101. Schreeg, L.A.; Kress, W.J.; Erickson, D.L.; Swenson, N.G. Phylogenetic Analysis of Local-Scale Tree Soil Associations in a Lowland Moist Tropical Forest. *PLoS ONE* **2010**, *5*, e13685. [CrossRef]
102. Westbrook, J.W.; Kitajima, K.; Burleigh, J.G.; Kress, W.J.; Erickson, D.L.; Wright, S.J. What Makes a Leaf Tough? Patterns of Correlated Evolutionbetween Leaf Toughness Traits and Demographic Rates among 197 Shade-Tolerant Woody Species in a Neotropical Forest. *Am. Nat.* **2011**, *177*, 800–811. [CrossRef]
103. Martin, A.R.; Erickson, D.L.; Kress, W.J.; Thomas, S.C. Wood Nitrogen Concentrations in Tropical Trees: Phylogenetic Patterns and Ecological Correlates. *New Phytol.* **2014**, *204*, 484–495. [CrossRef]
104. Mcmanus, K.M.; Asner, G.P.; Martin, R.E.; Dexter, K.G.; Kress, W.J.; Field, C.B.; Ustin, S.L.; Wynne, R.H.; Thenkabail, P.S. Phylogenetic Structure of Foliar Spectral Traits in Tropical Forest Canopies. *Remote Sens.* **2016**, *8*, 196. [CrossRef]
105. Chauvin, K.M.M.; Asner, G.P.; Martin, R.E.; Kress, W.J.; Wright, S.J.; Field, C.B. Decoupled Dimensions of Leaf Economic and Anti-Herbivore Defense Strategies in a Tropical Canopy Tree Community. *Oecologia* **2018**, *186*, 765–782. [CrossRef]
106. Swenson, N.G. The Assembly of Tropical Tree Communities—The Advances and Shortcomings of Phylogenetic and Functional Trait Analyses. *Ecography* **2013**, *36*, 264–276. [CrossRef]
107. Jones, F.A.; Erickson, D.L.; Bernal, M.A.; Bermingham, E.; Kress, W.J.; Herre, E.A.; Muller-Landau, H.C.; Turner, B.L. The Roots of Diversity: Below Ground Species Richness and Rooting Distributions in a Tropical Forest Revealed by DNA Barcodes and Inverse Modeling. *PLoS ONE* **2011**, *6*, e24506. [CrossRef] [PubMed]
108. Basset, Y.; Donoso, D.A.; Hajibabaei, M.; Wright, M.T.G.; Perez, K.H.J.; Lamarre, G.P.A.; De León, L.F.; Palacios-Vargas, J.G.; Castaño-Meneses, G.; Rivera, M.; et al. Methodological Considerations for Monitoring Soil/Litter Arthropods in Tropical Rainforests Using DNA Metabarcoding, with a Special Emphasis on Ants, Springtails. *Metabarcoding Metagenomics* **2020**, *4*, 151–163. [CrossRef]
109. Baker, T.R.; Pennington, R.T.; Dexter, K.G.; Fine, P.V.A.; Fortune-Hopkins, H.; Honorio, E.N.; Huamantupa-Chuquimaco, I.; Klitgård, B.B.; Lewis, G.P.; de Lima, H.C.; et al. Maximising Synergy among Tropical Plant Systematists, Ecologists, and Evolutionary Biologists. *Trends Ecol. Evol.* **2017**, *32*, 258–267. [CrossRef]
110. Pei, N.; Lian, J.Y.; Erickson, D.L.; Swenson, N.G.; Kress, W.J.; Ye, W.H.; Ge, X.J. Exploring Tree-Habitat Associations in a Chinese Subtropical Forest Plot Using a Molecular Phylogeny Generated from DNA Barcode Loci. *PLoS ONE* **2011**, *6*, e21273. [CrossRef]
111. de Oliveira, A.A.; Vicentini, A.; Chave, J.; Castanho, C.D.T.; Davies, S.J.; Martini, A.M.Z.; Lima, R.A.F.; Ribeiro, R.R.; Iribar, A.; Souza, V.C. Habitat Specialization and Phylogenetic Structure of Tree Species in a Coastal Brazilian White-Sand Forest. *J. Plant Ecol.* **2017**, *7*, 134–144. [CrossRef]
112. Mi, X.; Swenson, N.G.; Valencia, R.; John Kress, W.; Erickson, D.L.; Pérez, Á.J.; Ren, H.; Su, S.H.; Gunatilleke, N.; Gunatilleke, S.; et al. The Contribution of Rare Species to Community Phylogenetic Diversity across a Global Network of Forest Plots. *Am. Nat.* **2012**, *180*, E17–E30. [CrossRef]
113. Muscarella, R.; Uriarte, M.; Erickson, D.L.; Swenson, N.G.; Zimmerman, J.K.; Kress, W.J. A Well-Resolved Phylogeny of the Trees of Puerto Rico Based on DNA Barcode Sequence Data. *PLoS ONE* **2014**, *9*, e112843. [CrossRef]
114. Muscarella, R.; Uriarte, M.; Erickson, D.; Swenson, N.; Kress, W.; Zimmerman, J. *Variation of Tropical Forest Assembly Processes across Regional Environmental Gradients*; Elsevier: Amsterdam, The Netherlands, 2016. [CrossRef]
115. Wills, C.; Harms, K.E.; Wiegand, T.; Punchi-Manage, R.; Gilbert, G.S.; Erickson, D.; Kress, W.J.; Hubbell, S.P.; Gunatilleke, C.V.S.; Gunatilleke, I.A.U.N. Persistence of Neighborhood Demographic Influences over Long Phylogenetic Distances May Help Drive Post-Speciation Adaptation in Tropical Forests. *PLoS ONE* **2016**, *11*, e0156913. [CrossRef] [PubMed]
116. Smith, M.A.; Eveleigh, E.S.; McCann, K.S.; Merilo, M.T.; McCarthy, P.C.; van Rooyen, K.I. Barcoding a Quantified Food Web: Crypsis, Concepts, Ecology and Hypotheses. *PLoS ONE* **2011**, *6*, e14424. [CrossRef] [PubMed]
117. Navarro, S.P.; Jurado-Rivera, J.A.; GãMez-Zurita, J.; Lyal, C.H.C.; Vogler, A.P.; Jurado-Rivera, J.A.; Gómez-Zurita, J. DNA Profiling of Host–Herbivore Interactions in Tropical Forests. *Ecol. Èntomol.* **2010**, *35*, 18–32. [CrossRef]
118. Kishimoto-Yamada, K.; Kamiya, K.; Meleng, P.; Diway, B.; Kaliang, H.; Chong, L.; Itioka, T.; Sakai, S.; Ito, M. Wide Host Ranges of Herbivorous Beetles? Insights from DNA Bar Coding. *PLoS ONE* **2013**, *8*, e74426. [CrossRef] [PubMed]
119. García-Robledo, C.; Erickson, D.L.; Staines, C.L.; Erwin, T.L.; Kress, W.J. Tropical Plant-Herbivore Networks: Reconstructing Species Interactions Using DNA Barcodes. *PLoS ONE* **2013**, *8*, e52967. [CrossRef]
120. Strong, D.R., Jr. Rolled-Leaf Hispine Beetles (Chrysomelidae) and Their Zingiberales Host Plants in Middle America. *JSTOR* **1977**, *9*, 156–169. [CrossRef]
121. Palmer, C.M.; Wershoven, N.L.; Martinson, S.J.; ter Hofstede, H.M.; Kress, W.J.; Symes, L.B. Patterns of Herbivory in Neotropical Forest Katydids as Revealed by DNA Barcoding of Digestive Tract Contents. *Diversity* **2022**, *14*, 120. [CrossRef]
122. Colwell, R.K. Competition and Coexistence in a Simple Tropical Community. *Am. Nat.* **1973**, *107*, 737–760. [CrossRef]
123. Futuyma, D.J.; Moreno, G. The Evolution of Ecological Specialization. *Annu. Rev. Ecol. Syst.* **1988**, *19*, 207–233. [CrossRef]
124. Bizzarri, L.; Baer, C.S.; García-Robledo, C. DNA Barcoding Reveals Generalization and Host Overlap in Hummingbird Flower Mites: Implications for the Mating Rendezvous Hypothesis. *Am. Nat.* **2022**. *ahead of print*. [CrossRef]

125. Kartzinel, T.; Chen, P.; Coverdale, T. DNA Metabarcoding Illuminates Dietary Niche Partitioning by African Large Herbivores. *Proc. Natl. Acad. Sci. USA* **2015**, *112*, 8019–8024. [CrossRef] [PubMed]

126. Freeman, P.T.; Ang'ila, R.O.; Kimuyu, D.; Musili, P.M.; Ken-fack, D.; Lokeny Etelej, P.; Magid, M.; Gill, B.A.; Kartzinel, T.R. Opposing Gradients in the Abundance and Diversity of Plants and Large Herbivores Revealed with DNA Barcoding in a Semi-Arid African Savanna. *Diversity* **2022**, *14*.

127. González-Varo, J.P.; Arroyo, J.; Jordano, P. Who Dispersed the Seeds? The Use of DNA Barcoding in Frugivory and Seed Dispersal Studies. *Methods Ecol. Evol.* **2014**, *5*, 806–814. [CrossRef]

128. Baker, C.C.M.; Bittleston, L.S.; Sanders, J.G.; Pierce, N.E. Dissecting Host-Associated Communities with DNA Barcodes. *Philos. Trans. R. Soc. B Biol. Sci.* **2016**, *371*, 20150328. [CrossRef]

129. Schoch, C.L.; Seifert, K.A.; Huhndorf, S.; Robert, V.; Spouge, J.L.; Levesque, C.A.; Chen, W.; Fungal Barcoding Consortium, F.B.; Fungal Barcoding Consortium Author List, F.B.C.A.; Bolchacova, E.; et al. Nuclear Ribosomal Internal Transcribed Spacer (ITS) Region as a Universal DNA Barcode Marker for Fungi. *Proc. Natl. Acad. Sci. USA* **2012**, *109*, 6241–6246. [CrossRef]

130. Lücking, R.; Aime, M.C.; Robbertse, B.; Miller, A.N.; Ariyawansa, H.A.; Aoki, T.; Cardinali, G.; Crous, P.W.; Druzhinina, I.S.; Geiser, D.M.; et al. Unambiguous Identification of Fungi: Where Do We Stand and How Accurate and Precise Is Fungal DNA Barcoding? *IMA Fungus* **2020**, *11*, 14. [CrossRef]

131. Young, M.R.; Proctor, H.C.; Dewaard, J.R.; Hebert, P.D.N. DNA Barcodes Expose Unexpected Diversity in Canadian Mites. *Mol. Ecol.* **2019**, *28*, 5347–5359. [CrossRef]

132. Bonato, K.O.; Silva, P.C.; Malabarba, L.R. Unrevealing Parasitic Trophic Interactions-A Molecular Approach for Fluid-Feeding Fishes. *Front. Ecol. Evol.* **2018**, *6*, 1–8. [CrossRef]

133. Keet, J.-H.; Ellis, A.G.; Hui, C.; Roux, J.J.L. Legume–Rhizobium Symbiotic Promiscuity and Effectiveness Do Not Affect Plant Invasiveness. *Ann. Bot.* **2017**, *119*, 1319–1331. [CrossRef]

134. Smith, H.B.; Dal Grande, F.; Muggia, L.; Keuler, R.; Divakar, P.K.; Grewe, F.; Schmitt, I.; Lumbsch, H.T.; Leavitt, S.D. Metagenomic Data Reveal Diverse Fungal and Algal Communities Associated with the Lichen Symbiosis. *Symbiosis* **2020**, *82*, 133–147. [CrossRef]

135. Taberlet, P.; Coissac, E.; Pompanon, F. Towards Next-generation Biodiversity Assessment Using DNA Metabarcoding. *Molecular* **2012**, *21*, 2045–2050. [CrossRef] [PubMed]

136. Widmer, A.; Cozzolino, S.; Pellegrino, G.; Soliva, M.; Dafni, A. Molecular Analysis of Orchid Pollinaria and Pollinaria-Remains Found on Insects. *Mol. Ecol.* **2000**, *9*, 1911–1914. [CrossRef] [PubMed]

137. Sickel, W.; Ankenbrand, M.J.; Grimmer, G.; Holzschuh, A.; Härtel, S.; Lanzen, J.; Steffan-Dewenter, I.; Keller, A. Increased Efficiency in Identifying Mixed Pollen Samples by Meta-Barcoding with a Dual-Indexing Approach. *BMC Ecol.* **2015**, *15*, 20. [CrossRef] [PubMed]

138. Lowe, A.; Jones, L.; Witter, L.; Creer, S.; de Vere, N. Using DNA Metabarcoding to Identify Floral Visitation by Pollinators. *Diversity* **2022**, *14*.

139. Clare, E.L.; Schiestl, F.P.; Leitch, A.R.; Chittka, L. The Promise of Genomics in the Study of Plant-Pollinator Interactions. *Genome Biol.* **2013**, *14*, 1–11. [CrossRef]

140. Soininen, E.M.; Valentini, A.; Coissac, E.; Miquel, C.; Gielly, L.; Brochmann, C.; Brysting, A.K.; Sønstebø, J.H.; Ims, R.A.; Yoccoz, N.G.; et al. Analysing Diet of Small Herbivores: The Efficiency of DNA Barcoding Coupled with High-Throughput Pyrosequencing for Deciphering the Composition of Complex Plant Mixtures. *Front. Zool.* **2009**, *6*, 16. [CrossRef]

141. de Vere, N.; Rich, T.C.G.; Ford, C.R.; Trinder, S.A.; Long, C.; Moore, C.W.; Satterthwaite, D.; Davies, H.; Allainguillaume, J.; Ronca, S.; et al. DNA Barcoding the Native Flowering Plants and Conifers of Wales. *PLoS ONE* **2012**, *7*, e37945. [CrossRef]

142. Bell, K.L.; Batchelor, K.L.; Bradford, M.; McKeown, A.; MacDonald, S.L.; Westcott, D. Optimisation of a Pollen DNA Metabarcoding Method for Diet Analysis of Flying-Foxes (Pteropus Spp.). *Aust. J. Zool.* **2021**, *68*, 273–284. [CrossRef]

143. de Vere, N.; Jones, L.E.; Gilmore, T.; Moscrop, J.; Lowe, A.; Smith, D.; Hegarty, M.J.; Creer, S.; Ford, C.R. Using DNA Metabarcoding to Investigate Honey Bee Foraging Reveals Limited Flower Use despite High Floral Availability. *Sci. Rep.* **2017**, *7*, 42838. [CrossRef]

144. Jones, L.; Brennan, G.L.; Lowe, A.; Creer, S.; Ford, C.R.; de Vere, N. Shifts in Honeybee Foraging Reveal Historical Changes in Floral Resources. *Commun. Biol.* **2021**, *4*, 37. [CrossRef] [PubMed]

145. Richardson, R.T.; Lin, C.-H.; Sponsler, D.B.; Quijia, J.O.; Goodell, K.; Johnson, R.M. Application of ITS2 Metabarcoding to Determine the Provenance of Pollen Collected by Honey Bees in an Agroecosystem. *Appl. Plant Sci.* **2015**, *3*, 1400066. [CrossRef] [PubMed]

146. Bell, K.L.; Burgess, K.S.; Botsch, J.C.; Dobbs, E.K.; Read, T.D.; Brosi, B.J. Quantitative and Qualitative Assessment of Pollen DNA Metabarcoding Using Constructed Species Mixtures. *Mol. Ecol.* **2019**, *28*, 431–455. [CrossRef] [PubMed]

147. Harper, L.R.; Niemiller, M.L.; Benito, J.B.; Paddock, L.E. BeeDNA: Microfluidic Environmental DNA Metabarcoding as a Tool for Connecting Plant and Pollinator Communities. *bioRxiv* **2021**. [CrossRef]

148. Madden, M.J.L.; Young, R.G.; Brown, J.W.; Miller, S.E.; Frewin, A.J.; Hanner, R.H. Using DNA Barcoding to Improve Invasive Pest Identification at U.S. Ports-of-Entry. *PLoS ONE* **2019**, *14*, e0222291. [CrossRef]

149. Whitehurst, L.E.; Cunard, C.E.; Reed, J.N.; Worthy, S.J.; Marsico, T.D.; Lucardi, R.D.; Burgess, K.S. Preliminary Application of DNA Barcoding toward the Detection of Viable Plant Propagules at an Initial, International Point-of-Entry in Georgia, USA. *Biol. Invasions* **2020**, *22*, 1585–1606. [CrossRef]

150. Krishnamurthy, P.K.; Francis, R.A. A Critical Review on the Utility of DNA Barcoding in Biodiversity Conservation. *Springer* **2012**, *21*, 1901–1919. [CrossRef]

151. Liu, M.; Zhao, Y.; Sun, Y.; Li, Y.; Wu, P.; Zhou, S.; Ren, L. Comparative Study on Diatom Morphology and Molecular Identification in Drowning Cases. *Forensic Sci. Int.* **2020**, *317*, 110552. [CrossRef]
152. Bell, K.L.; Burgess, K.S.; Okamoto, K.C.; Aranda, R.; Brosi, B.J. Review and Future Prospects for DNA Barcoding Methods in Forensic Palynology. *Forensic Sci. Int. Genet.* **2016**, *21*, 110–116. [CrossRef]
153. Liu, J.; Milne, R.I.; Möller, M.; Zhu, G.F.; Ye, L.J.; Luo, Y.H.; Yang, J.B.; Wambulwa, M.C.; Wang, C.N.; Li, D.Z.; et al. Integrating a Comprehensive DNA Barcode Reference Library with a Global Map of Yews (Taxus L.) for Forensic Identification. *Mol. Ecol. Resour.* **2018**, *18*, 1115–1131. [CrossRef]
154. Lees, D.C.; Lack, H.W.; Rougerie, R.; Hernandez-Lopez, A.; Raus, T.; Avtzis, N.D.; Augustin, S.; Lopez-Vaamonde, C. Tracking Origins of Invasive Herbivores through Herbaria and Archival DNA: The Case of the Horse-Chestnut Leaf Miner. *Front. Ecol. Environ.* **2011**, *9*, 322–328. [CrossRef]
155. Veneza, I.; Silva, R.; Freitas, L.; Silva, S.; Martins, K.; Sampaio, I.; Schneider, H.; Gomes, G. Molecular Authentication of Pargo Fillets Lutjanus Purpureus (Perciformes: Lutjanidae) by DNA Barcoding Reveals Commercial Fraud. *Neotrop. Ichthyol.* **2018**, *16*, 1–6. [CrossRef]
156. Muellner, A.N.; Schaefer, H.; Lahaye, R. Evaluation of Candidate DNA Barcoding Loci for Economically Important Timber Species of the Mahogany Family (Meliaceae). *Mol. Ecol. Resour.* **2011**, *11*, 450–460. [CrossRef]
157. Nithaniyal, S.; Newmaster, S.G.; Ragupathy, S.; Krishnamoorthy, D.; Vassou, S.L.; Parani, M. DNA Barcode Authentication of Wood Samples of Threatened and Commercial Timber Trees within the Tropical Dry Evergreen Forest of India. *PLoS ONE* **2014**, *9*, e107669. [CrossRef] [PubMed]
158. Bolson, M.; De Camargo Smidt, E.; Brotto, M.L.; Silva-Pereira, V. ITS and TrnH-PsbA as Efficient DNA Barcodes to Identify Threatened Commercial Woody Angiosperms from Southern Brazilian Atlantic Rainforests. *PLoS ONE* **2015**, *10*, e0143049. [CrossRef] [PubMed]
159. Hassold, S.; Lowry, P.P.; Bauert, M.R.; Razafintsalama, A.; Ramamonjisoa, L.; Widmer, A. DNA Barcoding of Malagasy Rosewoods: Towards a Molecular Identification of CITES-Listed Dalbergia Species. *PLoS ONE* **2016**, *11*, e0157881. [CrossRef]
160. Lahaye, R.; Van Der Bank, M.; Bogarin, D.; Warner, J.; Pupulin, F.; Gigot, G.; Maurin, O.; Duthoit, S.; Barraclough, T.G.; Savolainen, V. DNA Barcoding the Floras of Biodiversity Hotspots. *Proc. Natl. Acad Sci. USA* **2008**, *105*, 2923–2928. [CrossRef]
161. Mitchell, A.; Rothbart, A.; Frankham, G.; Johnson, R.N.; Neaves, L.E. Could Do Better! A High School Market Survey of Fish Labelling in Sydney, Australia, Using DNA Barcodes. *PeerJ* **2019**, *7*, e7138. [CrossRef]
162. Siozios, S.; Massa, A.; Parr, C.L.; Verspoor, R.L.; Hurst, G.D.D. DNA Barcoding Reveals Incorrect Labelling of Insects Sold as Food in the UK. *PeerJ* **2020**, *8*, e8496. [CrossRef]
163. Thongkhao, K.; Pongkittiphan, V.; Phadungcharoen, T.; Tungphatthong, C.; Urumarudappa, S.K.J.; Pengsuparp, T.; Sutanthavibul, N.; Wiwatcharakornkul, W.; Kengtong, S.; Sukrong, S. Differentiation of Cyanthillium Cinereum, a Smoking Cessation Herb, from Its Adulterant Emilia Sonchifolia Using Macroscopic and Microscopic Examination, HPTLC Profiles and DNA Barcodes. *Sci. Rep.* **2020**, *10*, 14753. [CrossRef]
164. Staats, M.; Arulandhu, A.J.; Gravendeel, B.; Holst-Jensen, A.; Scholtens, I.; Peelen, T.; Prins, T.W.; Kok, E. Advances in DNA Metabarcoding for Food and Wildlife Forensic Species Identification. *Anal. Bioanal. Chem.* **2016**, *408*, 4615–4630. [CrossRef] [PubMed]
165. Jiao, L.; Yu, M.; Wiedenhoeft, A.C.; He, T.; Li, J.; Liu, B.; Jiang, X.; Yin, Y. DNA Barcode Authentication and Library Development for the Wood of Six Commercial Pterocarpus Species: The Critical Role of Xylarium Specimens. *Sci. Rep.* **2018**, *8*, 1945. [CrossRef] [PubMed]
166. Newmaster, S.G.; Grguric, M.; Shanmughanandhan, D.; Ramalingam, S.; Ragupathy, S. DNA Barcoding Detects Contamination and Substitution in North American Herbal Products. *BMC Med.* **2013**, *11*, 222. [CrossRef] [PubMed]
167. Yu, J.; Wu, X.; Liu, C.; Newmaster, S.; Ragupathy, S.; Kress, W.J. Progress in the Use of DNA Barcodes in the Identification and Classification of Medicinal Plants. *Ecotoxicol. Environ. Saf.* **2021**, *208*, 111691. [CrossRef] [PubMed]
168. Stoeckle, M.; Gamble, C.; Kirpekar, R.; Young, G.; Ahmed, S.; Little, D.P. Commercial Teas Highlight Plant DNA Barcode Identification Successes and Obstacles. *Sci. Rep.* **2011**, *1*, 42. [CrossRef] [PubMed]
169. Shapcott, A.; Liu, Y.; Howard, M.; Forster, P.; Kress, W.J.; Erickson, D.; Shimizu, Y.; McDonald, W. Comparing floristic diversity and conservation priorities across south east Queensland regional rain forest ecosystems using phylodiversity indexes. *Int. J. Plant Sci.* **2017**, *178*, 211–229. [CrossRef]
170. De Boer, H.J.; Ouarghidi, A.; Martin, G.; Abbad, A.; Kool, A. DNA Barcoding Reveals Limited Accuracy of Identifications Based on Folk Taxonomy. *PLoS ONE* **2014**, *9*, e84291. [CrossRef]
171. Pimentel, D.; Zuniga, R.; Morrison, D. Update on the Environmental and Economic Costs Associated with Alien-Invasive Species in the United States. *Ecol. Econ.* **2004**, *52*, 273–288. [CrossRef]
172. Comtet, T.; Sandionigi, A.; Viard, F.; Casiraghi, M. DNA (Meta)Barcoding of Biological Invasions: A Powerful Tool to Elucidate Invasion Processes and Help Managing Aliens. *Biol. Invasions* **2015**, *17*, 905–922. [CrossRef]
173. Ardura, A.; Zaiko, A.; Borrell, Y.J.; Samuiloviene, A.; Garcia-Vazquez, E. Novel Tools for Early Detection of a Global Aquatic Invasive, the Zebra Mussel Dreissena Polymorpha. *Aquat. Conserv. Mar. Freshw. Ecosyst.* **2017**, *27*, 165–176. [CrossRef]
174. Smith, M.A.; Fisher, B.L. Invasions, DNA Barcodes, and Rapid Biodiversity Assessment Using Ants of Mauritius. *Front. Zool.* **2009**, *6*, 31. [CrossRef]

175. Wang, A.; Wu, H.; Zhu, X.; Lin, J. Species Identification of Conyza Bonariensis Assisted by Chloroplast Genome Sequencing. *Front. Genet.* **2018**, *9*, 374. [CrossRef]

176. Agnarsson, I.; Biology, M.K.-S. Taxonomy in a Changing World: Seeking Solutions for a Science in Crisis. *Syst. Biol.* **2007**, *56*, 531–539. [CrossRef]

177. Gomes, A.C.; Andrade, A.; Barreto-Silva, J.S.; Brenes-Arguedas, T.; López, D.C.; Cde Freitas, C.; Lang, C.; de Oliveira, A.A.; Pérez, A.J.; Perez, R.; et al. Local Plant Species Delimitation in a Highly Diverse A Mazonian Forest: Do We All See the Same Species? *J. Veg. Sci.* **2013**, *24*, 70–79. [CrossRef]

178. Owen, N.R.; Gumbs, R.; Gray, C.L.; Faith, D.P. Phylogenetic Diversity and Conservation. *Nat. Commun.* **2019**, *10*, 859. [CrossRef]

179. Howard, M.G.; McDonald, W.J.F.; Forster, P.I.; Kress, W.J.; Erickson, D.; Faith, D.P.; Shapcott, A. Patterns of Phylogenetic Diversity of Subtropical Rainforest of the Great Sandy Region, Australia Indicate Long Term Climatic Refugia. *PLoS ONE* **2016**, *11*, e0153565. [CrossRef]

180. Humphreys, A.M.; Vorontsova, M.; Govaerts, R.; Ficinski, S.Z.; Lughadha, N.; Vorontsova, M.S. Global Dataset Shows Geography and Life Form Predict Modern Plant Extinction and Rediscovery. *Nat. Ecol. Evol.* **2019**, *3*, 1043–1047. [CrossRef]

181. Costion, C.M.; Kress, W.J.; Crayn, D.M. DNA Barcodes Confirm the Taxonomic and Conservation Status of a Species of Tree on the Brink of Extinction in the Pacific. *PLoS ONE* **2016**, *11*, e0155118. [CrossRef]

182. Ji, Y.; Ashton, L.; Pedley, S.M.; Edwards, D.P.; Tang, Y.; Nakamura, A.; Kitching, R.; Dolman, P.M.; Woodcock, P.; Edwards, F.A.; et al. Reliable, Verifiable and Efficient Monitoring of Biodiversity via Metabarcoding. *Ecol. Lett.* **2013**, *16*, 1245–1257. [CrossRef]

183. Procopio, N.; Ghignone, S.; Voyron, S.; Chiapello, M.; Williams, A.; Chamberlain, A.; Mello, A.; Buckley, M. Soil Fungal Communities Investigated by Metabarcoding Within Simulated Forensic Burial Contexts. *Front. Microbiol.* **2020**, *11*, 1686. [CrossRef]

184. Rivera, S.F.; Vasselon, V.; Jacquet, S.; Bouchez, A.; Ariztegui, D.; Rimet, F. Metabarcoding of Lake Benthic Diatoms: From Structure Assemblages to Ecological Assessment. *Hydrobiologia* **2018**, *807*, 37–51. [CrossRef]

185. Leray, M.; Knowlton, N. DNA Barcoding and Metabarcoding of Standardized Samples Reveal Patterns of Marine Benthic Diversity. *Proc. Natl. Acad. Sci. USA* **2015**, *112*, 2076–2081. [CrossRef]

186. Aguayo, J.; Fourrier-Jeandel, C.; Husson, C.; Loos, R. Assessment of Passive Traps Combined with High-Throughput. *Appl. Environ. Microbiol.* **2018**, *84*, e02637-17. [CrossRef]

187. Johnson, M.D.; Fokar, M.; Cox, R.D.; Barnes, M.A. Airborne Environmental DNA Metabarcoding Detects More Diversity, with Less Sampling Effort, than a Traditional Plant Community Survey. *BMC Ecol. Evol.* **2021**, *21*, 218. [CrossRef]

188. Cowart, D.A.; Pinheiro, M.; Mouchel, O.; Maguer, M.; Grall, J.; Miné, J.; Arnaud-Haond, S. Metabarcoding Is Powerful yet Still Blind: A Comparative Analysis of Morphological and Molecular Surveys of Seagrass Communities. *PLoS ONE* **2015**, *10*, e0117562. [CrossRef]

189. Pompanon, F.; Deagle, B.E.; Symondson, W.O.C.; Brown, D.S.; Jarman, S.N.; Taberlet, P. Who Is Eating What: Diet Assessment Using next Generation Sequencing. *Mol. Ecol.* **2012**, *21*, 1931–1950. [CrossRef]

190. Goldberg, A.R.; Conway, C.J.; Tank, D.C.; Andrews, K.R.; Gour, D.S.; Waits, L.P. Diet of a Rare Herbivore Based on DNA Metabarcoding of Feces: Selection, Seasonality, and Survival. *Ecol. Evol.* **2020**, *10*, 7627–7643. [CrossRef]

191. Chua, P.Y.S.; Crampton-Platt, A.; Lammers, Y.; Alsos, I.G.; Boessenkool, S.; Bohmann, K. Metagenomics: A Viable Tool for Reconstructing Herbivore Diet. *Mol. Ecol. Resour.* **2021**, *21*, 2249–2263. [CrossRef]

192. Vasquez, A.A.; Mohiddin, O.; Li, Z.; Bonnici, B.L.; Gurdziel, K.; Ram, J.L. Molecular Diet Studies of Water Mites Reveal Prey Biodiversity. *PLoS ONE* **2021**, *16*, e0254598. [CrossRef]

193. Leoro-Garzon, P.; Gonedes, A.J.; Olivera, I.E.; Tartar, A. Oomycete Metabarcoding Reveals the Presence of *Lagenidium* Spp. In Phytotelmata. *PeerJ* **2019**, *7*, e7903. [CrossRef]

194. Trzebny, A.; Slodkowicz-Kowalska, A.; Becnel, J.J.; Sanscrainte, N.; Dabert, M. A New Method of Metabarcoding Microsporidia and Their Hosts Reveals High Levels of Microsporidian Infections in Mosquitoes (Culicidae). *Mol. Ecol. Resour.* **2020**, *20*, 1486–1504. [CrossRef]

195. Hebert, P.D.N.; Braukmann, T.W.A.; Prosser, S.W.J.; Ratnasingham, S.; de Waard, J.R.; Ivanova, N.V.; Janzen, D.H.; Hallwachs, W.; Naik, S.; Sones, J.E.; et al. A Sequel to Sanger: Amplicon Sequencing That Scales. *BMC Genomics* **2018**, *19*, 219. [CrossRef]

196. Parks, M.; Cronn, R.; Liston, A. Increasing Phylogenetic Resolution at Low Taxonomic Levels Using Massively Parallel Sequencing of Chloroplast Genomes. *BMC Biol.* **2009**, *7*, 84. [CrossRef]

197. Kuo, L.-Y.; Li, F.-W.; Chiou, W.-L.; Wang, C.-N. First Insights into Fern MatK Phylogeny. *Mol. Phylogenetics Evol.* **2011**, *59*, 556–566. [CrossRef]

198. McKain, M.R.; Johnson, M.G.; Uribe-Convers, S.; Eaton, D.; Yang, Y. Practical Considerations for Plant Phylogenomics. *Appl. Plant Sci.* **2018**, *6*, e1038. [CrossRef]

199. Blanchet, S.; Prunier, J.G.; De Kort, H. Time to Go Bigger: Emerging Patterns in Macrogenetics. *Trends Genet.* **2017**, *33*, 579–580. [CrossRef]

200. Leigh, D.M.; van Rees, C.B.; Millette, K.L.; Breed, M.F.; Schmidt, C.; Bertola, L.D.; Hand, B.K.; Hunter, M.E.; Jensen, E.L.; Kershaw, F.; et al. Opportunities and Challenges of Macrogenetic Studies. *Nat. Rev. Genet.* **2021**, *22*, 791–807. [CrossRef]

201. Edwards, J.L. Research and Societal Benefits of the Global Biodiversity Information Facility. *BioScience* **2004**, *54*, 485–486. [CrossRef]

202. Hijmans, R.J.; Cameron, S.E.; Parra, J.L.; Jones, P.G.; Jarvis, A. Very High Resolution Interpolated Climate Surfaces for Global Land Areas. *Int. J. Climatol.* **2005**, *25*, 1965–1978. [CrossRef]
203. Vision, T. The Dryad Digital Repository: Published Evolutionary Data as Part of the Greater Data Ecosystem. *Nat. Preced.* **2010**. [CrossRef]
204. The International Nucleotide Sequence Database Collaboration. Available online: https://www.insdc.org (accessed on 26 January 2022).
205. Kress, W.J.; Soltis, D.E.; Kersey, P.J.; Wegrzyn, J.L.; Leebens-Mack, J.H.; Gostel, M.R.; Liu, X.; Soltis, P.S. Green Plant Genomes: What We Know in an Era of Rapidly Expanding Opportunities. *Proc. Natl. Acad Sci. USA* **2022**, *119*, 2115640118. [CrossRef]

MDPI

St. Alban-Anlage 66

4052 Basel

Switzerland

Tel. +41 61 683 77 34

Fax +41 61 302 89 18

www.mdpi.com

Diversity Editorial Office

E-mail: diversity@mdpi.com

www.mdpi.com/journal/diversity